CLIMATE CHANGE AND ITS CAUSES, EFFECTS AND PREDICTION

CLIMATE CHANGE

ICE SHEETS MELT AND CHANGES IN THE ARCTIC

Climate Change and its Causes, Effects and Prediction

Additional books and e-books in this series can be found on Nova's website under the Series tab.

CLIMATE CHANGE AND ITS CAUSES, EFFECTS AND PREDICTION

CLIMATE CHANGE

ICE SHEETS MELT AND CHANGES IN THE ARCTIC

GEORGE DE HAAS
EDITOR

snova
New York

Library of Congress Cataloging-in-Publication Data

ISBN: 978-1-53617-841-8

Published by Nova Science Publishers, Inc. † New York

CONTENTS

PREFACE

Chapter 1 focuses on the science to understand the physical processes and projections of mass loss of the major ice sheets in Greenland and Antarctica, as well as of mountain and other land-based glaciers. The chapter reports on current projections of glacier mass loss due to anthropogenic climate change, and in turn how that will affect sea level.

The diminishment of Arctic sea ice has led to increased human activities in the Arctic, and has heightened interest in, and concerns about, the region's future as reported in chapter 2. Issues such as Arctic territorial disputes; commercial shipping through the Arctic; Arctic oil, gas, and mineral exploration; endangered Arctic species; and increased military operations in the Arctic could cause the region in coming years to become an arena of international cooperation, tension, or competition.

In: Climate Change
Editor: George De Haas
ISBN: 978-1-53617-841-8

Chapter 1

EARTH'S THERMOMETERS: GLACIAL AND ICE SHEET MELT IN A CHANGING CLIMATE*

Committee on Science, Space, and Technology

Thursday, July 11, 2019

House of Representatives, Committee on Science, Space, and Technology. Washington, D.C.

The Committee met, pursuant to notice, at 10:01 a.m., in room 2318 of the Rayburn House Office Building, Hon. Eddie Bernice Johnson [Chairwoman of the Committee] presiding.

Committee On Science, Space, And Technology U.S. House Of Representatives, Hearing Charter. Earth's Thermometers: Glacial and Ice Sheet Melt in a Changing Climate.

Thursday, July 11, 2019, 10:00 a.m.

* This is an edited, reformatted and augmented version of Hearing Before the Committee on Science, Space, and Technology, House of Representatives, One Hundred sixteenth Congress, First Session, Serial No. 116–35, dated July 11, 2019.

PURPOSE

The purpose of this Full Committee hearing is to discuss the current state of the science on glacier and ice sheet melt. The hearing will focus on the science to understand the physical processes and projections of mass loss of the major ice sheets in Greenland and Antarctica, as well as of mountain and other land-based glaciers. The Committee will receive expert testimony on current projections of glacier mass loss due to anthropogenic climate change, and in turn how that will affect sea level. Additionally, this hearing will provide an opportunity to discuss the major sources of uncertainty related to glacial and ice sheet melt including research gaps, risks to communities from local glacier melt, as well as global risks from ice sheet instability and sea level rise, and the need for adaptation and mitigation.

WITNESSES

- Dr. Richard B. Alley, Evan Pugh Professor of Geosciences and Associate of the Earth and Environmental Systems Institute, Pennsylvania State University
- Dr. Robin E. Bell, Lamont Research Professor, Lamont-Doherty Earth Observatory, Columbia University
- Dr. Twila A. Moon, Research Scientist, National Snow and Ice Data Center's (NSIDC) Cooperative Institute for Research in Environmental Sciences
- Dr. Gabriel J. Wolkon, Research Scientist and Manager, Climate and Cryosphere Hazards Program, Division of Geological & Geophysical Surveys, Alaska Department of Natural Resources
- Dr. W. Tad Pfeffer, Fellow, Institute of Arctic and Alpine Research, University of Colorado Boulder

OVERARCHING QUESTIONS

- How is anthropogenic climate change, particularly rising temperatures, affecting glaciers and ice sheets?
- How do glaciers and ice sheets act as "Earth's thermometers" and what can past climates tell us about current rates of change?
- How much do melting glaciers and ice sheets contribute to sea level rise and what are the projections for future sea level changes?
- What methods do scientists use to study glaciers and ice sheets, and what are the major challenges and sources of uncertainty in understanding glacial and ice sheet melt?
- What is our understanding of "tipping points" or "thresholds" in ice sheet and glacial melt, such as at the Western Antarctic Ice Sheet?
- Globally, how does mountain glacial melt impact human society?

Background

Glaciers are defined as persistent, land-based, dense ice formations that form when accumulation of snow exceeds its ablation (melting and other forms of loss) over many years.[1] The world's approximately 198,000 glaciers cover less than one percent of Earth's land surface,[2] yet glacial ice is the largest reservoir of freshwater on Earth[3] and is an important source of water for plants, animals, and humans where they occur in temperate regions and release meltwater in the summer. Glaciers and ice sheets play a critical role in Earth's air and water cycles, ecosystem support through providing nutrients and shelter for plants and animals, and climate system.

---

[1] "What is a glacier?" National Snow & Ice Data Center. https://nsidc.org/cryosphere/glaciers/questions/what.html

[2] Davies, B. "Mapping the world's glaciers." 2017. http://www.antarcticglaciers.org/glaciers-and-climate/glacier- recession/mapping-worlds-glaciers/

[3] "Ice, Snow, and Glaciers and the Water Cycle." USGS. https://www.usgs.gov/special-topic/water-science- school/science/ice-snow-and-glaciers-and-water-cycle?qt-science_center_objects=0#qt-science_center_objects

Glaciers that are larger than 50,000 km^2 (20,000 mi^2) are called ice sheets, or continental glaciers.[4] The Earth's only two present day ice sheets are in Antarctica and Greenland.[5] Ninety percent of the Earth's ice mass is contained in the Antarctic ice sheet, the world's largest single ice mass, covering almost 14 million km^2 (5.4 million mi^2) and containing 30 million km^3 of ice.[6] The Greenland and Antarctic ice sheets hold enough water to raise sea levels by 65 m (over 213 ft); however, complete melting of the ice sheets is not expected to happen.[7]

Ninety-nine percent of glacial ice is contained in ice sheets in the polar regions, but mountain glaciers exist on every continent except Australia.[8] In the U.S., glaciers can be found in Washington, Oregon, California, Montana, Wyoming, Colorado, and Nevada, with the majority occurring in Alaska.[9] Glaciers are sometimes called "Earth's thermometers" because they are very sensitive to, and therefore indicators of, climatic changes.[10] Glaciers are also indicators of past climates because trapped air bubbles reveal past atmospheric conditions from thousands of years ago.[11] Glaciers and ice sheets have a slow response time to global warming, and glaciers have not yet caught up to the heat additions made in the past decades. Therefore, even if global carbon emissions stopped entirely today, glaciers are locked in to a certain amount of melt. Thousands of studies conducted by researchers around the world have documented melting glaciers, diminishing snow cover, and rising sea levels.[12]

---

4 "What is an ice sheet?" National Snow & Ice Data Center (NSIDC). https://nsidc.org/cryosphere/quickfacts/icesheets.html

5 Ibid.

6 Amos, Jonathan (2013-03-08). "*BBC News - Antarctic ice volume measured*". Bbc.co.uk.

7 Shepherd, A. et al. *Science*. 338: 1183-1189 (2012).

8 "State of the Cryosphere: Mountain Glaciers." NSIDC. https://nsidc.org/cryosphere/sotc/glacier_balance.html

9 USGS. "Where are glaciers found in continental North America?" https://www.usgs.gov/faqs/where-are-glaciers- found-continental-north-america?qt-news_science_products=0#qt-news_science_products

10 Moon, T. *et al.* 2018. "Rising oceans guaranteed: Arctic land ice loss and sea level rise." *Current Climate Change Reports*. https://doi.org/10.1007/s40641-018-0107-0

11 "Glaciers and climate change." NSIDC. https://nsidc.org/cryosphere/glaciers/ questions/climate.html

12 IPCC, 2014: Climate Change 2014: Synthesis Report. Contribution of Working Groups I, II and III to the Fifth Assessment Report of the Intergovernmental Panel on Climate Change

This hearing is focusing on land-based ice, as opposed to sea ice. Arctic sea ice is also diminishing at rapid rates due to anthropogenic warming,[13] but as it melts, it does not contribute to sea level changes unlike glacial and ice sheet melt.[14]

State of the Science on Glacial and Ice Sheet Melt

Globally, land-based ice is deteriorating. According to the Intergovernmental Panel on Climate Change Fifth Assessment Report (IPCC AR5), a diverse range of observational evidence from multiple data sources and independent analysis techniques provide consistent evidence of substantial retreat of mountain glaciers since the 1960s, and the increased surface melting of the Greenland ice sheet since 1993, due to warmer temperatures. According to satellite data, Greenland lost an average of 269 gigatons[15] of ice, equivalent to about 71 trillion gallons of water, per year between 2002 and 2016, with the pace accelerating in recent years (Figure 1).[16] Increased surface melt, runoff, and outlet glacier discharge from warmer air temperatures are the primary contributing factors. The portion of the Greenland Ice Sheet experiencing annual melt has increased since 1980, including through significant melting events. For example, an unprecedented 98.6% of the Greenland Ice Sheet surface experienced melt on a single day in July 2012.[17] While there are seasonal patterns of warm-weather ice melt and re-freezing in winter months, when seasonal melt outpaces re-freezing, there is net annual ice mass loss. Just last month, the National Snow and Ice Data Center (NSIDC) in Boulder, Colorado reported that the Greenland Ice Sheet appears to have experienced its biggest mid-June melt event on record.[18]

[Core Writing Team, R.K. Pachauri and L.A. Meyer (eds.)]. IPCC, Geneva, Switzerland, 151 pp (IPCC AR5)

[13] "Arctic sea ice minimum." NASA. https://climate.nasa.gov/vital-signs/arctic-sea-ice/

[14] Appell, D. "Loss of land ice (not sea ice) = more sea level rise." 2014. https://www.yaleclimateconnections.org/2014/11/loss-of-land-ice-not-sea-ice-more-sea-level-rise/

[15] 1 gigaton = 10^9 tons

[16] IPCC AR5

[17] Ibid.

[18] Samenow, Jason. June 14, 2018. "Temperatures leap 40 degrees above normal as the Arctic Ocean and Greenland ice sheet see record June melting." https://www.washingtonpost.com/

The Western Antarctic Ice Sheet (WAIS), the portion of the Antarctic ice sheet that covers the western part of the continent, is considered the most vulnerable ice sheet on Earth because its bed lies thousands of feet below sea level and is exposed to warm ocean currents.[19] In 2018, a major joint U.S.-UK research collaboration was initiated to study the possibility of "marine ice sheet instability" and "marine ice cliff instability" of the WAIS, focusing on the marine- terminating Thwaites Glacier.[20] Marine ice cliff instability is when a tall cliff that might form at the front of the glacier begins to calve and break in a runaway fashion. Marine ice sheet instability is an inherently unstable architecture caused by atmospheric and ocean warming, which could result in a positive feedback loop of rapid melting of the WAIS, triggering rapid sea level rise. The Thwaites Glacier has increased speed of movement and simultaneously experienced rapid ice thinning. Multiple studies indicate that this collapse is underway in the WAIS and may also be a cause of rapid ice front retreat occurring in Greenland.[21,22]

The WAIS has been experiencing mass loss since the early 1990s, and melt rates have more than tripled in the last 25 years.[23] Recent observed rapid mass loss from West Antarctica's floating ice shelves is attributed to increased glacial discharge rates due to diminishing ice shelves caused by the surrounding ocean becoming warmer.[24] Antarctica as a whole lost more

weather/2019/06/14/arctic-ocean-greenland- ice-sheet-have-seen-record-june-ice-loss/?utm_term=.2f92ed415b9d

[19] Fox, Douglas. "The West Antarctic Ice Sheet Seems to Be Good at Collapsing." *National Geographic*, National Geographic Society, 13 June 2018, www.news.nationalgeographic.com/2018/06/west-antarctic-ice-sheet-collapse- climate-change/

[20] The International Thwaites Glacier Collaboration, https://thwaitesglacier.org/about/itgc

[21] Moon, Twila. May 2017. "Saying goodbye to glaciers: Glacier volume is shrinking worldwide, with wide-ranging implications for society." Science. Vol. 356, Issue 6338.

[22] BBC 30 April 2018. "Thwaites Glacier: Biggest ever Antarctic field campaign." By Jonathan Amos https://www.bbc.com/news/science-environment-43936372

[23] Harvey, C. "Antarctic Melt Rate Has Tripled in the Last 25 Years." June 14, 2018. https://www.scientificamerican.com/article/antarctic-melt-rate-has-tripled-in-the-last-25-years/

[24] IPCC AR5

than 3 trillion tons of ice between 1992 and 2017.[25] More recent gravity data collected from space using NASA's Gravity Recovery and Climate Experiment (GRACE) satellites show that Antarctica has been losing more than one hundred km^3 (24 mi^3) of ice each year since 2002 (Figure 2).[26] A chunk of ice the size of Delaware broke off on July 12, 2019 from the Larsen C Ice Shelf of the WAIS, which might destabilize the entire ice shelf.[27]

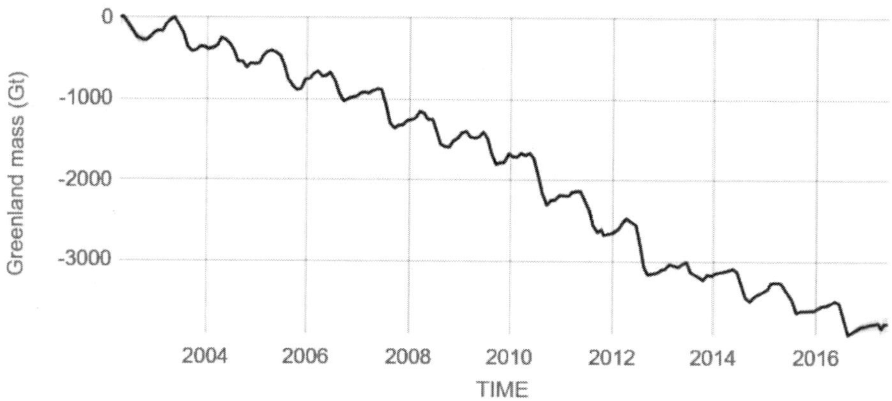

Source: climate.nasa.gov

Figure 1. Greenland ice sheet mass variation since 2002, showing loss of 286 gigatons ice per year. Data source: Ice mass measurement by NASA's Gravity Recovery and Climate Experiment (GRACE) satellites. Credit: NASA.

The IPCC AR5 predicts that global glacier volume, excluding glaciers on the periphery of Antarctica and the Greenland and Antarctic ice sheets, is projected to decrease by 15 to 55% if we limit global temperature rise to 2 degrees Celsius (RCP2.6) and by 35 to 85% under a high emissions scenario (RCP8.5). However, uncertainties remain in the quantification and

---

[25] Harvey, C. "Antarctic Melt Rate Has Tripled in the Last 25 Years." June 14, 2018. https://www.scientificamerican.com/article/antarctic-melt-rate-has-tripled-in-the-last-25-years/

[26] Conway, E. "Is Antarctica melting? – Climate Change: Vital Signs of the Planet." September 16, 2014. https://climate.nasa.gov/news/242/is-antarctica-melting

[27] Fox, Douglas. "The West Antarctic Ice Sheet Seems to Be Good at Collapsing." *National Geographic*, 13 June 2018, www.news.nationalgeographic.com/2018/06/west-antarctic-ice-sheet-collapse-climate-change/

modeling of key physical processes that contribute to the acceleration of land ice melting. Climate models are unable to capture the rapid pace of observed land ice melt over the last 15 years; a major factor is our inability to quantify and accurately model the physical processes driving the accelerated melting.[28]

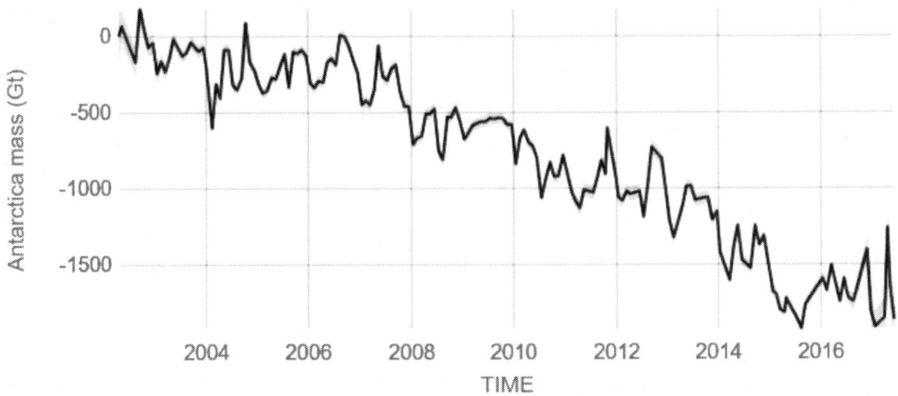

Source: climate.nasa.gov

Figure 2. Antarctic ice sheet mass variation since 2002, showing loss of 127 gigatons ice per year. Data source: Ice mass measurement by NASA's GRACE satellites. Credit: NASA.

The vast majority of global mountain glaciers are losing mass at significant rates. The annually averaged ice mass from 37 global reference glaciers has decreased every year since 1984. Annual average near-surface air temperatures across Alaska and the Arctic have increased over the last 50 years at a rate more than twice as fast as the global average temperature. Due to increasing temperatures, Alaska is losing about 75 billion tons of ice each year.[29]

A recent study provides the clearest picture of Himalayan glacier loss to date, using declassified U.S. spy satellite data from the past 40 years and

---

[28] IPCC AR5

[29] Fountain, H. "When the Glaciers Disappear, Those Species Will Go Extinct." April 17, 2019. https://www.nytimes.com/interactive/2019/04/16/climate/glaciers-melting-alaska-washington.html

combining it with contemporary satellite data.[30] The study found that the Himalayas lost 25% of their ice in the last 40 years, equivalent to eight billion tons of water each year. Glacial melt rates tracked temperature increases during the time period studied, meaning the melting can be attributed to warmer temperatures.

Methods for Studying Ice Loss

Measuring mass loss of glaciers is not straightforward but can be estimated used many types of observations, namely, the gravitational pull of the ice, surface elevation of ice sheets, and difference between ice accumulation and loss. The modern study of mass change in glaciers and ice sheets occurs over many spatiotemporal scales, from paleo-glaciological records dating back hundreds of thousands of years, to centimeter-scale in situ measurements, to global images using satellites.[31] Field studies enable detailed sampling and long-term monitoring and involve the use of various instruments. Global Positioning System (GPS), weather stations, seismometers, time- lapse cameras, and radar instruments elucidate glacier hydrology, subsurface environments, and glacier dynamics on time scales of minutes to months.[32] Challenges in observational work are that it is time-consuming, involves travel to remote and difficult to access locations, is expensive, and involves physically demanding work in harsh environments.

The advent of satellite monitoring in the 1990s was a major advancement for studying large ice sheets, allowing for improvements in the ability to estimate ice mass loss. NASA's GRACE satellite mission (2002-2017) and GRACE Follow On (launched May 2018)[33] help estimate ice mass variations, altimetry satellites (e.g., NASA's Ice, Cloud, and land

---

[30] Maurer, J.M. et al. 2019. "Acceleration of ice loss across the Himalayas over the past 40 years." Science Advances. Vol 5, no. 6. DOI: 10.1126/sciadv.aav7266

[31] Moon, Twila. May 2017. "Saying goodbye to glaciers: Glacier volume is shrinking worldwide, with wide-ranging implications for society." Science. Vol. 356, Issue 6338.

[32] Moon, T. *et al.* 2018. "Rising oceans guaranteed: Arctic land ice loss and sea level rise." *Current Climate Change Reports.*

[33] The NASA GRACE mission concluded in June 2017 and GRACE's successor mission, GRACE Follow-On, is launching in the summer of 2019.

Elevation Satellite, ICESat (2003-2009) and ICESat-2 (launched 2018)[34] and the European CryoSat (1999-2005) and CryoSat-2 (launched 2010))[35] detect changing surface elevations, and optical and radar imaging satellites measure ice motion, monitor glacier advance and retreat, and observe surface properties, including melt. Aerial surveys are able to cover inaccessible regions such as glacial crevasses and help with data collection in between satellite missions.[36]

Observations of ice sheets and glaciers and past records of glaciation are important for understanding Earth's climate system. Observational data is important in validating models in order to predict future changes. Ice sheets are important to the climate system, and incorporating ice sheet models into global climate models will improve projections. However, this is a challenge given ice sheet models are very high resolution and the climate models cannot currently accommodate that level of detail.[37]

Resulting Sea Level Rise

Sea levels have risen over eight inches (23 cm) since the Industrial Revolution and continue to rise 0.13 inches (3.2 mm) each year.[38] One third of current sea level rise is due to thermal expansion of seawater, one third is from ice sheet melt, and one third is from mountain glacial melt. The contribution to sea level rise from Antarctic and Greenland ice sheet melt has gone up from one-tenth just two decades ago, mostly due to Greenland ice losses.[39] The Fourth National Climate Assessment (NCA4) predicts global mean sea level will rise an additional 1.0-4.3 feet by 2100, depending on a low or high emissions scenario.[40] Even with a low emissions scenario and not considering additional contribution from ice

---

34 NASA ICESat and ICESat-2, https://icesat.gsfc.nasa.gov/

35 European Satellite Agency, CryoSat and CryoSat-2, https://www.esa.int/Our_Activities/Observing_the_Earth/CryoSat/Introducing_CryoSat

36 Moon, T. *et al.* 2018. "Rising oceans guaranteed: Arctic land ice loss and sea level rise." *Current Climate Change Reports.*

37 Fourth National Climate Assessment.

38 Nunez, C. 2017. "Sea level rise, explained." https://www.nationalgeographic.com/environment/global- warming/sea-level-rise/

39 Shepherd, A. et al. Science. 338: 1183-1189 (2012).

40 NCA4; Volume I; Ch. 12

sheet melt, we are locked in to approximately 1 foot of global sea level rise by the end of the century.[41]

Over the last two decades, significant progress has been made in understanding ice sheet dynamics through combined field and satellite observations and improving numerical models to capture responses of ice sheets to environmental change. However, there is major uncertainty in the amount of additional sea level rise that could occur due to melting of Greenland and Antarctic ice sheets, due to the possibility of marine ice sheet and marine ice cliff instability. Collapse of the WAIS could contribute an additional 11 feet of sea level rise.[42] Sea level rise threatens coastal communities in the U.S. and worldwide and will increase the frequency and extent of extreme flooding associated with coastal storms, such as hurricanes and nor'easters.[43]

Other Impacts of Glacial and Ice Sheet Melt

Glacial and ice sheet mass loss will also have other direct and indirect impacts on humans and ecosystems, including on the climate system and weather patterns, ocean circulation, Earth's rotation, drinking water, and certain fisheries. Mountain glaciers are an important drinking water supply for many people around the world, especially in India, Nepal, and some countries in South America. Approximately 800 million people depend on glacial meltwater from the high mountains of Asia alone.[44] Glacial melt leads to rapid, catastrophic floods and debris flows for these downstream communities.[45]

Recent research suggests ice sheet melt may substantially slow down the major ocean conveyor belt of heat, known as the Atlantic meridional overturning circulation, which helps regulate the climate and affects global

---

[41] Ibid

[42] Fox, Douglas. "The West Antarctic Ice Sheet Seems to Be Good at Collapsing." *National Geographic*, National Geographic Society, 13 June 2018, www.news.national geographic.com/2018/06/west-antarctic-ice-sheet-collapse- climate-change/

[43] IPCC AR5

[44] Pritchard, H.D. 2019. "Asia's shrinking glaciers protect large populations from drought stress." Nature, Vol 569. https://doi.org/10.1038/s41586-019-1240-1

[45] Ibid

weather patterns.[46] There is satellite evidence that ice sheet melt is also responsible for a slight decrease in the speed of the Earth's rotation. This is because ice sheets are at high latitudes, and when they melt, the water is redistributed toward lower latitudes in a phenomenon called "polar wander."[47]

Loss of glacial streams and meltwater will lead to extinction of small creatures that rely on them. If glacial meltwater continues to decline and stream temperatures rise, larger fish populations like salmon and similar fish may also be affected.[48]

Chairwoman JOHNSON. Good morning. This hearing will come to order. And without objection, the Chair is authorized to declare recess at any time.

I'd like to welcome our witnesses to the Science, Space, and Technology Committee's hearing entitled, "Earth's Thermometers: Glacial and Ice Sheet Melt in a Changing Climate." It seems as though we're bombarded on an almost daily basis with news articles and reports saying that the world's ice is melting faster than ever. As a matter of fact, I almost invited Mr. Young from Alaska, who moved to Alaska because it was too warm in the United States proper. Since I read about Alaska last week, I thought he might want to hear this.

Pictures show ice sheets in Greenland and Antarctica crashing into the oceans before our eyes. Just last month, a piece of ice the size of the State of Delaware broke off Antarctica, and Greenland was reported to have experienced the biggest June ice melt event on record with temperatures 40 degrees above normal.

The rate of change in the Arctic and Antarctic has been quickening in recent years, according to the Intergovernmental Panel on Climate Change (IPCC) and numerous other scientific bodies. For example, a study published in Nature in January that was led by an international team of

---

[46] Harvey, C. 2019. "Melting ice sheets could worsen extreme weather." https://www.scientificamerican.com/article/melting-ice-sheets-could-worsen-extreme-weather/

[47] Dunham, W. 2015. "Melting glaciers blamed for subtle slowing of Earth's rotation." https://www.reuters.com/article/us-science-rotation-idUSKBN0TU2F720151212

[48] Ibid

more than six dozen researchers tells us that melt rates have more than tripled in western Antarctica in the last 25 years.

Mountain glaciers are also experiencing rapid rates of change. Just a few weeks ago, declassified U.S. spy satellite data clearly showed that Himalayan glaciers lost 25 percent of their ice over the last 40 years. This is equivalent to 8 billion tons of water each year. This puts the hundreds of millions of people in that region who depend on glacial melt as a freshwater source at risk.

According to the 2014 IPCC Assessment Report, without significant reductions in global greenhouse gas emissions, mountain glaciers will lose 35 to 85 percent of their ice by the end of the century under a high emissions scenario. Newer reports indicate that IPCC estimates might even be conservative and that glacial and ice sheet melt rates could even be higher. We need to be listening to Earth's glaciers and ice sheets and what they're telling us about the changing climate.

Glacial and ice sheet melt is responsible for two-thirds of the 8 inches of sea-level rise that we've seen in the last 200 years from the anthropogenic warming, and that sea-level rise is only expected to continue. The western Antarctic ice sheet, which everyone is watching because it is thought to be the most unstable ice sheet, could add another 11 feet of additional sea-level rise if it collapses, which some experts expect could happen at some point. Such an increase would mean many coastal cities would be flooded, and the world as we know it would be different.

What's happening in Greenland, Antarctica, and the high mountain regions matters to us all. Glaciers and ice sheets play vital roles in regulating Earth's climate and weather, provide over two-thirds of the Earth's freshwater supply for drinking and agricultural uses, support fisheries and ecosystem health, and run hydro-power plants. I'm glad we have the opportunity to hear today from some of the Nation's leading glacial and ice sheet experts.

And I'd like to welcome Dr. Richard Alley, who last testified before this Committee in 2010. I also want to announce that later today we will be hosting a screening of the award-winning documentary "Chasing Ice" that documents changing ice in the Arctic. It will be followed by a question-and-answer session with two of our witnesses, Dr. Pfeffer, who was a scientific advisor to the film, and Dr. Moon. The screening is free and open to the public, and I hope all of you will join us.

This Committee plays an important role in authorizing both climate science and the research needed to better understand glaciers and ice sheets. Since the 1990s, NASA's (National Aeronautics and Space Administration's) ice-monitoring satellites have led to major discoveries of ice sheet dynamics and melt, while the National Science Foundation (NSF) has funded major field expeditions in ice sheets. I look forward to today's discussion with our distinguished panel to understand how Congress and the Committee in particular can address the critical research gaps in this field.

Thank you.

[The prepared statement of Chairwoman Johnson follows:]

Good morning. I would like to welcome our witnesses to the Science, Space, and Technology Committee's hearing entitled "Earth's Thermometers: Glacial and Ice Sheet Melt in a Changing Climate."

It seems as though we're bombarded on an almost daily basis with news articles and reports saying that the world's ice is melting faster than ever. Pictures show ice sheets in Greenland and Antarctica crashing into the oceans before our eyes. Just last month, a piece of ice the size of Delaware broke off of Antarctica, and Greenland was reported to have experienced the biggest June ice melt event on record with temperatures 40 degrees above normal.

The rate of change in the Arctic and Antarctic has been quickening in recent years, according to the Intergovernmental Panel on Climate Change and numerous other scientific bodies. For example, a study published in *Nature* in January that was led by an international team of more than six dozen researchers tells us that melt rates have more than tripled in Western Antarctica in the last 25 years.

Mountain glaciers are also experiencing rapid rates of change. Just a few weeks ago, declassified U.S. spy satellite data clearly showed that Himalayan glaciers lost 25% of their ice over the last 40 years. That is equivalent to eight billion tons of water each year. This puts the hundreds of millions of people in that region who depend on glacial melt as a fresh water source at risk.

According to the 2014 IPCC Assessment Report, without significant reductions in global greenhouse gas emissions, mountain glaciers will lose 35 to 85% of their ice by the end of the century under a high emissions scenario. Newer reports indicate that the IPCC estimates might even be conservative and that glacial and ice sheet melt rates could be even higher.

We need to be listening to Earth's glaciers and ice sheets and what they're telling us about the changing climate. Glacial and ice sheet melt is responsible for two-thirds of the 8 inches of sea level rise we've seen in the last 200 years from anthropogenic warming, and that sea level rise is only expected to continue. The Western Antarctic Ice Sheet, which everyone is watching because it is thought to be the most unstable ice sheet, could add another 11 feet of additional sea level rise if it collapses, which some experts expect could happen at some point. Such an increase would mean many coastal cities would be flooded and the world as we know it would be different.

What's happening in Greenland, Antarctica, and in high mountain regions matters to us all. Glaciers and ice sheets play vital roles in regulating Earth's climate and weather, provide over two-thirds of Earth's freshwater supply for drinking and agricultural uses, support fisheries and ecosystem health, and run hydropower plants. I'm glad we have the opportunity to hear today from some of the nation's leading glacial and ice sheet experts. We're lucky to have five distinguished glaciologists here today, and I would like to welcome back Dr. Richard Alley, who last testified before this Committee in 2010.

I also want to announce that later today we will be hosting a screening of the award-winning documentary *Chasing Ice* that documents changing ice in the Arctic. It will be followed by a question and answer session with two of our witnesses, Dr. Pfeffer (FEFF-er), who was a scientific advisor to the film, and Dr. Moon. The screening is free and open to the public, and I hope you can join us.

This Committee plays an important role in authorizing both climate science and the research needed to better understand glaciers and ice sheets. Since the 1990s, NASA's ice monitoring satellites have led to major discoveries of ice sheet dynamics and melt, while the National Science Foundation has funded major field expeditions to ice sheets. I look forward to today's discussion with our distinguished panel to understand how Congress, and this Committee in particular, can address the critical research gaps in this field. Thank you.

Chairwoman JOHNSON. And I now will offer our Ranking Member his opening statement time.

Mr. Lucas. Thank you, Chairwoman Johnson, for holding this hearing, which is another opportunity to examine the impacts of a changing climate on our country and the world at large. While today's hearing will examine the underlying science of this issue and concerns about climate change, I'd like for us to also focus on the agricultural, economic, and geopolitical consequences we can expect from glacial and sea ice melt and, more importantly, how we can address those.

For instance, polar ice sheets cool ocean currents, which affect global weather patterns. As I've mentioned a time or two, weather issues are of paramount importance to farmers and ranchers in Oklahoma and around the world. We do not have a firm grip on how these weather patterns will change due to melting and how we can prepare for these changes.

I also want to consider the economic and geopolitical consequences of glacial and sea ice melt. Five countries, including America and Russia, border the Arctic. Territorial disputes in this region will take on greater importance as resource-rich land and new shipping routes are revealed.

There are significant economic implications from the energy rights, mineral deposits, and tourism opportunities. For instance, Russia is claiming that some newly accessible routes should not be considered international waterways but a part of their sovereign territory. Better research will give us greater insights into how we can expect shipping routes to change so we can prepare to address these issues.

As the Science Committee, we have a responsibility to address our national research priorities, and those must be broader than just how the climate's changing. We need to understand the specific effects so we can adopt and continue our economic growth.

During our first full hearing of this Congress, Members of the Committee discussed how we could embrace a broader portfolio of basic research, energy innovation, and competitive technology to make energy production cleaner, more efficient, and less costly. I hope we can spend more time considering research into innovative technologies like nuclear reactors, battery storage, and carbon capture.

I'd like to thank our witnesses for being here today, and I look forward to our discussion. And I yield back, Madam Chair.

[The prepared statement of Mr. Lucas follows:]

Chairwoman Johnson, thank you for holding this hearing, which is another opportunity to examine the impacts of a changing climate on our country and the world at large.

While today's hearing will examine the underlying science of this issue and concerns about climate change, I'd like for us to also focus on the agricultural, economic, and geopolitical consequences we can expect from glacial and sea ice melt and—more importantly—how we can address those.

For instance, polar ice sheets cool ocean currents which affect global weather patterns. As I've mentioned once or twice, weather issues are of paramount importance to farmers and ranchers in Oklahoma and around the world. We do not have a firm grasp of how these weather patterns will change due to melting and how we can prepare for these changes.

I also want to consider the economic and geopolitical consequences of glacial and sea ice melt. Five countries, including America and Russia, border the Arctic. Territorial disputes in this region will take on greater importance as resource-rich land and new shipping routes are revealed.

There are significant economic implications from the energy rights, mineral deposits, and tourism opportunities. For instance, Russia is claiming that some newly accessible routes should not be considered international waterways but part of their sovereign territory. Better research will give us greater insight into how we can expect shipping routes to change so we can prepare to address these issues.

As the Science Committee, we have the responsibility to address our national research priorities and those must be broader than just how the climate is changing. We need to understand its specific effects so we can adapt and continue our economic growth.

During our first full committee hearing of this Congress, members of this Committee discussed how we must embrace a broad portfolio of basic research, energy innovation, and competitive technology to make energy production cleaner, more efficient, and less costly.

I hope we can spend more time considering research into innovative technologies like nuclear reactors, battery storage and carbon capture.

I'd like to thank our witnesses for being here today, and I look forward to our discussion.

Chairwoman Johnson. Thank you very much.

I'd like to extend a warm welcome to a guest in the audience, Maria, from Chandler, Arizona. Could you stand? We hear you're a rising senior in high school who's interested in studying engineering in college. And it's great to have the next generation of STEM (science, technology, engineering, and mathematics) professionals represented here today. And welcome to all the young people over here, too. Thank you for being here.

At this time I'd like to introduce our witnesses. Our first distinguished witness, Dr. Richard Alley, is the Evan Pugh Professor of Geosciences and Associate of the Earth and Environmental Systems Institute at the Pennsylvania State University. He has spent more than 40 years studying the great ice sheets to help predict future changes in climate and sea levels, and has made four trips to Antarctica, nine to Greenland, and additional expeditions to Alaska and elsewhere. He has authored or co-authored more than 300 scientific papers. He was involved in the IPCC group of contributors that won the 2007 Nobel Peace Prize. He won Pennsylvania State's highest teaching award, and has written a book on climate change and ice cores. He holds a Ph.D. in geology from the University of Wisconsin.

Our second witness, Dr. Robin Bell, is the PGI Lamont Research Professor at Lamont-Doherty Earth Observatory of Columbia University and a member of the faculty at Columbia Earth Institute. She directs programs in ice sheet dynamics, leads efforts to develop innovative technology, and works to improve the scientific culture, especially for women. She has led 10 major expeditions to the polar regions discovering an active volcano, large, deep lakes, and hidden mountain ranges buried by ice. She was instrumental in launching the International Polar Year in 2007 that brought together over 50,000 scientists. Currently, she is the President of the American Geophysical Union, the largest collection of Earth and space scientists in the world. And her PhD is in geophysics from Columbia University.

Our third witness is Dr. Twila Moon, who is a Research Scientist at the National Snow and Ice Data Center (NSIDC), part of the University of Colorado's Boulder Cooperative Institute for Research in Environmental Sciences. She studies modern changes in glaciers and ice sheets and the connection among ice, climate, ocean, and ecosystems. Her research focuses on the Greenland ice sheet and the Arctic and uses a variety of tools, including satellite remote sensing, fieldwork, and computer simulations. She also leads efforts to improve science and knowledge coproduction between scientists and stakeholders. Dr. Moon received her Ph.D. in Earth and space sciences from the University of Washington.

Our fourth witness, Dr. Gabriel Wolken, is a Research Scientist and Manager of the Climate and Cryosphere Hazards Program at the Alaska Division of Geological & Geophysical Surveys and a Research Assistant Professor at the International Research Center at the University of Alaska Fairbanks. There, he is a Senior Scientist in the Climate Adaptation Science Center. He studies snow and glacier change and their connection to climate and natural hazards through observations, remote sensing, and computer modeling. Dr. Wolken has a PhD in Earth and atmospheric sciences from the University of Alberta.

Our final witness, Dr. William Ted Pfeffer, is a Professor of Civil, Environmental, and Architectural Engineering and a Fellow at the Institute of Arctic and Alpine Research at the University of Colorado Boulder. He has been involved in glaciology research for 40 years, studying the world's mountain glaciers. He has conducted hundreds of field expeditions in the continental USA, Alaska, Canada, Norway, Greenland, Antarctica, the

Himalayas, and Africa. He has published over 60 peer-reviewed scientific papers and was a scientific advisor to the Emmy-winning film "Chasing Ice." Dr. Pfeffer earned his PhD in geophysics at the University of Washington.

As our witnesses should know, you will each have 5 minutes for your spoken testimony. Your written testimony will be included in the record of the hearing. When all of you have completed your spoken testimony, we will begin a round of questions. Each Member will have 5 minutes to question the panel. And so we will begin our witnesses now with Dr. Alley.

TESTIMONY OF DR. RICHARD B. ALLEY, EVAN PUGH PROFESSOR OF GEOSCIENCES AND ASSOCIATE OF THE EARTH AND ENVIRONMENTAL SYSTEMS INSTITUTE, PENNSYLVANIA STATE UNIVERSITY

Dr. Alley. Thank you, Madam Chairwoman, Ranking Member Lucas, distinguished Members, staff, and citizens, for this opportunity to address you.

We have high scientific confidence that the world is warming primarily because we burn fossil fuels and release CO2, and this is having broad-based impacts. You've asked us to tell you about changes in snow and ice of which we will get to some of them but not all.

We still have winter, we still have blizzards. Where and when snow and ice care about temperature we are seeing broad-based shrinkage, and this really is having impacts. Earlier spring snow melt means that you can lengthen the fire season. It affects eco-systems; it affects tourism. Loss of Arctic sea ice, as Representative Lucas mentioned, has national security implications, as well as weather implications. Glacier melt is changing streamflow in some of the most overused and politically sensitive rivers on Earth.

I will focus particularly on sea level, which is the biggest global footprint of melting ice. Sea level is rising. Recently, it's been about 1 inch per 8 years. It is rising not because of natural cycles but because of warming. The ocean expands as it warms. The mountain glaciers are melting. The edges of Greenland are melting and putting extra water into the ocean. And there's faster flow of non-floating ice into the ocean from parts of Greenland and Antarctica.

We are committed to some additional sea-level rise. Just as if you drop an ice cube into your tea, it is committed to melting, but it takes a while to melt. The ice has not caught up with the warming we have already caused. But by the time our students are getting old, the decisions that we humans make now and in the future will grow to be the dominant control on how much sea-level rise we experience.

This sea-level rise is already having implications. You can Google the picture of the octopus in the parking garage in Miami on a high tide, not a storm. But the impacts could become much larger. The general projections are that if we don't change our energy system, we will get something like 3 feet of sea-level rise by 2100 above the natural level, the pre-industrial level.

And I'd like to speak about the uncertainties in that, right? So I'd like to do an analogy first. I ride my bicycle to work at Penn State. My wife drives our car. But I drove down here. I saw commuters in the D.C. area. My impression is that a commuter in D.C. expects to spend half an hour stuck in traffic. The best thing that can happen to a commuter is no traffic, but they might spend an hour, and they might get run over by a drunk driver and be in the hospital or worse. What they expect, the most likely future, is well on the good end of the possible futures when you get in that car.

When we look at the sea-level rise, it is similar. Three feet if we don't change our energy system, maybe 2, maybe 4, maybe 5, 10. We're not sure. It could be much worse. And there isn't much better to offset the much worse. There are drunk drivers in the climate system.

I'd like to explain one of them. If you ever get the chance to go to Glacier Bay National Park and Preserve in Alaska, it is a gloriously beautiful place. You can cruise 65 miles up the Bay and see little glaciers breaking off little icebergs in shallow water, and it's still spectacular.

When Vancouver was on his cruise in 1796, there was no Glacier Bay. It was entirely full of ice up to a mile thick. When John Muir went by, less than a century later, the Bay was mostly open because icebergs had been breaking off the front of the glacier like dominoes at a rate of up to 7 miles a year, falling over.

That process has happened to other glaciers in Alaska. You have world experts on that process here. It has happened in Chile, in Svalbard. It's happening in Greenland and the Antarctic Peninsula.

It happened to ice sheets in the past. And it's well-known that this happens when it gets too warm where ice flows into the ocean. So far, those have been in narrow valleys. They're spectacular locally but one collapse doesn't raise global sea level a lot. If this starts to happen in parts of Antarctica rather than a narrow valley, it will open into a broad embayment. If that breaks as rapidly as we have seen elsewhere, in the next century you might get 10 feet or so of extra sea-level rise. It could be faster than that.

It is very clear that the uncertainties can be reduced if you fund bright young people to work with the co-panelists up here. That's self-serving, but it's correct. But there may be a little irreducible uncertainty in the same way that you can never predict where every drunk driver might be out on the highway. If we raise temperature, we raise sea level with high confidence, and the uncertainties are it could be a little better, a little worse, or a lot worse. Thank you.

[The prepared statement of Dr. Alley follows:]

Antarctic Contributions to Future Sea-Level Rise: Possible Tipping Points, Testimony of Dr. Richard B. Alley,* Pennsylvania State University

For the hearing entitled: Earth's Thermometers: Glacial and Ice Sheet Melt in a Changing Climate Before the Committee on Science, Space and Technology, United States House of Representatives.

10:00 a.m., July 11, 2019

* Any opinions, findings, conclusions, or recommendations expressed in this publication are those of the author and do not necessarily reflect those of the Pennsylvania State University, the Intergovernmental Panel on Climate Change, the National Academy of Sciences, or other organizations.

Personal Introduction and Background

My name is Richard Alley. I am an Evan Pugh University Professor of Geosciences and Associate of the Earth and Environmental Systems Institute at the Pennsylvania State University. I have authored or coauthored more than 300 refereed scientific papers, and I have made more than 1000 public presentations concerning my areas of expertise. I have been recognized with awards for research, teaching and service, including election to the US National Academy of Sciences and foreign membership in the Royal Society; as noted above, my comments are my own and do not represent these bodies or any other bodies, but the recognition by these bodies may help establish my credentials as a witness.

My research is especially focused on the great ice sheets of Antarctica and Greenland, their potential for causing major changes in sea level, the climate records they contain, and their other interactions with the environment; I also study mountain glaciers, and ice sheets of the past. I have served with distinguished national and international teams on major scientific assessment bodies, including chairing the U.S. National Research Council's Panel on Abrupt Climate Change (report published in 2002), and serving the U.S. Climate Change Science Program, and the Nobel-Peace-Prize-Winning Intergovernmental Panel on Climate Change (IPCC) in various ways on their Second (1995), Third (2001) and especially Fourth (2007) Assessment Reports. I have had the honor on several occasions of providing requested testimony and briefings to high government officials at the federal as well as state levels, including to legislative committees chaired by members of both major political parties, and to executive officials in administrations of both major political parties, drawing on my expertise to provide scientific information to those working for the public good. Additional information is given in the short biography at the end of this document.

My testimony here is updated from my testimony of November 17, 2010 to the Subcommittee on Energy and Environment of the House Committee on Science and Technology of the United States House of Representatives; the consistency of this testimony reflects the consistency of the scientific understanding, which continues to strengthen without fundamentally changing.

Background on Climate Change, Global Warming and Sea-Level Rise

Scientific assessments such as those of the National Academy of Sciences of the United States (e.g., National Research Council, 1975; 1979; 2001; 2006; 2008; 2010a; 2010b), the U.S. Climate Change Science Program (CCSP), and the Intergovernmental Panel on Climate Change (IPCC) have for decades consistently found with increasingly high scientific confidence that human activities are raising the concentration of carbon dioxide and other greenhouse gases in the atmosphere, that this has a warming effect on the climate, that the climate is warming as expected, and that the changes so far are small compared to those projected if humans burn much of the fossil fuel on the planet.

The basis for expecting and understanding warming from carbon dioxide is the fundamental physics of how energy interacts with gases in the atmosphere. This knowledge has been available for over a century, was greatly refined by military research after World War II, and is directly confirmed by satellite measurements and other data (e.g., American Institute of Physics, 2008; Harries et al., 2001; Griggs and Harries, 2007).

With very high scientific confidence, this warming is causing sea level to rise. Since 1993, when high-quality satellite altimetry data have been available, sea-level rise has averaged approximately 3.1 mm/yr, or about 1 inch per 8 years. (Of many sources for this general information on sea-level change, Church et al., 2013 and Lemke et al., 2007 are good starting points, and important information is available at http://sealevel.colorado.edu/ and in Nerem et al., 2018.) The rate of sea-level rise has accelerated. The rise comes from several sources. Much of the energy added to the Earth system because of the rise in atmospheric greenhouse gases has heated the ocean, causing the water to expand and raise sea level. Warming is causing melting of glaciers, shifting water from land to the ocean, with local impacts on water availability and hazards as well as global implications through sea level. Melting has increased mass loss from the surface of the Greenland ice sheet. Faster flow of non-floating ice into the ocean to float and melt has caused mass loss from some parts of the Greenland and Antarctic ice sheets. In addition, groundwater "mining"—removing well-water from the ground more rapidly than replenished—may be contributing a little to sea-level rise. (Note that the IPCC found with high confidence that warming is causing loss of other temperature- sensitive snow and ice, including springtime snow cover and Arctic sea ice, but that these do not contribute in any significant way to sea-level rise; e.g., Lemke et al., 2007.)

Many strands of mutually supporting evidence are woven into the confident knowledge that loss of land ice and warming of the ocean are driving sea-level rise. A large and consistent scientific literature exists on this topic (e.g., The IMBIE Team, 2018), and the synopsis of techniques in Lemke et al. (2007; section 4.6.2.1) provides a broad overview. As explained there, for large ice sheets the gain or loss of mass contributing to sea-level change is measured in three ways: by repeatedly "weighing" the ice sheet using satellites that measure Earth's gravity field, supplemented by data from aircraft and surface measurements; by repeatedly measuring the surface elevation of the ice sheet using altimeters on satellites or aircraft supplemented by surface measurements; and, by measuring and modeling the addition of snowfall to the ice sheet, and the loss by runoff of meltwater or by flow of nonfloating ice into the ocean to float and then melt. Measurements of ocean temperature provide data to estimate the expansion of ocean water, and measurements of changes in mountain glaciers similar to those for ice sheets but involving more on-the-glacier work and more analysis of size changes from satellite imagery provide additional constraints. Sea-level measurements from satellites and tide gauges monitor the rise of the ocean. Scientists then assess the consistency among the various lines of evidence, and find agreement. Additional tests are also applied. Melting of ice from near the poles and motion of the water into the ocean actually slows Earth's rotation very slightly, something like a spinning ice skater sticking out her arms. And, while

Greenland is far north, it is not all the way to the North Pole, so melting of Greenland's ice causes Earth to wobble a bit, something like the skater moving one arm but not the other. These changes in Earth's rotation are tiny, they are not important to most people most of the time, but they test and confirm the other measurements.

Insights from Climate History

The hills around Los Angeles have always burned, so we worry about people shooting illegal fireworks during dry summers. People have always died, so we worry about murder. An arson investigator must understand natural fires, and a homicide investigator must understand how people live and die naturally.

The science of climate and sea level includes a "CSI" component (see, for example, CCSP, 2009; Masson-Delmotte et al., 2013). Summarizing those summaries and many other studies, climate has always changed, which shows that climate is changeable. Climate has changed for many reasons—dust from single large volcanic eruptions blocking the sun and causing occasional cold years, features of Earth's orbit slowly shifting sunshine around over tens of thousands of years, small changes in the brightness of the sun, shifts in ocean circulation, the very slow drifting of continents—but naturally caused changes in greenhouse gases and especially carbon dioxide have been very important over Earth's climate history. When warming has occurred, ice has melted and sea level has risen. Careful study of these natural changes and their causes contributes to the strong knowledge that the changes now occurring are not primarily the continuation of some natural cycle, but are instead caused primarily by the human-driven rise in greenhouse gases, mostly carbon dioxide.

Past warming and cooling in response to changing greenhouse-gas concentrations and other forcings including features of Earth's orbit have caused much larger past changes in ice and sea level than have occurred in the last century or that are projected for the next century. But, human burning of fossil fuels could cause a climate change that rivals or exceeds those of the past in combined size and speed (White et al., 2010), raising questions about the future that are discussed next.

Looking to the Future

Some additional sea-level rise is already committed, under the future emissions pathways considered by the IPCC. Just as an ice cube placed in a glass of tea takes a while to melt, the warming of the ocean and mass loss from land ice have not yet "caught up" with the warming that has already been caused, and would continue to contribute to some additional sea-level rise even if temperature were stabilized today. Looking toward the time when today's students are old, toward the end of this century and beyond, future human decisions become increasingly important and then dominant in controlling projected sea-level rise (e.g., Church et al., 2013, Fig. 13.27). Under strong warming (RCP8.5), projected sea- level rise in 2100 compared to preindustrial is slightly more than 3 feet, with an uncertainty that the IPCC considered likely to be no more than about 1 foot (details and more-precise numbers available in the source).

One occasionally hears the unsubstantiated claim that the IPCC projections are overly alarmist. In fact, several studies have suggested that, if the IPCC is open to criticism on this subject, it is overly conservative. Rahmstorf et al. (2007), for example, compared observed sea-level rise at the time to projections from earlier IPCC work, and found that the sea was rising faster than the most-likely projection and near the projected likely upper limit. The Third Assessment Report of the IPCC projected a most-likely future in which climate change would cause growth of the Antarctic ice sheet over this century (Church et al., 2001), but Antarctic ice is shrinking (The IMBIE Team, 2018); more-recent central projections from the IPCC indicate less ice loss from Antarctica than observed (Slater and Shepherd, 2018). Garner et al. (2018) compared the history of IPCC projections to other science-based projections, many of which were available to the IPCC assessment teams; Figure 2a in Garner et al. (2018) shows no tendency for the IPCC projections to be higher than in the underlying literature, and some tendency for the IPCC projections to be lower. The new expert elicitation from Bamber et al. (2019) includes information obtained since the IPCC Fifth Assessment Report (Church et al., 2013), and yields potential for higher sea-level rise than generally projected. Quoting from the Abstract, with English units added:

> "For a +2°C (+3.6 °F) temperature scenario consistent with the Paris Agreement, we obtain a median estimate of a 26 cm (0.85 ft) SLR contribution by 2100, with a 95th percentile value of 81 cm (2.7 ft). For a +5°C (+9°F) temperature scenario more consistent with unchecked emissions growth, the corresponding values are 51 and 178 cm (1.7 and 5.8 ft), respectively. Inclusion of thermal expansion and glacier contributions results in a global total SLR estimate that exceeds 2 m (6.6 ft) at the 95th percentile. Our findings support the use of scenarios of 21st century global total SLR exceeding 2 m (6.6 ft) for planning purposes. Beyond 2100, uncertainty and projected SLR increase rapidly. The 95th percentile ice sheet contribution by 2200, for the +5°C (+9°F) scenario, is 7.5 m (24.6 ft) as a result of instabilities coming into play in both West and East Antarctica. Introducing process correlations and tail dependences increases estimates by roughly 15%."

Focus on Ice-Sheet Changes

The large ice sheets of Greenland and Antarctica are of special interest, because they are so big and thus could affect sea level so much. Expansion of the ocean as it warms gives just under 1 foot of rise per degree F of warming (0.4 m/1°C) (Levermann et al., 2013), but the ~1000-year time for heat to mix into the ocean makes the resulting sea-level rise relatively slow. Melting of all of the world's mountain glaciers and small ice caps might raise sea level by about 1 foot (0.3 m), but melting of the great ice sheets would raise sea level by just over 200 feet (more than 60 m), with ~23 feet from Greenland (7.3 m) and the rest from Antarctica (Lemke et al., 2007). We do not expect to see melting of most of that ice over the next century or centuries, but even a relatively small change in the ice sheets could matter to the world's coasts; roughly 10% of the world's population lives within 33 feet (10 m) of sea level (McGranahan et al., 2007). I thus next consider the possibility of rapid changes (National Research Council, 2002; 2013; CCSP, 2008; 2009).

Ice-Sheet Behavior

An ice-sheet is a two-mile-thick, continent-wide pile of snow that has been squeezed to ice under the weight of more snowfall. (For a more-detailed background, see Cuffey and Paterson, 2010.) Growth or shrinkage of an ice sheet depends on the balance between snowfall and melting on top, and on flow taking accumulated ice to lower elevation, often to the ocean to make icebergs.

The balance between snowfall and melting is of great importance. I will not treat this aspect in detail, however; additional information is available from the IPCC reports cited herein. Generally, warming of subfreezing conditions is expected to increase snowfall because warmer air can carry more water, although additional processes are important. For Greenland and mountain glaciers, the increase in melting from warming is expected to exceed the increase in snowfall. For Antarctica, fairly large warming would be required for mass loss by surface melting to become important, so the surface is expected to gain mass with moderate warming; however, flow increases have recently exceeded changing snowfall leading to mass loss, and this is expected to continue. I thus focus next on the flow characteristics. A review of much of this information is available in Alley et al. (2015) and Scambos et al. (2017) as well as in the IPCC reports and in Cuffey and Paterson (2010). The literature is large and increasing rapidly.

All piles, including glaciers and ice sheets, tend to spread under their own weight, restrained by their own strength (which is why spilled coffee spreads on a table top but the stronger table beneath does not spread), by friction beneath (so pancake batter spreads faster on a greased griddle than on a dry waffle iron), or by "buttressing" from the sides (so an appropriately placed spatula will slow the spreading of the pancake batter).

Some early gothic cathedrals suffered from the "spreading-pile" problem, in which the sides tended to bulge out while the roof sagged down, with potentially unpleasant consequences. The beautiful solution was the flying buttress, which transfers some of the spreading tendency to the strong earth beyond the cathedral. Ice sheets also have "flying buttresses", called ice shelves. In the coldest regions, the ice reaching the ocean usually does not immediately break off to form icebergs, but remains attached to the ice sheet while spreading over the ocean. The friction of these ice shelves with local high spots in the sea floor, or with the sides of embayments, helps restrain the spreading of the ice sheet much as a flying buttress supports a cathedral. The ice shelves are at the melting point where they contact water below, and are relatively low in elevation hence warm above. Ice shelves thus are much more easily affected by climatic warming than are the thick, cold central regions of ice sheets. Rapid melting or collapse of several ice shelves has occurred recently, in response to both atmospheric warming and to intrusion of warmer ocean water, allowing the "gothic cathedrals" behind to spread faster, contributing to sea-level rise. Many additional ice shelves remain that have not changed notably, and these contribute to buttressing of much more ice than was supported by those ice shelves that experienced the large recent changes, so the potential for similar changes contributing to sea-level rise in the future is large.

There are no large ice shelves fed by glaciers in warmer areas, including Alaska, southern Greenland, and elsewhere. Warming of air or water beyond some threshold leads to ice-shelf loss, leaving a calving cliff of the sort that tourists visit on Alaskan cruises or that produce Greenland icebergs that have been filmed and viewed tens of millions of times on video- sharing sites.

Many tourists on Alaskan cruises visit Glacier Bay National Park and Preserve, Alaska. When George Vancouver visited the area in 1794, the bay, which is now about 65 miles long, was completely filled with ice. Less than a century later, when John Muir studied there, most of the bay had lost its ice, as icebergs broke off the front of the glacier at a rate of up to 7 miles per year (Meier and Post, 1987; Post, 1975), and the ice thinned by as much as 1 mile (Larsen et al., 2005). As described by Meier and Post (1987) and many other researchers, ice that ends in the ocean often stabilizes on a local high point or a narrowing of a fjord. It then may remain there for some time despite small climate changes. Under sufficiently large forcing, though, it may retreat rapidly through the deeper or wider part of its valley to the next point of stability, losing ice at a rate that tends to increase with water depth and valley width, but that also depends on other controls. Such behavior has been observed from other Alaskan glaciers, in Greenland, and along the Antarctic Peninsula and elsewhere, and geological evidence indicates that such processes contributed to loss of older ice masses during past, natural warmings.

For Glacier Bay, and many others, the glacier flowed in a relatively narrow fjord. When the ice retreated rapidly from one point of stability to the next, the local changes were spectacular, but the global significance was relatively small. Even in Greenland, the beds of the deep fjords rise inland to near or above sea level, and rapid iceberg calving in deep water cannot discharge most of the ice sheet. Some parts of the Antarctic ice sheet, however, have glaciers that drain large basins rather than narrow fjords, but with ice that is much too thick to float and thus that can raise sea level. Attention is especially, but not uniquely, focused on Thwaites Glacier in West Antarctica, where sufficient retreat could trigger loss of the marine ice of West Antarctica that would add about 11 feet to the total sea-level rise (Scambos et al., 2017). One attempt to put this process into a model and simulate the future found that, once rapid retreat was initiated, most of West Antarctica's marine ice would be lost over the next century or so (DeConto and Pollard, 2016). That model, though, did not allow icebergs to calve faster than a rate that was based on previous observations in Greenland; Thwaites could retreat to regions where the water is much deeper and the valley much wider than in Greenland, and the tendency for deeper water and wider valleys to have faster calving then allows the possibility of even faster sea-level rise.

The uncertainties about this topic are very large. Thwaites Glacier, and other Antarctic outlets, may resist retreat, or they may retreat with intact ice shelves that limit iceberg calving and thus slow the resulting sea-level rise, and additional stabilizing influences may resist collapse. But, Thwaites may retreat rapidly, and East Antarctic ice may join, with even greater potential to cause rise than has been modeled by, e.g., DeConto and Pollard (2016). The basic physical processes involved have been known for a long time, but properly modeling them is at a very early stage, subject to deep uncertainty.

Uncertainty and Commuting

I am fortunate to ride my bicycle or jog to work; my wife and I share one car, which she mostly drives. But, I have on occasion driven to Washington, DC, and observed automobile commuting. In my experience, a typical commuter on the Beltway heading downtown at rush hour expects notable delays—say, half an hour. The best commute has no delays. Some commuters may waste an hour in traffic. But, occasionally, a commuter is run over by a drunk driver, and ends up in the hospital or worse. When a commuter sets off on a trip, the most-likely future (half an hour in traffic) is close to the "good" end of the possible futures (no traffic, versus run over by drunk driver).

The uncertainties about sea-level rise have a vaguely similar distribution. With warming, ocean water expands, mountain glaciers and the edges of Greenland's ice sheet melt more, and the ocean rises in response. The relevant science, as assessed by the IPCC or other authoritative bodies, gives very high confidence that some additional sea-level rise is already committed as the ice and ocean "catch up" with the warming already caused. Furthermore, additional warming will cause more sea-level rise. The uncertainties include a little less or a little more sea-level rise. But, those projections do not include major instability of Antarctic ice.

If "collapse" is triggered in West or East Antarctica, a natural "drunk driver", then the sea-level rise could be much larger and faster than in the most-likely projections, with no real agreement on what might be the worst-case scenario.

Dedicated police officers, watchful bartenders, and other community members help reduce drunk driving, while automotive engineers, highway designers, and additional experts help improve safety in the event of an accident. Society thus has considerable knowledge of ways to reduce risks to commuters. But, we still cannot predict the where and when of every drunk driver. Please note that my next statement is potentially self-serving, because colleagues, students, and I personally enjoy conducting research and may receive funding to do research; but, further research can greatly reduce uncertainties on ice sheets and sea-level rise, providing guidance to policy-makers. Nonetheless, even with well-supported, vigorous research, including ongoing efforts by exceptional colleagues, there may remain some irreducible uncertainty.

Synopsis

With high scientific confidence, human actions are raising the greenhouse-gas concentration of the atmosphere especially by releasing carbon dioxide from fossil-fuel burning, this is having a warming influence on the climate, and the resulting rise in temperature is contributing to sea-level rise by expanding ocean water, melting ice, and changing ice flow. Some additional sea-level rise is already committed, but human decisions will become increasingly important in determining how much the sea rises over the coming decades and centuries. If Antarctic ice avoids rapid collapse, then uncertainties in projections of sea-level rise are important but not huge, are being addressed by ongoing research, and can be reduced further by planned research. Rapid iceberg calving under too much warming is a well-known but poorly modeled process; if this becomes active in large, deep Antarctic basins, then sea-level rise could be much larger than generally projected, with much greater uncertainties.

References

Alley, R. B., Anandakrishnan, S., Christianson, K., Horgan, H. J., Muto, A., Parizek, B. R., Pollard, D. & Walker, R. T. (2015). Oceanic forcing of ice-sheet retreat: West Antarctica and more. *Annual Reviews of Earth and Planetary Sciences, 43*, 207–231.

American Institute of Physics. (2008). *The Discovery of Global Warming*, by Spencer Weart, http://www.aip.org/history/climate.

Bamber, J. L., Oppenheinmer, M., Kopp, R. E., Aspinall, W. P. & Cooke, R. M. (2019). Ice sheet contributions to future sea-level rise from structured expert judgement. *Proceedings of the National Academy of Sciences of the United States of America, 116*(23), 11195-11200.

CCSP. (2008). *Abrupt Climate Change*. A report by the U.S. Climate Change Science Program and the Subcommittee on Global Change Research [Clark, P.U., A.J. Weaver (coordinating lead authors), E. Brook, E.R. Cook, T.L. Delworth, and K. Steffen (chapter lead authors)]. U.S. Geological Survey, Reston, VA, 244 pp.

CCSP. (2009). *Past Climate Variability and Change in the Arctic and at High Latitude.* A report by the U.S. Climate Change Program and Subcommittee on Global Change Research [Alley, R.B., Brigham-Grette, J., Miller, G.H., Polyak, L., and White, J.W.C. (coordinating lead authors). U.S. Geological Survey, Reston, VA, 461 pp.

Church, J. A., Gregory, J. M., Huybrechts, P., Kuhn, M., Lambeck, C., Nhuan, M. T., Qin, D. & Woodworth, P. L. (2001). Changes in sea level. in: J.T Houghton, Y. Ding, D.J.Griggs, M. Noguer, P.J. Van der Linden, X. Dai, K. Maskell, and C.A. Johnson(eds.): *Climate Change 2001: The Scientific Basis: Contribution of Working Group Ito the Third Assessment Report of the Intergovernmental Panel on Climate Change*, Cambridge University Press (Cambridge, New York), 639-694.

Church, J. A., Clark, P. U., Cazenave, A., Gregory, J. M., Jevrejeva, S., Levermann, A., Merrifield, M. A., Milne, G. A., Nerem, R. S., Nunn, P. D., Payne, A. J., Pfeffer, W. T., Stammer, D. & Unnikrishnan, A. S. (2013). Sea Level Change. In: *Climate Change 2013: The Physical*

Science Basis. Contribution of Working Group I to the Fifth Assessment Report of the Intergovernmental Panel on Climate Change [Stocker, T.F., D. Qin, G.-K. Plattner, M. Tignor, S.K. Allen, J. Boschung, A. Nauels, Y. Xia, V. Bex and P.M. Midgley (eds.)]. Cambridge University Press, Cambridge, United Kingdom and New York, NY, USA.

Cuffey, K. M. & Paterson, W. S. B. (2010). *The Physics of Glaciers, Fourth edition.* Oxford. DeConto, R.M and D. Pollard. 2016. Contribution of Antarctica to past and future sea-level rise. *Nature, 531*(7596), 591–597. doi:10.1038/nature17145.

Garner, A. J., Weiss, J. L., Parris, A., Kopp, R. E., Horton, R. M., Overpeck, J. T. & Horton, B. P. (2018). Evolution of 21st century sea level rise projections. *Earth's Future*, 6, 1603–1615.

Griggs, J. A. & Harries, J. E. (2007). Comparison of spectrally resolved outgoing longwave radiation over the tropical Pacific between 1970 and 2003 using IRIS, IMG, and AIRS. *Journal of Climate*, *20*, 3982–4001.

Harries, J. E., Brindley, H. E., Sagoo, P. J. & Bantges, R. J. (2001). Increases in greenhouse forcing inferred from the outgoing longwave radiation spectra of the Earth in 1970 and 1997. *Nature*, *410*, 355–57.

The IMBIE Team. (2018). Mass balance of the Antarctic ice sheet from 1992 to 2017. *Nature*, *558*, 219-222. doi: 10.1038/s41586-018-0179-y.

Larsen, C. F., Motyka, R. J., Freymuller, J. T., Echelmeyer, K. A. & Ivins, E. R. (2005). Rapid viscoelastic uplift in southeast Alaska caused by post-Little Ice Age glacial retreat. *Earth and Planetary Science Letters*, *237*(3-4), 548-560, doi: 10.1016/j.epsl.2005.06.032.

Lemke, P., Ren, J., Alley, R. B., Allison, I., Carrasco, J., Flato, G., Fujii, Y., Kaser, G., Mote, P., Thomas, R. H. & Zhang, T. (2007). Observations: Changes in Snow, Ice and Frozen Ground, in Solomon, S., D. Qin, M. Manning, Z. Chen, M. Marquis, K. B. Averyt, M. Tignor and H. L. Miller [eds.], *Climate Change 2007: The Physical Science Basis. Contribution of Working Group I to the Fourth Assessment Report of the Intergovernmental Panel on Climate Change* [Cambridge University Press, New York.

Levermann, A., Clark, P. U., Marzeion, B., Milne, G. A., Pollard, D., Radic, V. & Robinson, A. (2013). The multimillennial sea-level commitment of global warming. *Proceedings of the National Academy of Sciences of the United States of America*, *110*(34), 13745-13750, https://doi.org/10.1073/pnas.1219414110.

Masson-Delmotte, V., Schulz, M., Abe-Ouchi, A., Beer, J., Ganopolski, A., González Rouco, J. F., Jansen, E., Lambeck, K., Luterbacher, J., Naish, T., Osborn, T., Otto-Bliesner, B., Quinn, T., Ramesh, R., Rojas, M., Shao, X. & Timmermann, A. (2013). Information from Paleoclimate Archives. In: *Climate Change 2013: The Physical Science Basis. Contribution of Working Group I to the Fifth Assessment Report of the Intergovernmental Panel on Climate Change* [Stocker, T.F., D. Qin, G.-K. Plattner, M. Tignor, S.K. Allen, J. Boschung, A. Nauels, Y. Xia, V. Bex and P.M. Midgley (eds.)]. Cambridge University Press, Cambridge, United Kingdom and New York, NY, USA.

McGranahan, G., Balk, D. & Anderson, B. (2007). The rising tide: Assessing the risks of climate change and human settlements in low elevation coastal zones. *Environment and Urbanization*, *19*, 17–37.

Meier, M. F. & Post, A. (1987). Fast tidewater glaciers. *Journal of Geophysical Research*, *92*(B9), 9051-9058.

National Research Council, National Academy of Sciences, National Academies Press, Washington, DC:

1975: United States Committee for the Global Atmospheric Research Program, *Understanding Climate Change: A Program for Action*

1979: Ad Hoc Study Group on Carbon Dioxide and Climate, *Carbon Dioxide and Climate: A Scientific Assessment*

2001: Committee on the Science of Climate Change, *Climate Change Science: An Analysis of Some Key Questions*

2002: Committee on Abrupt Climate Change, *Abrupt Climate Change: Inevitable Surprises*

2006: Committee on Surface Temperature Reconstructions for the Last 2,000 Years, *Surface Temperature Reconstructions for the Last 2,000 Years*

2008: *Understanding and Responding to Climate Change: Highlights of National Academies Reports*
2010a: Committee on Stabilization Targets for Atmospheric Greenhouse Gas Concentrations, *Climate Stabilization Targets: Emissions, Concentrations, and Impacts over Decades to Millennia*
2010b: Panel on Advancing the Science of Climate Change, *Advancing the Science of Climate Change*
2013: Committee on Understanding and Monitoring Abrupt Climate Change and Its Impacts, *Abrupt Impacts of Climate Change: Anticipating Surprises*

Nerem, R. S., Beckley, B. D., Fasullo, J. T., Hamlington, B. D., Masters, D. & Mitchum, G. T. (2018). Climate-change-driven accelerated sea-level rise detected in the altimeter era. *Proceedings of the National Academy of Sciences of the United States of America*, *115*(9), 2022-2025.

Post, A. (1975). *Preliminary hydrology and historic terminal changes of Columbia Glacier, Alaska.* U.S. Geological Survey Hydrological Investigations Atlas.

Rahmstorf, S., Cazenave, A., Church, J. A., Hansen, J. E., Keeling, R. F., Parker, D. E. & Somerville, R. C. J. (2007). Recent climate observations compared to projections. *Science*, *316*, p. 709.

Scambos, T., Bell, R. E., Alley, R. B., Anandakrishnan, S., Bromwich, D. H., Brunt, K., Christianson, K., Creyts, T., Das, S., DeConto, R., Dutrieux, P., Fricker, H. A., Holland, D., MacGregor, J., Medley, B., Nicholas, J. P., Pollard, D., Siegfried, M. R., Smith, A. M., Steig, E. J., Trusel, L. D., Vaughan, D. G. & Yager, P. L. (2017). How Much, How Fast?: A Review and Science Plan for Research on the Instability of Antarctica's Thwaites Glacier in the 21st century. *Global and Planetary Change*, *153*, 16-34.

Slater, T. & Shepherd, A. (2018). Antarctic ice losses tracking high. *Nature Climate Change*, *8*, 1024-1026.

White, J. W. C., Alley, R. B., Brigham-Grette, J., Fitzpatrick, J. J., Jennings, A. E., Johnsen, S. J., Miller, G. H., Nerem, R. S. & Polyak, L. (2010). Past rates of climate change in the Arctic. *Quaternary Science Reviews*, *29*(15-16), 1716-1727.

Biography

Dr. Richard Alley (PhD 1987 University of Wisconsin; MSc 1983 and 1980 The Ohio State University, all in Geology) is Evan Pugh University Professor of Geosciences and Associate of the Earth and Environmental Systems Institute at the Pennsylvania State University. He has spent more than 40 years studying the great ice sheets to help predict future changes in climate and sea level, and has made four trips to Antarctica, nine to Greenland, and additional expeditions to Alaska and elsewhere. He has been honored for research (including election to the US National Academy of Sciences and Foreign Membership in the Royal Society), teaching (including Penn State's highest teaching award), and service. Dr. Alley participated in the UN Intergovernmental Panel on Climate Change (co-recipient of the 2007 Nobel Peace Prize), and has provided requested advice to numerous government officials in multiple administrations including a US Vice President, Presidental Science Advisors, and committees and individual members of the US Senate and House of Representatives. He has authored or coauthored more than 300 refereed scientific papers. He was presenter for the PBS TV miniseries on climate and energy *Earth: The Operators' Manual*, and author of the book. His popular account of climate change and ice cores, *The Two-Mile Time Machine*, was Phi Beta Kappa's science book of the year. Dr. Alley is happily married with two grown daughters, two stay-at-home cats, a bicycle, and a pair of soccer cleats.

Dr. Richard B. Alley
Evan Pugh University Professor
Department of Geosciences, and Earth and Environmental Systems Institute, The Pennsylvania State University.

Chairwoman Johnson. Thank you very much. Dr. Bell.

TESTIMONY OF DR. ROBIN E. BELL, LAMONT RESEARCH PROFESSOR, LAMONT-DOHERTY EARTH OBSERVATORY, COLUMBIA UNIVERSITY

Dr. BELL. Thank you very much. Chairwoman Johnson, Ranking Member Lucas, Members of the Committee, I'm very pleased to be here today. I'm going to take you on a visual tour because the ice sheets are beautiful. I think that's why we all study them. And we want to share a little bit of that beauty with you, so this is a picture of what Antarctica looks like. And just to give you a sense of scale, this is a huge iceberg with tiny scientists in front of it.

What I'm going to show you today is the evidence for change. I will tell anyone this who stops me anywhere on the street—you stop me on my electric motorcycle, I'm going to tell you this story. There are three signs of change—three really clear signs that the ice sheets are speeding up and changing. One is they're moving faster. In the 1990s, they were moving 1 mile a year. In the 2000s, they're moving 2 miles a year. They've doubled in speed.

Ice happens to be like the mozzarella cheese on top of your pizza, so when you bite into the cheese and stretch it, it gets thinner. So the second measurements we've made is by zapping the ice sheet with a laser, and that's what you see forming is that yellow on the surface is actually where the elevation, just like the cheese is getting stretched, the ice sheet is getting stretched, that's more than half a football field of stretching where the ice sheet is getting lower, second measurement.

Our third measurement is one we make and NASA makes with partners—makes from space. Can we turn the video on—animation on, please? You will see that this is Antarctica again—now we're looking at a whole map of Antarctica, and you're going to see a red dot develop. And what that red dot is showing we're actually losing mass. And remember I showed you it sped up, it lowered. This is a different measurement. This is basically the ice sheet on the bathroom scale. And what you can see is the ice sheet is losing mass predominantly in that place that Richard referred to, the place that's furthest north and exposed to the warming ocean. The ice sheet is losing mass.

We could show you the same things for Greenland, three very clear signals, kind of the scientific gold standard. We like to make independent measurements. This is the evidence that the ice sheets are changing.

What does it mean? We go next to NOAA (National Oceanic and Atmospheric Administration) and we look at NOAA's global collection of tide gauges—so these are really high-tech instruments. They're like pipes stuck in the water, OK? But they measure the tides going up and down and up and down, and they measure storms—the tide levels go way up, 12 feet in New York during Sandy. But you can see most of those are going up. Sea level almost everywhere on the planet is going up except where the planet is still recovering from the ice sheet that was more than 20,000 years ago and it's bouncing back up like a mattress. But this predominant signal globally is it's going up.

There's even one of those fancy pipes right here in Southeast, Washington, and that record goes back to about when my dad was born. And so since my dad was born right here in Washington, D.C., sea level has gone up a foot. And we're using Beth for scale here. Beth is, for today's purposes, 2 meters or 6 feet roughly. And you can see, sea level has risen almost a foot, almost to my knee— I like to think of it—I put my hand on my leg because then I realize what it really means. That's how far sea-level has come up since my dad was born.

So what does that mean? We are working on this problem—I'm back to the uncertainty question. Can I tell you how much sea-level is going to go up in the next hundred years? We are working on it as hard as we can. This is just a range of forecasts published this year. You can see the results—they are spread. This is again Beth for scale, about 6 feet. They range a lot. But when we looked— that's what we're working on is how to be able to tell our communities how much is sea level to go up in the next hundred years because that's what we're building infrastructure. The big bridge we just spent $4 billion on close to my house needs to know what we're going to plan for sea level. Are we going to plan for a couple feet or a lot more?

So when we look at the glacier melt budget altogether, Antarctica is in the next hundred years is on the order of maybe over our knees, maybe a little bit more. Greenland is going to be in there, too. We're going to have warming oceans, and we're going to have mountain glaciers. And while I have this as roughly 4 feet, 3 feet, we don't know. This is cutting-edge research.

And what can we do to improve it? There's a priority of three ideas in my mind—there's three important things to do. One is get up close and personal to the ice sheets. We need to understand better how the ice sheets work so we can improve our models. We used to not be able to have very good models of weather. We do much better now. So number one is get up close and personal.

Second is we need to invest in the workforce. Right now, there are 1,400 scientists at the AGU who are affiliated with ice. Do you know there are 140,000 people enrolled in law school every year? We just don't have enough people working on this. We need more scientists, engineers, educators, creative minds like Maria over there. We need to talk her into studying ice somehow.

And we also need to look at how convergent science works. We need to figure out how to pull together the work that we do, which is on the polar caps, to what's happening at the coastlines around the planet because we kind of need an ice sheet person in every community because we need to understand what the community needs to respond to.

So am I hopeful? Yes, I am hopeful because we are in a unique place as a species that we know how the ice sheet works, we know how sea level rises—we are understanding how our planet works. And we, as scientists, we're all members of the American Geophysical Union. We're actually putting our money where our mouth is. We have a building here in Washington that we just renovated, so it is the first net-zero renovation building in Washington. That means we're taking less energy than we are generating, more energy than we are using to run this building. We'd love to have you come visit. And we're also very happy to look forward and that this is a time for action among all of us, and we need to bring everybody to the table. Thank you very much.

[The prepared statement of Dr. Bell follows:]

Written Testimony of Professor Robin Elizabeth Bell, PGI Lamont Research Professor, Lamont-Doherty Earth Observatory of Columbia University, President American Geophysical Union. Earth's Thermometers: Glacial and Ice Sheet Melt in a Changing Climate For Committee on Science Space and Technology U.S. House of Representatives. July 11, 2019

Introduction

Chairwoman Johnson, Ranking Member Lucas, Members of the House Science Committee, thank you for inviting me and my esteemed colleagues to speak today on this important topic- the melting ice on our home planet. This change is happening at the ends of our planet but is lapping at our doorsteps now. I am Robin Elizabeth Bell, and I am the PGI Lamont Research Professor at Lamont-Doherty Earth Observatory of Columbia University and a member of the Earth Institute faculty. At Lamont, I direct programs in ice sheet dynamics, lead efforts to develop innovative technology and work to improve the scientific culture especially for women. I have led ten major expeditions to the polar regions, to both Greenland and Antarctica resulting in discoveries ranging from active volcanism beneath the West Antarctic Ice sheet, to large deep lakes encased by two miles of ice to hidden mountain ranges buried by ice where water under the pressure of thick ice is forced uphill. I was the first women to chair the National Academy of Science's Polar Research Board (2002-2008) where I was instrumental in launching the International Polar Year 2007-9 that brought together over 50,000 scientists from around the globe. The International Polar Year fostered major expeditions, new international collaborations and discoveries that were only possible because of the partnerships between 60 nations. The polar regions are still a challenging place to work and science remains an international team sport. I also co-chaired the recent National Academy report *A Strategic Vision for NSF Investments in Antarctic and Southern Ocean Research* (2015) that set the priorities for Antarctic science. This report identified changing ice as the highest priority science in Antarctica. Recently, I chaired the National Academy of Science Review of the Draft Fourth National Climate Assessment – a comprehensive undertaking by U.S. scientists and citizens documenting the impacts of climate change around the country (USGCRP 2018). Currently, I have the great honor of serving as the President of the American Geophysical Union, or AGU as we all refer to it. Formed 100 years ago, AGU is the society of over 60,000 Earth and space scientists from around the globe who together promote discovery for the benefit of humanity. Today, I am honored to speak about the changing ice sheets with you. Thirty years ago when I first flew over Antarctica in a Naval Research Laboratory P-3 it seemed unimaginable to me that the vast ice sheet below could change. Now we know those white expanses are changing and these changes matter to our homes and communities around the globe. The changing polar ice is tightly linked to the changing coastlines. Although I speak to you today in my capacity as a private citizen, my testimony is based on my decades of experience studying our planet's ice.

Evidence for Changing Ice

The surprising wakeup call for the polar science community came in early 2002. This buzzing alarm came from the Antarctic Peninsula, the part of Antarctica that is the furthest north, jutting towards South America. This is also the destination of Antarctic cruises which over 14,000 Americans visit each year The Antarctic Peninsula is where global temperatures have risen the most - more than 7°F over 50 years. We had thought ice sheets and the state-sized ¼-1/2 mile thick pieces of floating ice that pin them in place changed really slowly. These floating extensions of continental ice sheets are called ice shelves. But by 2002, warming temperatures had started to produce more meltwater on top of the ice. The floating Larsen B Ice Shelf, the size of Rhode Island, developed hundreds of lakes. Suddenly the ice shelf disintegrated into thousands of icebergs over the course of two weeks (Scambos, Hulbe et al. 2003). The change occurred before our very eyes. The Larsen B ice shelf had been in place for over 10,000 years (Domack, Duran et al. 2005). Once the floating ice shelf disintegrated, the glaciers that flowed into the ice shelf sped up, pushing more ice into the ocean (Rignot, Casassa et al. 2004, Scambos, Bohlander et al. 2004). Glaciers are the earth's conveyor belts delivering ice to the ocean and an ice shelf controls the speed– if an ice shelf collapses, the conveyor belt speeds up. The satellite images of this collapse were printed in major newspapers around the globe (for example: The Charleston Gazette, St Louis Post, The Gazette – Ft. Wayne Indiana, The Patriot – Harrisburg, PA, Chicago Tribune, Rocky Mountain News, The Economist, The Wall Street Journal, The New York Times, Toronto Star, Calgary Herald, The Press - New Zealand, Belfast Telegraph, The Australian, China Daily, The Statesman- India). Suddenly, changing ice was newsworthy. Together scientists and the public from Harrisburg to India learned Antarctic ice could change faster than we imagined. The Antarctic conveyor belt had sped up. For the first time many around the world saw the link between blue meltwater on the ice shelf surface, the glacier conveyor belt speeding up and sea level rising.

Over the ensuing decades, the evidence for the changing ice on our planet has become very clear. I will focus on the grounded ice, the large ice sheets in Antarctica and Greenland where thick ice, in places over two miles thick, rests on solid ground although the ground may be well below sea level. Melting these ice sheets will raise sea level around the globe. Antarctica holds 200 feet of potential sea level rise and Greenland 20 feet of sea level rise– although no scientists are suggesting they will completely disappear any time soon. These very thick ice sheets are distinct from the relatively thin floating sea ice (around 10 feet) that covers much of the Arctic Ocean and rings the Southern Ocean close to Antarctica. Sea ice is like the layer of ice cubes floating in a punch bowl. The Arctic sea ice has been steadily shrinking over the past two decades and recently the Antarctic sea ice has begun to retreat. Changing sea ice shifts the Earth's albedo and weather patterns, impacts food available to wildlife from penguins to polar bears, and opens new shipping routes. But shrinking sea ice itself will not cause sea level to rise, since sea ice is already floating in the water. The major source of future sea level rise are the grounded ice sheets. Melting ice sheets are like the kid with a new twenty-pound bag of ice at the picnic who pours the entire bag into the bowl without thinking. The glaciers are conveyor belts of ice being delivered to the ocean, and we see them speeding up.

I often get asked "do you believe the ice is changing?" My response is – changing ice is not a belief but knowledge that emerges from three independent observations. These independent observations are primarily based on satellite measurements enabled by NASA working with other space agencies around the globe. The first measurement is how fast the ice moves. Several parts of the Antarctic Ice Sheet (key parts of the conveyor belts) have doubled their speed in the past two decades, showing that the ice is speeding up (Rignot, Mouginot et al. 2011). The second measurement is the height of the ice surface, and is made using laser and radar instruments from a satellite or aircraft. In the same places where the ice is speeding up the ice surface is getting lower. Ice, like mozzarella cheese atop a pizza, is getting thinner and lower because it is stretching. The third measurement is ice sheet mass, or weight, which is calculated from observations from a pair of identical satellites chasing each other and measuring changes in the gravity field (Velicogna, Sutterley et al. 2014, Harig and Simons 2015). In the same places that the ice is speeding up and lowering, it is losing mass. These three measurements together demonstrate in more detail than ever before how the ice in Greenland and Antarctica is changing.

Scientists from around the globe have used these three key observations to quantify how fast the ice sheets are changing. To quantify the change over a large continent like Antarctica, the size of the lower 48 states, requires careful examination of each measurement and resolving issues such as how the snow that falls on Antarctica turns into ice. After much lively debate and testing of assumptions by a team of 77 scientists from around the world, the clear signal is that Antarctica is losing ice, as is Greenland. The current mass loss from the ice sheets is contributing one millimeter of sea level rise globally each year (Shepherd, Ivins et al. 2018) although this rise is not evenly distributed around the globe. Antarctica is now losing mass at twice the rate it was in the 1990s. For these calculations, the team broke Antarctica up into three parts, the Peninsula where the Larsen B Ice Shelf was; West Antarctica, the ice sheet that rests on low-lying topography and is exposed to changes in the ocean temperature and East Antarctica, the large ice sheet where the South Pole is that sits on higher topography. Each region stores different amounts of ice, has a different history and a different susceptibility to a warming world. The West Antarctic Ice Sheet is the most susceptible to warming oceans and atmosphere as it sits lower and is in direct contact with the ocean. West Antarctica is where the greatest changes have occurred over the past decade. Most of the 0.3 inches (8 mm) of sea level rise from Antarctica in the last decade has come from West Antarctica. This region was the highest priority in the 2015 National Academy report *A Strategic Vision for NSF Investments in Antarctic and Southern Ocean (National Academies of Sciences and Medicine 2015).*

Evidence for Changing Coastlines

Why did the changing ice emerge as the highest priority in the National Academy Report? We are beginning to see the melting ice, including from Antarctica, at the tide gauges along our coastline. Globally, average sea level has risen 8-9 inches since 1880, with the global rise since 1993 being 3 inches (Hay, Morrow et al. 2015, Nerem, Beckley et al. 2018). Right here at the dock along the bike path in Southeast Washington sea level has risen a foot in since 1919 (https://tidesandcurrents.noaa.gov/sltrends). I put my hand on my leg just below my knee and realize the water level has risen that far since my father was born.

At most locations around the globe sea level is rising now, although the ocean turns out to be more complicated than the punch in the punchbowl. At a few locations, sea level is actually falling. Three major components make up the change at an individual coastal city: the change in ocean temperature, the melting ice and whether the land the city rests on is rising or sinking. Up to now the warming of the ocean waters by 1.3°F since 1960 is the major signal that has appeared at our coasts. But, melting ice has the greatest potential for new rapid sea level rise globally. To complicate things further, the melting ice contribution to changes sea level is modulated by the self-gravitation of the ice sheets. Already the modulation of the impact of melting ice in Greenland by the self-gravitation is apparent in the tide gauges along the east coast of the United States. Because of this gravitational effect, sea level is rising faster in the Southeastern US than in New England. Atop these signals are local impacts. The land cities and towns are built on can be rising or sinking, impacting local sea level. In cities like Juneau and Stockholm (Milne, Davis et al. 2001) the land is rising due to the loss of ice 20,000 years ago while cities like Norfolk, Virginia and New Orleans are sinking due to removal of groundwater (Sweet, Kopp et al. 2017). Every community is going to see a different future sea level depending the ocean temperature, the changing ice, and whether the land is rising or falling. Linking the changing ice to the changing coastlines is a challenge that will require collaboration from the ice to the shorelines.

Impacts of Changing Coastlines Now and Looking Ahead

So we have begun to witness the melting ice and see the impact along our shorelines. The higher sea level made the impact of recent major storms like Maria, Harvey, Irma and Sandy more devastating. For example, close to my home 30 miles from the Atlantic Ocean they used bulldozers to clear boats from the roads after Superstorm Sandy. Because of the sea level rise over the past century, 45,000 more people were impacted by Sandy's flooding. The impact of rising sea level is not just during major storms. All around the US we are seeing increased nuisance flooding. Nuisance flooding is called sunny day flooding where high tides in fair weather make it difficult to get home because the roads are flooded. Miami and Norfolk are both experiencing this and are working to adapt to this.

Scientists are working to provide these cities with the forecasts of future sea level they need to adapt.

Looking Ahead: Current Ice Sheet Change Projections

Looking ahead the scientific community is scrambling to provide answers to how fast and how much will sea level rise in each community from ice sheet melt. Suddenly city managers, architects, reinsurance companies and resiliency officers care about Antarctic ice. The efforts to answer the how-fast-how-much question range from simple exercises to frame the problem to quizzing experts locked in a room (Bamber and Aspinall 2013) to probabilistic projections (Edwards, Brandon et al. 2019) and full-blown ice sheet models (Feldmann and Levermann 2015, DeConto and Pollard 2016). These models are like weather models only for ice. In contrast to weather and hurricane models, these models are still in the early stages of development. Ice sheet modeling scientists have made big advances in these efforts, such as figuring out how to capture mathematically the changing forces when ice goes afloat and using the latest supercomputing resources to allow the models to include many of the important stresses at play within the ice. The ice sheet models are now linked to different futures, whether temperatures go up a little, a lot or a huge amount. These different futures will be determined by how much CO2 we release into the atmosphere. The science community is working through this collaboratively and through peer review, the way good science happens. An idea is published, the community tests it and new ideas are advanced. Since scientists have never watched an ice sheet disappear, we use records from the past. We know sea level rises when temperature rise --- Miami is built on rocks formed in a shallow sea very similar to the Bahamas today. The hills of Miami formed 120,000 years ago when the planet was warm and sea level was 19-30 feet higher than it is now. The other point we use to calibrate our models is from three million years ago, the last time CO_2 was as high as it is now sea level was 19-65 feet higher than it is now.

The challenges the scientists working on the models face include that is that we are still learning so much about how ice sheets work. For example, while we are all familiar with how water flows across our familiar landscape, we are just now working to understand what happens when water collects on Antarctica. Greenland wears a necklace of blue ponds every summer and has water hidden in crevasses and in the snow. What happens if Antarctica warms until it looks like Greenland (Bell, Banwell et al. 2018)? Will all the new water make the remaining ice shelves disintegrate like the Larsen B, triggering more glaciers/conveyor belts to accelerate, or will rivers form atop the ice (Kingslake, Ely et al. 2017)? Will we will see giant ice cliffs that become unstable causing a sudden runaway collapse of the ice switching the glacier conveyor belts to hyper-fast? These are the ice processes that might produce drastically accelerate sea level rise. Models with lots of meltwater and collapsing cliffs predict close to six feet of sea level rise from Antarctica by 2100. More recent publications suggest that the number might be closer to 1-1.5 feet (45 cm). As we discover new important processes and discover more, these numbers will change. Our knowledge-base and our models are evolving. My family has a boat on the Hudson and we worry about hurricanes every summer. Thirty years ago the hurricane models could not tell us whether the hurricane was going to hit Maine or our New York home, now we can plan much better. We knew Sandy was possibly coming ten days out and were able to prepare. The improvement in hurricane prediction illustrates that the ice predictions can improve if we work on it by building our knowledge base, deepening the bench of scientists and fostering interdisciplinary and international collaborations.

Three Essentials to Improve Ice Sheet Melt Projections

The Antarctic melt projections for 2100 range from just below my knee or over my head, or, quantitatively 1-6 feet. How can we narrow down this answer about how Antarctica will melt in the coming decades? My neighbors are asking me. There are three critical things essential to improving the predictions: knowledge of processes (or how ice sheets work), people (to explore, discover, model and communicate, and fostering collaboration: 1) Processes: We have never witnessed an ice sheet collapse and improving our predictions requires getting up close and personal with the ice sheets to better understand how ice sheets work and intense efforts to decide how best to describe these processes in ice-sheet models. 2) People: The community studying ice around the world has grown but the community is still really small. 3) Collaboration: Because changing ice is controlled by the ocean, the atmosphere, the underlying geology and ice physics and Antarctica are huge, this work requires collaboration across disciplines and nations.

Our understanding of the process of how ice sheets work has made huge advances. Prior to the International Geophysical Year in 1958, we did not even know how much ice there was in Antarctica. By the 1980s we began to understand why those giant conveyor belts of ice can deliver so much ice to the ocean (Alley 1986). These conveyor belts can be over 60 miles wide and in Antarctica move up to about 1.5 miles per year. In Greenland the conveyor belts move even faster – more than 7.5 miles a year. In the 1990s we began both to drill through the ice sheet and to study extensive regions with aircraft and we discovered that the geology underneath matters. In the 2000s we realized there were extensive networks of water beneath the ice including large lakes, one the size of New Jersey (Kapitsa, Ridley et al. 1996), smaller lakes that will slowly fill and drain (Fricker, Scambos et al. 2007), and water networks that move the water. Where the water goes matters because the water is part of the basal lubrication system. Some of the big unknowns include: what is happening in this hidden environment beneath the ice, how will the warming ocean and atmosphere attack the ice sheet and will surface water trigger collapse of all the major ice shelves?

We have discovered a lot but there remains a lot to be learned. We as a species have lived with changing weather and have a deep knowledge of weather systems behave. Our grandmothers understood the wispy angular clouds they called mare's tails meant rain soon, but we can now predict to the hour when the rain will arrive. We as a species have far less experience with collapsing ice sheets. To improve our models, we must get up close and personal with the ice sheets. The satellite record has clearly shown us that change is happening but it is the work in Antarctica from surface ships and aircraft that is essential to foster the advances in understanding of how ice sheets work that will improve our projections. NASA's Operation Icebridge is an example of the importance of comprehensive imaging of the ice sheets that fostered a new norm of freely available open data. The National Science Foundation has responded to the 2015 National Academy report by launching a major program collaboratively with the United Kingdom's Natural Environment Research Council (NERC), the International Thwaites Glacier Collaboration. Thwaites Glacier, one of the largest conveyor belts, is considered one of the most unstable pieces of ice on the planet. Thwaites Glacier is wide and is perched on a topographic ridge where the warming ocean is known to be thinning the ice. Because this glacier can deliver a lot of ice to the ocean fast and because it is already showing signs of thinning and shrinking it is a major threat and a high priority (NAS). The major NSF/NERC initiative (Scambos, Bell et al. 2017) this as an example of the type of work that is essential to launch around all of Antarctica. Advancing the basic understanding of how the ice sheets work and the processes that control their melting, will improve our predictions. Think of Antarctic scientists as the hurricane hunters for sea level.

The second critical need to improve our projections is people. As President of AGU and as former President of the Cryosphere Section (the best job title ever), I am acutely aware of how small our community is. Now, the AGU Cryosphere section has 1,492 members. This number includes scientists from around the globe studying ice, snow and sea ice. To put that in perspective in 2010, there were about 140,000 people enrolled in law school in the US. In a single year, 100 times more people were studying law than the entire global community studying changing ice. There is an acute personnel problem. We need more scientists working on this problem if we are to improve our projections. Science and the science of melting ice from the Arctic to the South Pole must be an open welcoming community. The science is remarkable and the discoveries to be made remarkable. We have barely started to scratch the surface of the ice sheets.

The third need is to fully embrace ice as part of the changing earth and enable truly convergent work. When your child is in the hospital with a sudden ailment you really want the specialists to be working together to provide the best care. The ice community is coming to the realization that we need to take a similar approach. We recently completed ROSETTA, a study of the largest ice shelf in Antarctica, the Ross, just a little smaller than Texas. Using Recovery Act funding, in partnership with the New York Air National Guard, we repurposed military imaging technology for ice studies. After three years of flying the IcePod over the Ross Ice Shelf we realized that it was impossible to understand how the ice will melt without bringing all the specialists to the table. We learned that the geology is in essence protecting that sector of West Antarctica from the warming global ocean but the vulnerability is to heat pumped under the ice shelf from the shallow ocean waters by strong winds (Tinto, Padman et al. 2019). It took scientists from many disciplines working together on the same data sets to converge on these complex processes. We were acting like that team of specialists working together for the good for a patient. It is essential to foster this convergent work for the planet and our species. To move the Antarctic work forward will require interdisciplinary and international collaboration as fostered by NSF in the ITGC program but on a larger scale. Ice science must also be more tightly linked to our changing coastlines so each community will know how to respond and adapt. I am hopeful. With investment the hurricane forecasts have improved. We can improve the melt forecasts and provide better information to our neighbors.

Thank you for inviting me to speak today. I am heartened that the House Science Committee is considering this very important issue. We are very fortunate as a species to have the capacity to see how our home planet works and have the capacity to address this issue both scientifically and technically. If we continue to foster collaborative science across disciplines, the science community will be able to provide our communities with accurate projections of how sea level will rise. If ice scientists work with coastal scientists, we can develop tailored projections for each community. I am also heartened as I see individuals, communities, state governments and professional societies taking action to reduce the underlying cause of the changing ice – our greenhouse gas emissions. The AGU community is very proud of our headquarters on Florida Avenue. This building, long known for the planets in the sidewalk, is Washington DC's first net-zero emissions building renovation. Reaching net zero required multiple technologies from solar panels to heat exchange with the sewer system to green walls. Similarly, using the same multi-pronged strategy, we as a species can address the issue of climate change and ice melt with broad concerted efforts, from individuals, communities and governments.

References

Alley, R. B. (1986). "Deformation of till beneath ice stream B, West Antarctica." *Nature*, *322*, 57-59.

Bamber, J. L. & Aspinall, W. (2013). "An expert judgement assessment of future sea level rise from the ice sheets." *Nature Climate Change*, *3*(4), 424.

Bell, R. E., Banwell, A. F., Trusel, L. D. & Kingslake, J. (2018). "Antarctic surface hydrology and impacts on ice- sheet mass balance." *Nature Climate Change*, *8*(12), 1044-1052.

DeConto, R. M. & Pollard, D. (2016). "Contribution of Antarctica to past and future sea-level rise." *Nature*, *531*, 591–597.

Domack, E., Duran, D., Leventer, A., Ishman, S., Doane, S., McCallum, S., Amblas, D., Ring, J., Gilbert, R. & Prentice, M. (2005). "Stability of the Larsen B ice shelf on the Antarctic Peninsula during the Holocene epoch." *Nature*, *436*(7051), 681-685.

Edwards, T. L., Brandon, M. A., Durand, G., Edwards, N. R., Golledge, N. R., Holden, P. B., Nias, I. J., Payne, A. J., Ritz, C. & Wernecke, A. (2019). "Revisiting Antarctic ice loss due to marine ice-cliff instability." *Nature*, *566*(7742), 58.

Feldmann, J. & Levermann, A. (2015). "Collapse of the West Antarctic Ice Sheet after local destabilization of the Amundsen Basin." *Proceedings of the National Academy of Sciences*, *112*(46), 14191-14196.

Fricker, H. A., Scambos, T., Bindschadler, R. & Padman, L. (2007). "An active subglacial water system in West Antarctica mapped from space." *Science*, *315*, 1544-1548.

Harig, C. & Simons, F. J. (2015). "Accelerated West Antarctic ice mass loss continues to outpace East Antarctic gains." *Earth and Planetary Science Letters*, *415*, 134-141.

Hay, C. C., Morrow, E., Kopp, R. E. & Mitrovica, J. X. (2015). "Probabilistic reanalysis of twentieth-century sea- level rise." *Nature*, *517*(7535), 481.

Kapitsa, A., Ridley, J. K., Robin, G. d. Q., Siegert, M. J. & Zotikov, I. (1996). "Large deep freshwater lake beneath the ice of central East Antarctica." *Nature*, *381*, 684-686.

Kingslake, J., Ely, J. C., Das, I. & Bell, R. E. (2017). "Widespread movement of meltwater onto and across Antarctic ice shelves." *Nature*, *544*(7650), 349.

Milne, G. A., Davis, J. L., Mitrovica, J. X., Scherneck, H. G., Johansson, J. M., Vermeer, M. & Koivula, H. (2001). "Space-geodetic constraints on glacial isostatic adjustment in Fennoscandia." *Science*, *291*(5512), 2381-2385.

National Academies of Sciences E. and Medicine. (2015). A Strategic Vision for NSF Investments in Antarctic and Southern Ocean Research. Washington, DC, The National Academies Press.

Nerem, R., Beckley, B., Fasullo, J., Hamlington, B., Masters, D. & Mitchum, G. (2018). "Climate-change–driven accelerated sea-level rise detected in the altimeter era." *Proceedings of the National Academy of Sciences*, 201717312.

Rignot, E., Casassa, G., Gogineni, P., Krabill, W., Rivera, A. & Thomas, R. (2004). "Accelerated ice discharge from the Antarctic Peninsula following the collapse of Larsen B ice shelf." *Geophysical Research Letters*, *31*(18).

Rignot, E., Mouginot, J. & Scheuchl, B. (2011). "Ice flow of the Antarctic ice sheet." *Science, 333*(6048), 1427-1430.

Scambos, T., Hulbe, C. & Fahnestock, M. (2003). "Climate-induced ice shelf disintegration in the Antarctic Peninsula." *Antarctic Peninsula Climate Variability: Historical and Paleoenvironmental Perspectives, Antarct. Res. Ser, 79*, 79-92.

Scambos, T. A., Bell, R. E., Alley, R. B., Anandakrishnan, S., Bromwich, D., Brunt, K., Christianson, K., Creyts, T., Das, S. & DeConto, R. (2017). "How much, how fast?: A science review and outlook for research on the instability of Antarctica's Thwaites Glacier in the 21st century." *Global and Planetary Change, 153*, 16-34.

Scambos, T. A., Bohlander, J. A., Shuman, C. A. & Skvarca, P. (2004). "Glacier acceleration and thinning after ice shelf collapse in the Larsen B embayment, Antarctica." *Geophys. Res. Lett., 31*.

Shepherd, A., Ivins, E., Rignot, E., Smith, B., van den Broeke, M., Velicogna, I., Whitehouse, P., Briggs, K., Joughin, I., Krinner, G., Nowicki, S., Payne, T., Scambos, T., Schlegel, N., Agosta, G. A. C., Ahlstrøm, A., Babonis, G., Barletta, V., Blazquez, A., Bonin, J., Csatho, B., Cullather, R., Felikson, D., Fettweis, X., Forsberg, R., Gallee, H., Gardner, A., Gilbert, L., Groh, A., Gunter, B., Hanna, E., Harig, C., Helm, V., Horvath, A., Horwath, M., Khan, S., Kjeldsen, K. K., Konrad, H., Langen, P., Lecavalier, B., Loomis, B., Luthcke, S., McMillan, M., Melini, D., Mernild, S., Mohajerani, Y., Moore, P., Mouginot, J., Moyano, G., Muir, A., Nagler, T., Nield, G., Nilsson, J., Noel, B., Otosaka, I., Pattle, M. E., Peltier, W. R., Pie, N., Rietbroek, R., Rott, H., Sandberg-Sørensen, L., Sasgen, I., Save, H., Scheuchl, B., Schrama, E., Schröder, L., Seo, K. W., Simonsen, S., Slater, T., Spada, G., Sutterley, T., Talpe, M., Tarasov, L., van de Berg, W. J., van der Wal, W., van Wessem, M., Vishwakarma, B. D., Wiese, D., Wouters, B. & The, I. t. (2018). "Mass balance of the Antarctic Ice Sheet from 1992 to 2017." *Nature, 558*(7709), 219-222.

Sweet, W. V., Kopp, R. E., Weaver, C. P., Obeysekera, J., Horton, R. M., Thieler, E. R. & Zervas, C. (2017). *Global and regional sea level rise scenarios for the United States.*

Tinto, K., Padman, L., Siddoway, C., Springer, S., Fricker, H., Das, I., Tontini, F. C., Porter, D., Frearson, N. & Howard, S. (2019). "Ross Ice Shelf response to climate driven by the tectonic imprint on seafloor bathymetry." *Nature Geoscience*, *1*.

USGCRP, Ed. (2018). *Impacts, Risks, and Adaptation in the United States: Fourth National Climate Assessment*, Volume II Washington, DC, USA, U.S. Global Change Research Program.

Velicogna, I., Sutterley, T. & Van Den Broeke, M. (2014). "Regional acceleration in ice mass loss from Greenland and Antarctica using GRACE time-variable gravity data." *Geophysical Research Letters*, *41*(22), 8130-8137.

Biography

I am Robin Elizabeth Bell, and I am the PGI Lamont Research Professor at Lamont-Doherty Earth Observatory of Columbia University and a member of the Earth Institute faculty. At Lamont, I direct programs in ice sheet dynamics, lead efforts to develop innovative technology and work to improve the scientific culture especially for women. I have led ten major expeditions to the polar regions, to both Greenland and Antarctica resulting in discoveries ranging from active volcanism beneath the West Antarctic Ice sheet, to large deep lakes encased by two miles of ice to hidden mountain ranges buried by ice where water under the pressure of thick ice is forced uphill. I was the first women to chair the National Academy of Science's Polar Research Board (2002-2008) where I was instrumental in launching the International Polar Year 2007-9 that brought together over 50,000 scientists from around the globe. I also co-chaired the recent National Academy report A Strategic Vision for NSF Investments in Antarctic and Southern Ocean Research (2015) that set the priorities for Antarctic science. Recently, I chaired the National Academy of Science Review of the Draft Fourth National Climate Assessment – a comprehensive undertaking by U.S. scientists and citizens documenting the impacts of climate change around the country (USGCRP 2018). Currently,

I have the great honor of serving as the President of the American Geophysical Union, or AGU as we all refer to it.

Chairwoman Johnson. Thank you very much. Dr. Moon.

TESTIMONY OF DR. TWILA A. MOON, RESEARCH SCIENTIST, NATIONAL SNOW AND ICE DATA CENTER'S COOPERATIVE INSTITUTE FOR RESEARCH IN ENVIRONMENTAL SCIENCES

Dr. MOON. Chairwoman Johnson, Ranking Member Lucas, Members of the Committee, thank you for the opportunity to testify today.

Land ice loss has serious consequences within the United States and across the globe, and I'm honored to share my scientific expertise with the Committee.

Glaciers and ice sheets are Earth's water towers. Only 2.5 percent of the world's water is fresh water, and most of that fresh water is contained within glaciers and ice sheets or land ice. As Earth's water towers, glaciers are valuable sources of drinking water, irrigation water, and hydropower. But land ice is now melting at a rapid and accelerating pace, increasing risks for hundreds of millions of people who depend on them for survival and prosperity. And it is raising sea levels across the globe.

Today, land ice loss is the biggest contributor to sea-level rise. Sea-level rise can contaminate drinking water, erode coasts, over-whelm stormwater and wastewater systems, and cause increased or permanent flooding. Over just the last 25 years, average sea level around the globe has already risen 3 inches. But because sea-level rise is not evenly distributed, some areas like regions of the U.S. Gulf Coast and eastern seaboard are already dealing with more than double this amount.

The impacts we are facing today, however, may pale in comparison to the changes we could experience in the future. If we continue on our current path of high greenhouse gas emissions, it's reasonable to expect 2.5

feet or more of sea-level rise in the next 80 years. In regions of the Gulf Coast and the eastern seaboard, that number will be significantly higher.

The Greenland ice sheet, which is more than 2 miles thick in its center and covers an area the size of Texas, California, Arizona, and Nevada combined is an important player in sea-level rise. Since the early 2000s, ice loss from Greenland has increased rapidly, and Greenland is now a primary player in land ice contribution to sea-level rise.

The cause of ice loss is clear. Greenland and glaciers around the world are melting and more rapidly spilling their ice into the sea as a direct result of warming air and warming ocean water due to manmade greenhouse gas emissions. During the last 2 decades, the science community has made substantial strides in understanding Greenland ice sheet behavior and projecting future ice loss. But for any given future greenhouse gas emissions pathway, there is still a large range in projections for how much ice Greenland will lose.

Narrowing the range of future possibilities and our projections of them is possible. The United States can lead by supporting targeted research on the physical processes that control ice sheet behavior by developing systems to collect long-term observations and by fostering iterative research that connects observations and computer models. Science will also advance more quickly and better serve the public good if strong connections are fostered among scientific disciplines and between scientists and stakeholders. You can ensure this happens by increasing coordinated opportunities for interagency funding and actively funding activities that bring together scientists and decisionmakers.

Finally, I want to emphasize a critical difference in the roles of science and policy in addressing land ice loss and its impacts. Increasing scientific knowledge is essential to more accurately project what the future is likely to bring given that we are on a particular emissions pathway. But policy has the power to determine which emissions pathway we take. Embarking on a lower-emissions strategy will make a fundamental difference in how much and how quickly land ice disappears.

U.S. leadership on mitigating green-house gas emissions within our lifetimes will reverberate to positively impact the world for millennia.

Thank you for giving attention to this important topic. You have the power to make a difference between a manageable future and a painful one. I look forward to supporting you with complete and accurate science and to answering your questions.

[The prepared statement of Dr. Moon follows:]

Rapid Land Ice Loss and Impacts for the United States Statement of Twila A. Moon, Research Scientist, National Snow and Ice Data Center, Cooperative Institute for Research in Environmental Sciences, University of Colorado, Boulder before the Committee on Science, Space, and Technology, U.S. House of Representatives for the hearing Earth's Thermometers: Glacial and Ice Sheet Melt in a Changing Climate. 11 July 2019.

Chairwoman Johnson, Ranking Member Lucas, and members of the House Committee on Science, Space, and Technology, thank you for the opportunity to testify on the critical issue of rapid global land ice loss, its implications, and the challenges and opportunities for moving forward. I am heartened to see the Committee taking up this topic; sea-level rise and other impacts from land ice loss have serious consequences within the United States and across the globe. I am honored to inform the Committee's knowledge and actions via my testimony. I am a Research Scientist at the National Snow and Ice Data Center[49] and the Cooperative Institute for Research in Environmental Sciences[50] at the University of Colorado Boulder, and my testimony today is as an expert in glacier and ice sheet science. My comments represent the views of a scientific expert, not those of the University of Colorado. Since I aim to synthesize science information within this testimony, I have provided academic paper references sparingly, but would be happy to provide additional resources on any topics or statements contained herein.

[49] For more information: nsidc.org

[50] For more information: cires.colorado.edu

The Importance of Land Ice

The title of this hearing refers to glaciers and ice sheets - Earth's land ice - as "Earth's thermometers". It may be more accurate to describe glaciers and ice sheets as Earth's water towers. Only 2.5% of the world's water is fresh water, and more than two thirds of this fresh water is contained in glaciers and ice sheets[51]. These global water towers are critical as sources of fresh water and as long-term reservoirs to store it.

Glaciers around the World Provide Water for Drinking, Irrigation, Energy, and Other Uses

And because warm, sunny weather produces ice melt, glacier-produced fresh water is often an abundant and vital resource during dry seasons or drought. For example, roughly 800 million people in Asia, including less geopolitically stable regions across Pakistan, Afghanistan, and India, depend on glacier melt as an important water source. During average years, land ice in the region provides a small percentage of the total water supply, but in drought years glacier melt becomes much more important, sustaining millions of people and the economies they depend on. Glacier melt is important in the production of hydropower as well[52]. However, ice in the region is melting more quickly than it is being resupplied. This provides a temporary bump in water availability, but future water and energy shortages due to glacier loss are certain.

As a source of cold water, glacier melt helps to regulate stream temperatures and sustain ecosystems. In Glacier National Park, for example, glacier melt is critical to maintaining the cooler water temperatures needed for native fish species and for a cascade of invertebrates that form a fundamental portion of the food web[53]. Glacier melt also helps to cycle nutrients by eroding rocks and transporting

---

[51] Shiklomanov, I. (1993), World fresh water resources in 'Water in crisis: A guide to the world's fresh water resources'. P. H. Gleick (ed.).

[52] Case study details: Pritchard, H. D. (2019), Asia's shrinking glaciers protect large populations from drought stress, *Nature*, 1–20, doi:10.1038/s41586-019-1240-1.

[53] For example: Clark, A. M., J. T. Harper, and D. B. Fagre (2018), Glacier-Derived August Runoff in Northwest Montana, *Arct Antarct Alp Res*, *47*(1), 1–16, doi:10.1657/AAAR0014-033.

sediments underneath the ice to river and ocean ecosystems. In Alaska, the roughly $1B salmon fishing industry depends strongly on the nutrients and water properties of the Alaska Coastal Current. Roughly half of the Alaska Coastal Current waters come from seasonal glacier melt[54]. But Alaska is losing ice; the region is the second largest contributor to ice loss in the Arctic (after Greenland). Ongoing ice loss is likely to affect the future of the salmon industry and the many people who depend on it.

As the world's water towers, Earth's glaciers and ice sheets protect our coastal communities and economies from a myriad of impacts related to sea-level rise, including coastal erosion, frequent or permanent flooding, saltwater inundation of fresh water resources, disruption of storm water systems, and destruction of infrastructure including schools, homes, superfund sites, and military bases[55]. These impacts are already evident within the U.S. and across the globe as a result of almost 3 inches of sea-level rise during the last 25 years. But additional sea-level rise from ice contained within Greenland, Antarctica, and the rest of the Earth's glaciers will likely dwarf what is currently being experienced. Midrange projections for additional sea level rise by 2100 are 17 inches for IPCC RCP2.6, 21 inches for RCP4.5, and 29 inches for RCP8.5, often referred to as the 'business as usual' pathway[56]. Note that these projections do not include mechanisms for potential accelerated ice loss that scientists are currently studying, particularly in Antarctica.

---

[54] Case study details: O'Neel, S. et al. (2015), Icefield-to-Ocean Linkages across the Northern Pacific Coastal Temperate Rainforest Ecosystem, *BioScience*, *65*(5), 499–512, doi:10.1093/biosci/biv027. For a more accessible summary: https://alaska.usgs.gov/products/pubs/2014/2014_Timm_ONeel_etal_ACCC_website_factsheet.pdf

[55] For more information on the military and sea level rise: climateandsecurity.org/militaryexpertpanel

[56] IPCC (2013), *Climate Change 2013: The Physical Science Basis. Contribution of Working Group I to the Fifth Assessment Report of the Intergovernmental Panel on Climate Change*, edited by T. F. Stocker, D. Qin, G.-K. Plattner, M. Tignor, S. K. Allen, J. Boschung, A. Nauels, Y. Xia, V. Bex, and P. M. Midgley, Cambridge University Press, Cambridge, United Kingdom and New York, NY, USA. p. 1180.

Recent & Rapid Global Land Ice Loss

Creation of the world-leading American economy has come about, in part, because of a stable and reliable climate. Development has depended upon and reinforced well defined coastal margins, allowing 39% of the U.S. population to thrive and prosper in shoreline counties[57], driving the U.S. economy and providing stable locations to base infrastructure for shipping, military activities, and other functions vital in today's connected global economy. This stability, however, is at risk. Since 1993, global average sea-level has already risen almost 3 inches. But sea-level rise is not evenly distributed around the world, and some communities have already experienced much higher sea-level rise than the global average. In the United States, for example, coastal erosion related to sea-level rise is displacing communities in Louisiana, drinking water problems are affecting California, and there is increased regular flooding across the Gulf Coast and Eastern Seaboard[58] (Figure 3). Similar impacts are being felt around the world.

As noted, sea-level rise is not the same everywhere. Local sea-level rise is due to the combined effects of local vertical land motion (some land is sinking, some is rising, depending on various factors), ocean and atmospheric currents moving ocean waters towards or away from the coast, additions of groundwater to the ocean, expansion of ocean water as it warms, and added ocean water due to loss of land ice. Today, the largest of these contributors is land ice loss. Land ice loss has increased quickly and across the globe since the mid-20th Century, with contemporary rates - and extents - of ice loss unprecedented over human history.

Satellites are a critical tool for studying contemporary changes in ice mass. The NASA GRACE and GRACE-FO (Gravity Recovery and Climate Experiment / -Follow On) satellites directly measure ice loss via changes created in the gravitational pull of the ice mass. The NASA ICESat and ICESat-2 (Ice, Cloud, and land Elevation) satellites use lasers

---

[57] oceanservice.noaa.gov/facts/population.html

[58] Moon, T. A. et al. (2019), The expanding footprint of rapid Arctic change, *Earth's Future*, 1–13, doi:10.1029/2018EF001088. And references therein.

reflected off of the Earth surface to measure the changing surface elevations of glaciers and ice sheets.

Figure 3. Flooding and storm damage in the US are connected to rapid global land ice loss. a) South Ponte Vedra Beach, Florida, b) Black Creek area near Jacksonville, Florida, c) Charleston, South Carolina, and d) Lumberton, North Carolina. Images: James Balog/Earth Vision Institute[59].

Data from the joint NASA/USGS Landsat satellites are used to measure the surface speed of land ice and to map its extent. Combining these speed measurements with observed and modeled precipitation and ice thickness information tells us how much snow is accumulated on an ice sheet versus how much ice is lost from melting and calving of icebergs into the ocean. Subtracting the losses from the gains tells us how much the ice mass has changed. These techniques alone provide three methods to independently and accurately measure ice mass changes, and results from these techniques give a consistent picture of rapid worldwide ice loss that has accelerated. This conclusion is supported by other data, including long-term measurements of glacier advance/retreat from committed individuals

[59] Figure from: Moon, T. A. et al. (2019), The expanding footprint of rapid Arctic change, Earth's Future, 1–13, doi:10.1029/2018EF001088.

and institutions, examination of historical photographs, and deduction from geologic records. It is now unequivocal that contemporary ice loss is a direct result of warming air and warming ocean water due to human-caused climate change.

Role of the Greenland Ice Sheet

The Greenland Ice Sheet reaches more than two miles thick at its center, with hundreds of fast-moving outlet glaciers along its edges, which act as conveyor belts to move ice from the ice sheet interior to the ocean. Smaller glaciers and ice caps (regions of land ice that are smaller than ice sheets, but include areas of ice connecting multiple glaciers) began losing ice due to climate change earlier in the 20th century. The Greenland Ice Sheet was mostly in balance through the 1980s, but that changed in the late 1990s[60]. During the 21st century, Greenland has lost ice at an increasing tempo.

Greenland ice is being lost through melt on the ice sheet surface and where the ice contacts the oceans along its edge. Ice is also being lost as large and small icebergs that break or 'calve', into the ocean. While glaciers and ice caps continued to add the most water to the world's ocean up into the early 21st century, ice loss from the Greenland Ice Sheet has recently reached the same level[61], with a reservoir of potential future sea-level rise that eclipses small glaciers and ice caps.

There are several indicators of current rapid ice loss in Greenland. Widespread surface thinning and ice edge retreat is observed around the entire ice sheet, exposing new land and ocean water. While personally conducting ship-based field work in northwestern Greenland in August 2018, our ship's navigation map did not show the elevation of the ocean seafloor underneath us. Instead, because the ice sheet edge in the area had

---

[60] For example: Mouginot, J. et al. (2019), Forty-six years of Greenland Ice Sheet mass balance from 1972 to 2018, *Proc National Acad Sciences*, doi:10.7280/D1MM37.

[61] For example: Chen, X., X. Zhang, J. A. Church, C. S. Watson, M. A. King, D. Monselesan, B. Legresy, and C. Harig (2017), The increasing rate of global mean sea-level rise during 1993–2014, *Nature Climate change*, 7(7), 492–495, doi:10.1038/nclimate3325.

retreated almost two miles since 2000, the map showed our ship ostensibly motoring across the ice sheet itself (Figure 4). Individual glacier retreat of multiple miles since 2000 is common[62]. Along with adding ice to the ocean, the speed at which the Greenland coastal environment is changing may increase risk across valuable industries like tourism and resource extraction.

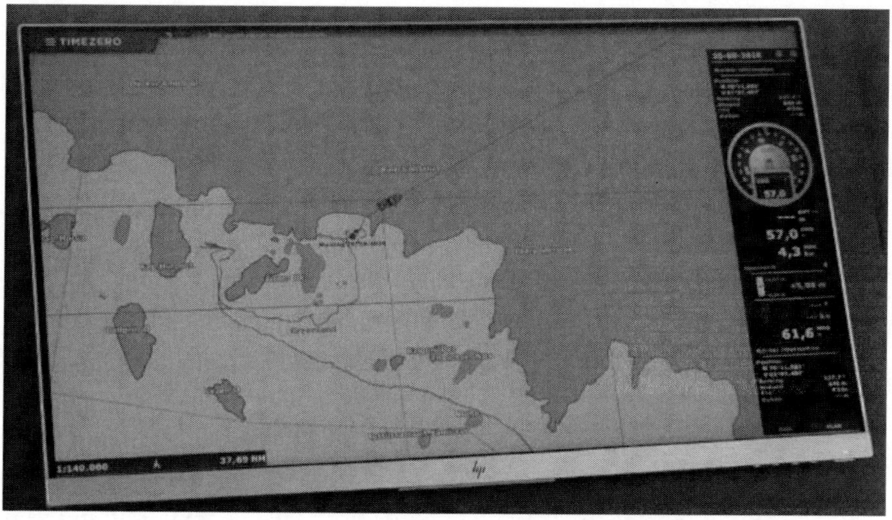

Figure 4. Rapid retreat of the Greenland ice sheet edge is changing coastal boundaries in Greenland, and in the U.S., where ice loss creates rising sea levels. Here, a science research ship appears to be traveling across the ice sheet itself because navigational maps are not keeping pace with the fast ice edge retreat. Image: Twila Moon.

The rapid ice loss in Greenland is akin to the rapid melt you see in your kitchen when placing an ice cube in a water glass or leaving it lying on the countertop. The ice sheet is melting in response to warming air temperatures and ocean temperatures that are warming at depths that matter to ice (~600-1300 ft). This handy analogy has its limits, however, as

---

[62] The NASA MEaSUREs (Making Earth System Data Records for Use in Research Environments) Program provides valuable data on a variety of Greenland Ice Sheet metrics, including glacier advance/retreat. To overview NASA MEaSUREs data at NSIDC: nsidc.org/data/measures.

the Greenland Ice Sheet system is much more complex than a household block of ice and understanding its behavior requires continued research:

- Depending on the season, the ice sheet surface is a mix of bright, reflective new snow; darker and often dirty bare glacier ice; and very dark melt lakes fed by an extensive system of surface streams. The character of the surface, which changes from region to region and over time, helps to determine how much energy, or heat, the ice sheet absorbs to melt more ice or reflects, minimizing the amount of melt. Research indicates that there will be more surface melt in the future, but also that year-to-year variations in the quantity of surface melt will span a wider range than in the past. How will the ice sheet surface transform in coming decades and respond to year-to-year variations in weather?
- Underneath the ice sheet is an entire landscape of mountains and valleys. In some areas, the ice sheet is frozen to the land, or 'bed', beneath it while in other areas liquid water and water-saturated sediments sit between the ice and the bed, affecting how easily the ice moves (or slides) over the land. Understanding the properties, and even the shape, of the bed beneath the ice is challenging, but it also vital to determining how the ice sheet moves. Will small topographic features not yet mapped speed up or slow down ice loss? How will the ice sheet be affected by increasingly large amounts of meltwater flowing underneath it?
- Some areas of the ice sheet edge end on land, but there are also hundreds of fast- moving outlet glaciers that connect directly to the ocean, each with their own unique width, depth, slope, and flow path. These areas where the ice sheet interacts with the ocean are hot spots for activity. This is where melt water discharges from underneath the glaciers into the ocean, warm ocean water at depth contacts the glacier edge, and icebergs calve into the ocean. The ocean plays a substantial role in determining how quickly an outlet glacier can retreat and flush its ice into the sea. If the elevation of the land underneath the glacier drops further inland, retreat can

create a cascade of glacier speedup and thinning that drives further retreat and may create a self-reinforcing cycle of ice loss. How will interactions between the ice sheet and ocean change the speed of ice loss? How will the fresh water from Greenland transform the ocean water, ocean currents, and marine ecosystems?

Ice loss from Greenland has impacts in addition to sea-level rise. The additional fresh water added to the ocean is changing the temperature and salinity of ocean water, and also changing the quantity and cycling of important nutrients[63]. It is likely that these changes are influencing plants and animals around Greenland, including in areas supporting important commercial fisheries. The fresher and colder water that Greenland adds to the ocean may also alter ocean currents in the North Atlantic region. North Atlantic ocean currents help to determine the climate in North America and Europe, and research is ongoing to understand how quickly and to what degree ocean currents may change, and the contributing role of Greenland Ice Sheet melt.

The past two decades have seen strong advances in characterizing, understanding, and predicting changes in the Greenland Ice Sheet. In the early 1990s our knowledge of the large ice sheets was so limited that the science community did not consider them major players in climate change on decadal timescales. Today, we understand that ice sheets are major players, and changes in Greenland are already impacting U.S. communities and the U.S. economy. The Greenland Ice Sheet, however, remains a remote and difficult place to study, and increasing our ability to predict what will happen to the Greenland Ice Sheet in coming decades will require investment in observing systems, understanding the physical processes controlling ice sheet change, and building scientific capacity and stakeholder connections.

---

[63] For example: Cape, M. R., F. Straneo, N. Beaird, R. M. Bundy, and M. A. Charette (2018), Nutrient release to oceans from buoyancy-driven upwelling at Greenland tidewater glaciers, *Nat Geosci*, 1–8, doi:10.1038/s41561-018-0268-4.

Understanding the Future of Greenland Ice Loss: The Powerful Impact of Science

Satellite observations in the early 2000s revealed that the Greenland Ice Sheet was starting to respond to climate change. Since then, the science community has made substantial progress in understanding the behavior of the ice sheet and projecting future changes, and watched as the initial changes became much larger and more rapid. The United States has played a key role in these advances, supported via investment in satellites, in airplane- and ship-based surveys, and through advances in modeling. It is clear that while there may be short-term (less than a decade) departures from the current trend of ice loss, there is no expectation of a long-term (multi-decadal) re-stabilization under current greenhouse gas emission rates.

There remains, however, a large range in projections of ice loss for each individual future greenhouse gas emissions scenario. This spread is due to remaining limitations in our knowledge of ice sheet physics and ice sheet - atmosphere - ocean interaction, and the challenges in fully incorporating newly discovered physical relationships into computer models. The spread in the range of projected ice loss adds to the difficulty in pinpointing how much sea-level rise our nation, states, counties, and cities should plan for when making infrastructure investments. Knowing when two feet of sea-level rise will arrive to the Texas coast, for example, is essential for planning for improvements to coastal infrastructure or adjusting regional flood mitigation and planning.

Glaciology (the study of land ice) and the general study of the Earth's cryosphere (all things frozen) is a fairly young field of research, given the historic difficulty of access and vast expanse of land ice. Current resources support only a limited number of cryosphere scientists. Yet, we are trying to understand one of the most rapidly changing elements on our planet. The Arctic is warming at more than double the speed of the global average.

Additional glaciological research is necessary if we want to narrow the range of future ice loss projections and inform our Nation and the world about the ramifications for sea-level rise, water resources, etc., and the associated implications for decision-making. The United States can lead this effort via focused support to: 1) conduct research in challenging locations and understand the physics of Earth's land ice and its interactions with the atmosphere and ocean, 2) foster iterative research to connect observations and computer models to improve projections, and 3) create new and stronger connections across scientific disciplines and with decision makers through knowledge co-production so that science can advance quickly while serving societal needs.

Critical Research in Challenging Locations

Harsh weather, remote locations, large distances, and sparse infrastructure conspire to make studying Earth's land ice challenging and expensive. Yet understanding how massive ice sheets behave and how physical processes work at critical interfaces, like the ice-ocean boundary, is imperative for improving our projections of ice loss and impacts such as sea- level rise. Strengthening funding for focused research on processes controlling ice sheet behavior is vital. Through these efforts, we will increase understanding of ice-ocean interactions, ice sheet surface properties, ice sheet hydrology, and iceberg calving to improve future projections. Increased support must also include developing long-term observations that span all seasons over multiple decades. The U.S. Arctic Observing Network (AON) and international Sustaining Arctic Observing Networks (SAON) can play a valuable role in supporting these observations for the Arctic.

Connecting Current Observations to Future Projections

Observations fundamentally inform scientific knowledge about land ice processes, and are critical in developing better computer models to help scientists explore the causes of observed changes and project what will happen in coming decades. Not all physics can be seamlessly included in computer models - some processes must be captured using approximations, or 'parameterizations', that are created through observational analysis. Consistent and efficient information sharing between modeling research and observational research can fast track scientific understanding. Achieving this will require sustained support and encouragement of science that integrates these tools.

Building Collaborations to Lead

Understanding the expansive and compounding impacts of land ice loss require that new collaborations be created across different research disciplines, including areas of Earth science, social sciences, and infrastructure and economy research. Old approaches that silo science within specific disciplines must be broken down. The United States has the opportunity to be a world leader in creating strong collaborations between scientists and stakeholders. These collaborations need to be fostered via long-term, repeat interactions so that the United States can form and champion co-production of scientific knowledge and rapid information sharing. This should include support to coordinate science and stakeholder communities so that decisions at all levels are informed by the most robust science, not simply the latest headline. U.S. federal action that can support these advances by increasing coordinated opportunities for interagency funding and actively funding activities that integrate scientists with decision makers and planners. In my experience, the most effective stakeholder-scientist partnerships fully support participation across all groups.

International collaboration is also key. Land ice loss and sea-level rise are not just U.S. concerns, and coordinating research in the U.S. with international efforts will serve all. The current 17-nation MOSAiC mission[64] and the U.S. - U.K. International Thwaites Glacier Collaboration[65] are excellent examples. Research in Greenland is also

[64] For more information: mosaic-expedition.org
[65] For more information: thwaitesglacier.org

enhanced via strong partnerships with Greenland, Denmark, and other Arctic nations.

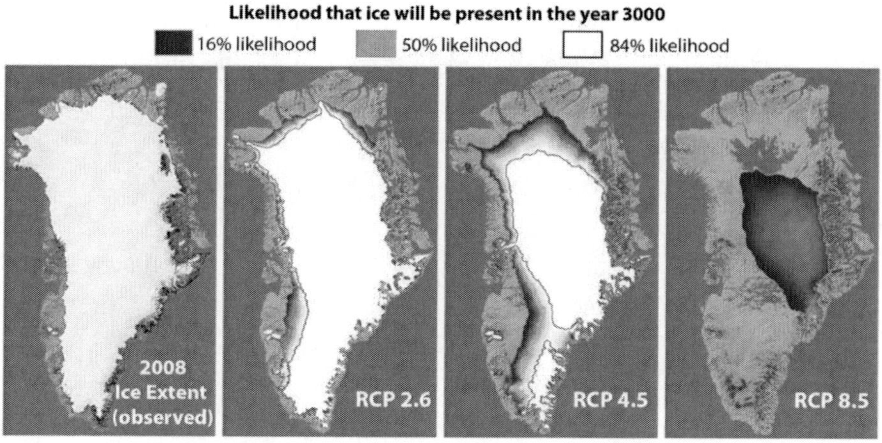

Figure 5. Computer simulations show that greenhouse gas emissions pathways followed this century will result in dramatically different Greenland Ice Sheet extents in the year 3000. All ice lost will contribute to global sea-level rise[66].

Creating the Future of Greenland Ice Loss: The Powerful Impact of Mitigation

The impacts of land ice loss are visible today. More difficult to comprehend is that our climate has not caught up with the greenhouse gases that have already been added to the atmosphere, meaning that more warming and additional ice loss is certain. Continued sea- level rise over the next several decades is guaranteed and will require adaptation. If greenhouse gas emissions are reduced, however, sea-level rise can still be kept to a more manageable level.

The full power of mitigation to create the future of Earth's land ice is underappreciated. U.S. leaders, including Congress, must think beyond the

[66] Figure modified from: Aschwanden, A., M. A. Fahnestock, M. Truffer, D. J. Brinkerhoff, R. Hock, C. Khroulev, R. Mottram, and S. A. Khan (2019), Contribution of the Greenland Ice Sheet to sea level over the next millennium, Sci Adv, 5(6), eaav9396, doi:10.1126/sciadv.aav9396.

short time scales of politics and realize the power they now hold for fundamentally creating the world our grandchildren and their grandchildren will live in. Recent research using a cutting-edge Greenland Ice Sheet computer simulation tells us that choosing to lead the world on a path of reduced greenhouse gas emissions can be the difference between maintaining most of the Greenland Ice Sheet or allowing it to melt completely within the next 1000 years (Figure 3).

Let me be clear here about how I see the various roles of science, policy, and leadership. Increasing scientific knowledge is essential to more accurately project what the future is likely to bring given that we are on a particular greenhouse gas emissions pathway. But policy has the power to determine which emissions pathway we take. Embarking on a lower emissions strategy will make a fundamental difference in how quickly - and how much - land ice disappears. U.S. leadership on mitigation within our lifetimes will reverberate to positively impact the world for millennia.

Communicating Scientific Urgency

Challenges remain for communicating with decision makers, planners, and the public about land ice loss and its impacts, including within the United States. The language of science can be difficult to understand. For example, the public meaning of 'positive feedback' is a good response or praise, while the scientific meaning of 'positive feedback' is a self-reinforcing cycle, often with negative consequences[67]. This disconnect is a problem. Fostering interactive and sustained connections between scientists and decision making and public communities via intentional funding can create a shared language for clear understanding and informed planning.

It is also critical that decision makers and the public understand that land ice loss at the poles has strong impacts on low and mid-latitude countries, including the United States. These impacts are also not confined to coastal regions. For example, if limited federal resources are being used

[67] Somerville and Hassol, Physics Today 64, 10, 48 (2011); https://doi.org/10.1063/PT.3.1296.

to respond to sea-level rise in coastal states, interior states may find it more difficult to secure resources to address other priorities or emergencies.

That mitigation today will not immediately stop ice loss and sea-level rise poses a challenge for maintaining momentum on mitigation while at the same time adapting to change. But strong mitigation today can reduce the worst outcomes of ice loss. America cannot only lead the world in understanding and projecting ice loss. America can also lead on actually determining the future of land ice and the extent to which ice loss will impact our citizens and the citizens of the world.

Thank you for the opportunity to testify today on glacier and ice sheet melt in Greenland and across the globe. I look forward to answering your questions and ensuring that the Committee has the most complete and up-to-date information on the science of ice loss and its impacts in the United States and around the world.

Biography

Dr. Twila Moon is a Research Scientist at the National Snow and Ice Data Center, part of the University of Colorado Boulder's Cooperative Institute for Research in Environmental Sciences, a world leader in Earth science. Dr. Moon's expertise is in contemporary changes in glaciers and ice sheets, and the connections among ice, climate, ocean, and ecosystem. While she has conducted work across the globe, her primary focus is on the Greenland Ice Sheet and the Arctic.

Dr. Moon's research uses tools ranging from satellite remote sensing to field work to computer simulations. Her work has been published in high-impact journals such as *Science* and *Nature*, and received extensive media coverage; for example, National Public Radio, the Associated Press, and the BBC. She is an accomplished science communicator and is leading efforts to improve science and knowledge co-production between scientists and stakeholders.

Dr. Moon has a BS in Geological and Environmental Sciences (Stanford) and an MS and PhD in Earth and Space Sciences (University of

Washington). Subsequently, she was a U.S. National Science Foundation Postdoctoral Fellow (Ocean Sciences Division) at the University of Oregon and a Cooperative Institute for Research in Environmental Sciences Postdoctoral Fellow at the University of Colorado Boulder. She then held the position of Lecturer in Cryospheric Sciences at the University of Bristol, England, before returning to the National Snow and Ice Data Center. She serves on the Acting Committee for the Greenland Ice Sheet - Ocean Interactions Science Network. Additional information on her work is available at www.changingice.com.

Chairwoman Johnson. Thank you very much. Dr. Wolken.

TESTIMONY OF DR. GABRIEL J. WOLKEN, RESEARCH SCIENTIST AND MANAGER, CLIMATE AND CRYOSPHERE HAZARDS PROGRAM, DIVISION OF GEOLOGICAL & GEOPHYSICAL SURVEYS, ALASKA DEPARTMENT OF NATURAL RESOURCES

Dr. Wolken. Good morning. Chairwoman Johnson, Ranking Member Lucas, staff, and Members of the Committee on Science, Space, and Technology, thank you very much for the invitation to come speak to you today.

As a citizen, I'm very pleased to be here. And I congratulate you on selecting this topic to consider. As a scientist, it means very much to me to be here to speak to you about evidence-based decision making, the data that we have to talk to you about today on glaciers and ice sheet change.

I live in Alaska, and Alaskans are very in touch with their surroundings. The cryosphere is that place on Earth where water is in its solid form, so snow, ice, and permafrost. Recently, while doing some fieldwork near Valdez, Alaska, it looks much like what you're seeing today. And so Valdez is in a fjord. It used to be covered by ice. Now the ice is melting quickly.

Upon completing a bathymetric survey or mapping the lake surface below the water near Valdez glacier, we were at the shoreline and reviewing our data and very happy about what we discovered because now we can start to find out how much water in the lake has contributed to the melting of the glacier that terminates into it. A woman and her dog named Elvis, a slobbering basset hound, came up to us and she says, what are you doing? And I said, well, we're trying to find out how deep the lake is. She said it's 600 feet deep. We looked at each other and said you're absolutely right. We just used $25,000 in equipment to figure that out. What did you do? She and her friend went out in a canoe and lowered a rope. And they discovered that the rope wasn't long enough, so they paddled back to shore and grabbed the rope. And then they tied the extra rope onto it, lowered it down, and they discovered that it was 600 feet deep near the glacier.

Now, she is a Valdez resident for 30 years. She said this glacier is melting faster than anything I've seen in the area. Where does all this water go? Well, the answer to that is in the oceans. And so Alaskans are keenly aware of their environment. They're keenly aware of the changes.

This same woman lives in an area where outburst floods impact her house every single year. The glacier releases tremendous amounts of water, rips out the dike, challenges the bridge, and gives them an opportunity to see the power of change. So the cryosphere is changing in Alaska, and glaciers are a part of that. It's very important for us to understand what is happening.

In Alaska we have a very large State. It's one-fifth the size of the rest of the United States. It's huge. We have thousands and thousands of glaciers. We know changes physically on three of those glaciers. We have mass balance data that began back in 1966. And with those data, we are able to understand how glacier change is happening over long-term. That is incredibly valuable to us.

So most of the information that we have today is built on the shoulders of giants and the data that they were able to start collecting a long time ago. It's important that we start that process now. So collecting data now in various places in state means that we can evaluate and quantify the amount of change that we have between now and whenever we're worried about the change. We do this so that we can build better computer simulations so that we can plan.

As policymakers and decisionmakers, it is imperative to have the right scientific information, and we cannot provide that without the money, without the funding, without the students, without the resources to be able to provide the information that is necessary for local stakeholders such as the woman in Valdez and her dog, as well as important federally mandated decisions that have to be made in this country. So evidence-based decisionmaking is what we are after in order to have sound change and be able to communicate to the local residents such as those in Valdez and Alaska so that we can actually start planning for some of these changes. Thank you.

[The prepared statement of Dr. Wolken follows:]

The Impacts of Glacier Change in Alaska, Testimony of Gabriel J. Wolken, Alaska Division of Geological & Geophysical Surveys and University of Alaska Fairbanks For the hearing entitled Earth's Thermometers: Glacial and Ice Sheet Melt in a Changing Climate Before the Committee on Science, Space, and Technology, United States House of Representatives, 11 July 2019.

Chairwoman Johnson, Ranking Member Lucas, and members of the House Committee on Science, Space, and Technology, thank you for the opportunity to testify on the critical issue of glacier and ice sheet melt. As a citizen of this country I am grateful that you consider this issue relevant and important, and as a scientist I am honored to be in a position to provide evidence-based testimony to the committee to aid your decision making process. I am a Research Scientist and Manager of the Climate and Cryosphere Hazards Program at the Alaska Division of Geological & Geophysical Surveys and a Research Assistant Professor in the International Arctic Research Center at the University of Alaska Fairbanks. I provide testimony today as an expert in glacier science. My comments represent my views as a scientist and private citizen, and not those of the State of Alaska or the University of Alaska Fairbanks.

Key Points

- Glaciers in Alaska are in steep decline, and are among the fastest-melting glaciers on Earth.
- The rapid melting of glaciers in Alaska is a result of rising air temperatures associated with global climate change.
- Glaciers in Alaska are projected to continue to melt and contribute to global sea level rise.
- Hazards associated with rapid glacier retreat are impacting infrastructure, and threatening public safety and resource security.
- Narrowing the range in mass loss estimates requires a comprehensive and coordinated multi-agency collaboration, and a sustained long-term funding structure.

Glaciers and Ice Sheets

A glacier is a large mound of ice, snow, water, rock and sediment that forms on land and flows outward or down slope under its own weight. A glacier forms when snow that has accumulated doesn't melt and is buried by subsequent years of snow with the same fate. As more snow is added to the mound over many years, the snow near the bottom gets compressed and forms ice. In Alaska, glaciers form in high- mountain areas where they are constrained by topography, and flow down to lower elevation.

Glaciers gain mass by snow accumulation and lose mass by surface melt and runoff, and by iceberg separation and submarine melting at the glacier-water interface. During periods of net accumulation, glaciers gain mass and advance, whereas during periods of net loss, glaciers melt and retreat. Similar to an accountant who balances the credits and debits from a bank account, we monitor mass gains and losses of a glacier over time by computing mass balance.

Ice sheets, located in Antarctica and Greenland are massive expanses of glacier ice and snow that span entire continents. These masses of ice subsume the terrain and flow outward from a high point. At their margins ice sheets becomes thinner, and interact with the terrain.

The Role and Significance of Glaciers and Ice Sheets

Glaciers and ice sheets are important components of the cryosphere, or the parts of earth where water is in its solid form. Glaciers and ice sheets cover about 10% of Earth's land area and represent the majority of perennial land ice. Land ice is distributed as glaciers across high-mountain regions all over the world, but the majority of this ice is concentrated near Earth's poles. There glaciers and ice sheets play a vital role as regulators of global weather and climate, reflecting radiation to cool the planet and providing long-term stability for atmospheric and oceanic currents.

Glaciers and ice sheets are also an integral part of the hydrologic cycle. Together with other forms of perennial land ice, they hold about 69% of the freshwater available on the planet. They temporarily store freshwater at high altitudes and latitudes, serving as frozen water reservoirs that help to maintain a relatively stable sea level and providing consistent water resources. Runoff from glaciers provides critical baseline flow during warm and dry periods when other water sources are unavailable. This runoff is crucial to billions of people around the world, particularly those in arid regions who rely on glacier-sourced streams for drinking water and irrigation for agriculture. Runoff from glaciers also regulates stream temperature and provides nutrients for plants, insects, fish, and other animals. Glacier runoff is also important for hydropower production in many parts of the world.

Observed Glacier Change

As ice melt outpaces snow accumulation worldwide, glacier change has become one of the most important and widely recognized indicators of climate change. Historically unprecedented mountain glacier recession has proceeded on a global scale for the last thirty years (Zemp et al., 2015). In the Arctic, glaciers and ice caps have been experiencing increasingly negative cumulative mass balances since the early 1990s (Wolken et al., 2017), and are a leading contributor to global sea level change despite their relatively small size compared to ice sheets in Antarctica and Greenland (Gardner et al., 2011, 2013; Jacob et al., 2012).

Over the last several decades, Alaska has warmed twice as fast as the rest of the United States. Glaciers in Alaska (including northwest Canada) represent about 22% of the Arctic land ice area and currently have one of the highest glacier melt rates in the world (Gardner et al., 2013; Pfeffer et al., 2014), with annual thinning rates that reach several meters (tens of feet) per year for some glaciers terminating near sea level (VanLooy et al., 2006; Larsen et al., 2007; Larsen et al., 2015). This rapid and sustained melting and retreat of glaciers in Alaska is expected to continue as a consequence of climate warming, as predicted unequivocally by all current climate models (IPCC AR5, 2013; Huss and Hock, 2018).

Impacts of Glacier and Cryosphere Change

The rapid and sustained melting of glaciers in Alaska has led to a disproportionately large contribution to global sea level rise. This trend is projected to continue with inevitable consequences including lowland flooding and the displacement of communities, increased coastal erosion, degradation of water quality and agricultural soils, and loss of ecological habitats among other effects already impacting communities in the United States and elsewhere around the world.

As high glacier mass loss rates continue, the storage capacity of snow and ice will diminish, which is projected to have significant impacts on future water resource availability, even in basins with low glacier coverage (Huss and Hock, 2018). As a result, river discharge volume and timing in seasonal river runoff will change, and the ability of glaciers to sustain this flow during warm and dry periods will be reduced or lost. In Alaska, this will continue to produce clear and profound impacts on marine and terrestrial ecosystems at multiple levels, negatively influencing key species, including salmon, as well as the billion dollar fishing industry on which people depend. These changes will also heavily impact hydropower production, tourism, socio-economics, and the livelihoods and lifestyles of many people.

Rapid changes in climate can have a major effect on the cryosphere system and cause an increase in hazards. This is because changes in climate can modify natural physical processes and increase the magnitude and frequency of certain cryosphere hazards (e.g., avalanches, floods, erosion, slope instability, glacier collapses, and glacier lake outburst floods), which if not properly addressed, may have a damaging effect on Alaska's communities and infrastructure, as well as on the security and livelihoods of Alaskans. Rapid glacier retreat in recent decades has caused new ice- dammed lakes to form in valleys that were formerly occupied by glaciers. These new lakes are inherently unstable, with many resulting in glacial lake outburst floods that threaten communities and damage infrastructure (as in Juneau and Valdez). Rising air temperature and an increase in melt for Alaska's glaciers has also caused slopes surrounding glaciers to become unstable, which has led to an increase in the magnitude and frequency of landslides, and placed Southeast Alaska at the epicenter of these catastrophic events (Coe et al., 2016).

Improving our Understanding of Future Glacier Melt

Fifty-three years ago, a small team of U.S. Geological Survey scientists ventured onto two glaciers in Alaska to collect data to help understand snowpack variations in high-mountain settings. Those measurements on the Gulkana and Wolverine glaciers became the first data entries into Alaska's Benchmark Glacier program, which has provided some of the longest continuous data for evaluating the health of glaciers in Alaska. Although many advances in glacier monitoring have been made since the beginning of the Alaska Benchmark Glacier program, the observational data from this program are still used every year and represent one of the three important tools that support modern glacier science.

Modern glacier science relies on observations, remote sensing, and modeling to measure glacier melt and simulate future changes. In Alaska, the observational record is long, but there are only three long-term, continuous records of glacier mass balance for the entire state, which currently hosts many thousands of glaciers. Because of the lack of ground-based observational data, remote sensing and computer modeling are used to broaden the scope and extend estimates of glacier melt into the future based on a range of emission scenarios. But the current range in glacier ice loss projections can be narrowed. This can happen by increasing the number and distribution of long-term, ground-based observational data, increasing the quality and coverage of remote sensing products, and improving melt models and climate data products so that they more accurately represent the variables and processes involved in glacier melt and runoff in climatically diverse and topographically complex areas of the state. To achieve this, a comprehensive and coordinated multi-agency collaboration and a sustained long-term funding structure are required.

References

Coe, J. A., Bessette-Kirton, E. K. & Geertsema, M. (2018). Increasing rock-avalanche size and mobility in Glacier Bay National Park and Preserve, Alaska detected from 1984 to 2016 Landsat imagery. *Landslides*, *15*(3), 393-407.

Gardner, A. S., Moholdt, G., Wouters, B., Wolken, G. J., Burgess, D. O., Sharp, M. J., Cogley, G. J., Braun, C. & Labine, C. (2011). Sharp acceleration of mass loss from Canadian Arctic Archipelago glaciers and ice caps. *Nature*, *473*, 357-360.

Gardner, A. S., Moholdt, G., Cogley, J. G., Wouters, B., Arendt, A. A., Wahr, J., Berthier, E., Hock, R., Pfeffer, W. T., Kaser, G., Ligtenberg, S. R. M., Bolch, T., Sharp, M. J., Hagen, J. O., van den Broeke, M. R.

& Paul, F. (2013). A consensus estimate of glacier contributions to sea level rise: 2003 to 2009. *Science*, *340*, 852-857.

Huss, M. & Hock, R. (2018). Global-scale hydrological response to future glacier mass loss. *Nat. Clim. Chang*, *8*(2), 135–140 doi:10.1038/ s41558-017-0049-x.

IPCC. (2013). *Climate Change 2013: The Physical Science Basis. Contribution of Working Group I to the Fifth Assessment Report of the Intergovernmental Panel on Climate Change*. Cambridge University Press, Cambridge, United Kingdom and New York, USA doi: 10.1017/CBO9781107415324.

Jacob, T., Wahr, J., Pfeffer, W. T. & Swenson, S. (2012). Recent contributions of glaciers and ice caps to sea level rise. *Nature*, *482*, 514–518.

Larsen, C. F., Motyka, R. J., Arendt, A. A., Echelmeyer, K. A. & Geissler, P. E. (2007). Glacier changes in southeast Alaska and northwest British Columbia and contribution to sea level rise. *Journal of Geophysical Research: Earth Surface*. *112*(F1), doi: 10.1029/ 2006JF000586.

Larsen, C., Burgess, E., Arendt, A., O'Neel, S., Johnson, A. & Kienholz, C. (2015). Surface melt dominates Alaska glacier mass balance. *Geophysical Research Letters*. doi: 10.1002/2015GL064349.

Pfeffer, W. T., et al. (2014). The Randolph Glacier Inventory: a globally complete inventory of glaciers. *Journal of Glaciology*, *60* (221), 537-552.

VanLooy, J., Forster, R. & Ford, A. (2006). Accelerating thinning of Kenai Peninsula glaciers, Alaska. *Geophysical Research Letters*., *33*(21), doi: 10.1029/2006GL028060.

Wolken, G., Sharp, M., Andreassen, L. M., Burgess, D., Cogley, J. G., Copland, L., Kohler, J., O'Neel, S., Pelto, M., Thomson, L. & Wouters, B. (2017). Glaciers and ice caps outside Greenland (in "State of the Climate in 2016"). *Bulletin of the American Meteorological Society*, *98* (8), 140-143, doi:10.1175/2017BAMSStateoftheClimate.1.

Zemp, M., Frey, H., Gärtner-Roer, I., Nussbaumer, S., Hoelzle, M., Paul, F. & Vincent, C. (2015). Historically unprecedented global glacier decline in the early 21st century. *Journal of Glaciology*, *61*(228), 745-762. doi:10.3189/2015JoG15J017.

Biography

Dr. Gabriel Wolken is a research scientist and manager of the Climate and Cryosphere Hazards Program at the Alaska Division of Geological & Geophysical Surveys, and a research assistant professor at the International Arctic Research Center, University of Alaska Fairbanks, where he is a senior scientist in the Climate Adaptation Science Center.

Dr. Wolken's expertise is in snow and glacier change, and the connection these changes have to climate and natural hazards. His research uses a combination of field-based observations, remote sensing, and computer modeling to monitor, simulate, and assess the impacts of snow and ice changes on communities, infrastructure, and resources. Through his affiliations, Dr. Wolken is dedicated to fostering scientific partnerships that increase the collection of critical data and improve the co-production of scientific products that lead to informed decision- making.

Dr. Wolken has a PhD in Earth and Atmospheric Sciences from the University of Alberta, an MA from the University of Wyoming, and a BSc from Creighton University. He was a Postdoctoral Fellow in the Arctic and Alpine Research Group and lecturer at the University of Alberta. Over the course of his career, he has authored numerous papers, reports and data products, and he is a regular contributor to governmental climate reports.

Chairwoman Johnson. Thank you very much. Dr. Pfeffer.

TESTIMONY OF DR. W. TAD PFEFFER, FELLOW, INSTITUTE OF ARCTIC AND ALPINE RESEARCH, UNIVERSITY OF COLORADO BOULDER

Dr. Pfeffer. Along with my colleagues, I'd like to thank you all, Chairwoman Johnson, Ranking Member Lucas, all the Members of the Committee and staff.

Like my colleagues, I was pleased, surprised, jumped at the opportunity to come and talk to you today about subjects that I've spent two-thirds of my life on. I've spent a long time living on glaciers and have had a good opportunity to see the changes and study them.

As Chair Johnson mentioned, I'm a glaciologist. I've done this for 40 years, and I've had opportunities to work in landscapes that have changed dramatically over time mostly in Alaska. I work mostly on the small glaciers of the world, the 200,000 glaciers other than the Greenland and Arctic ice sheets. And I want to talk mostly about them, and I really want to come back and focus on Alaska, which is one of the hotspots in the world both literally and figuratively in sea-level rise but also in fresh water flowing into the ocean, in fires and environmental change in the coastal regions.

These small glaciers matter for a wide variety of reasons, and I also want to try to concentrate today on the reasons that have direct ties to the United States. There are a number of global issues. Water resources, water availability from the Himalayas, for example, is going to be critical for Nepal, India, Pakistan, Bhutan, places like that. They also produce significant geo-hazards of landslides, flooding, what we call outburst flooding as glaciers retreat and leave behind very unstable steep slopes. When this happens in places like Nepal, these are very unstable landscapes in the same valleys where a lot of people live. It's one of the reasons that these hazards are as great as they are in the Himalayas. It's because we've got the mountains there, glaciers changing, and also people living in that landscape. That's one of the reasons that that's not quite so much of a problem in the United States because we are not obliged to live right next door to glaciers in most places, not at all.

They also have significant environmental impacts by changing the temperature of the waters the glaciers drain into and by changing the salinity of the water. One of the effects of Alaska that we don't understand particularly well yet but we know it's there is the fact that the ice sheet or the glacier runoff from Alaska that flows into the Gulf of Alaska and the Pacific, works its way up through a gap in the Aleutian Islands and enters the Arctic basin. And it turns out that that's quite a large chunk of the fresh water entering the Arctic basin, and that fresh water influences, among other things, the extent of sea ice in the Arctic.

We don't have a good handle on how much that flow is in part because we're not making comprehensive measurements of the water flow into the Gulf of Alaska from glaciers. As my colleague Dr. Wolken mentioned, we're not monitoring the glaciers in Alaska very well. We're not really keeping track of them. So while we can see that they're melting, we can measure their height change or we could up until recently anyway, we don't have good observations of where the water is going. We don't have good gauges measuring that flow.

One of the last things that I want to come back to in my statement again, though, is sea-level rise. As Dr. Alley pointed out, the ice sheets contain virtually all of the fresh water that's locked up on land in ice. You take all the other glaciers, about 200,000 glaciers, you only get about a foot of sea-level rise out of them if you put them all into the ocean. But they're like a big bucket with a little tiny—sorry, they're like a small bucket with a big hole in it. That water is leaving the small reservoir very fast. In effect, if you look at the combined most recent measurements of where new water coming into the ocean is coming from, more than 50 percent of it is coming from these small glaciers, and the remaining smaller percentage is coming from the ice sheets.

Now, that's right now. That's in the short term. The longer term, the ice sheets are certainly going to take over. But in the short term—and this is a term, say, on the order of 30, 40, 50 years where decisionmakers, planners, policymakers really need to have the most robust information, and they need the greatest handle on uncertainties. We have to look at the entire picture, the ice sheets and the glaciers and all of their consequences of which sea-level rise is just one.

So I'll stop there for now and be happy to continue and answer your questions.

[The prepared statement of Dr. Pfeffer follows:]

Role of Global Glaciers in Environmental Change and Sea Level Rise, Testimony of W. Tad Pfeffer, University of Colorado at Boulder For the hearing entitled Earth's Thermometers: Glacial and Ice Sheet Melt in a Changing Climate Before the Committee on Science, Space and Technology, United States House of Representatives, 10:00 a.m., July 11, 2019.

Personal Introduction and Background

My name is William Tad Pfeffer. I am a glaciologist employed by the University of Colorado at Boulder, where I am a Professor of Civil, Environmental, and Architectural Engineering, and a Fellow of the University's Institute of Arctic and Alpine Research (INSTAAR). I have been at UC Boulder for 31 years, and have been an active glaciological researcher for 40 years. My particular sphere of expertise is in the study of the world's "small" glaciers – meaning all of the world's ca. 200,000 glaciers exclusive of the two ice sheets covering Greenland and Antarctica. I have worked extensively in glaciological laboratory experiments, numerical modeling, and theoretical analysis, and have conducted hundreds of field expeditions over 35 years in the Continental USA, Alaska, Canadian Arctic, Svalbard, Greenland, Antarctica, the Himalayas, and Africa. I have published over 60 papers in the refereed scientific literature, including several seminal and highly-cited studies of glacier physics and of global glacier contributions to sea level rise. I served as a co-author of the 2012 National Research Council Report "Sea Level Rise for the Coasts of California, Washington, and Oregon: Past, Present, and Future."

I was also a Lead Author for Chapter 13 (Sea Level Change) of the IPCC Fifth Assessment (AR5), Working Group 1, in 2013. Most recently, I have shifted my focus to science planning and policy and to the historical development of glaciological and sea level research. Starting in 2013, I was a founding editor of the Oxford University Press Handbook Series on Planning for Climate Change Hazards. I also served in 2015-16 as a National Academy of Sciences Jefferson Fellow; in this capacity, I worked at USAID in Washington DC as a senior science advisor in the Office of Energy and Infrastructure, Europe and Eurasia.

My testimony reflects my own views and scientific judgement, and does not represent the views or positions of any institution or agency, including the University of Colorado.

While the potential for rapid sea level rise from the earth's two major ice sheets tends to be the most visible sphere of snow and ice research, there are other issues of concern to glaciologists involving ice in its many forms at the earth's surface. Ranging from permafrost in Siberia and Alaska to seasonal river ice in New England, from seasonal snow in Colorado's ski country to glacier lake outburst floods (or "GLOFs") in the Himalayas, and from glacier-fed rivers in India to sea ice blocking shipping routes across the Canadian Northwest Passage, ice is a crucial element of our environment anywhere on earth where freezing occurs. As temperatures rise, melting ice mobilizes liquid water, weakens previously strong frozen materials, increases the permeability of thawing soils, speeds aqueous chemical reactions, and drives a multitude of other processes, all with the potential to dramatically alter our environment. In this testimony I will focus in problems directly involving glaciers (leaving aside some equally important issues involving permafrost, river ice, and sea ice) and briefly summarize a few of what I view as the most important outstanding environmental problems. I will also concentrate on those problems that affect the United States directly, or indirectly through economic and political reactions to environmental changes elsewhere in the world.

- Seasonal snow and glacier runoff as a water resource. Society everywhere in earth depends critically on freshwater for domestic use (cooking, cleaning, washing, etc.) as well as for agricultural irrigation, industrial use, and hydropower generation. All fresh water moving on the earth's surface starts as rain or snow, but that fraction falling at high elevations as snow will remain in place (either seasonally as snow or for many years as ice) until melting conditions at the surface allow the water to move downslope. Water stored in the mountains as snow and ice acts as a reservoir, delaying the drainage of precipitation, which may arrive in very imbalanced "wet season/dry season" cycles, until later in the season. This benefits users of the water by spreading the downstream arrival of water throughout the year and storing water as glacier ice during wet years to be released during dry years. This is a critically important benefit for agriculture anywhere (including the Western US), but no more so than in the Indian Subcontinent, where very large populations depend upon crops irrigated by runoff from the Himalayas. Recent research (Maurer et al, 2019) indicates that the glaciers of the Himalayas are experiencing losses at rates that have doubled over the past 40 years. The economic and political effects of major seasonal water shortages in India and neighboring countries – a probable consequence of continued snowpack depletion and glacier losses – could be profound and global in its indirect consequences.
- Glacier recession and geohazards. People and infrastructure living in the immediate vicinity of glaciers are exposed to natural hazards including flooding, landslides, and rockfalls, all associated with slopes destabilized by the removal of glacier ice (Richardson and Reynolds, 2000) Such risks are global in extent but are particularly concentrated in parts of the world with high population densities in mountain regions, and specifically on the south side of the Himalayas (Bhutan, India, Nepal, Pakistan) and in the Andes on the west coast of South America (Harrison et al, 2018). These regions (along with virtually all of the earth's mountain regions)

are subject to landslide and rockfall hazards, but glacier retreat dramatically magnifies these hazards. Advancing glaciers disaggregate rocks and soil at their base and margins and plow this material forward and the margins of the glacier, creating *moraines* that surround the glacier terminus and valley sides. When a glacier retreats, the moraines are left behind, and "proglacial" lakes frequently form in the enclosed depression formed between the retreating terminus and the inner side of the moraine wall. Moraines are intrinsically weak materials, being composed of an incohesive mixture of soil and rocks of many sizes; they also typically have very steep slopes. These factors all favor the incidence of slope failures and landslides, and when proglacial lakes are formed, additional hazards are created due to the easily eroded moraine dams.

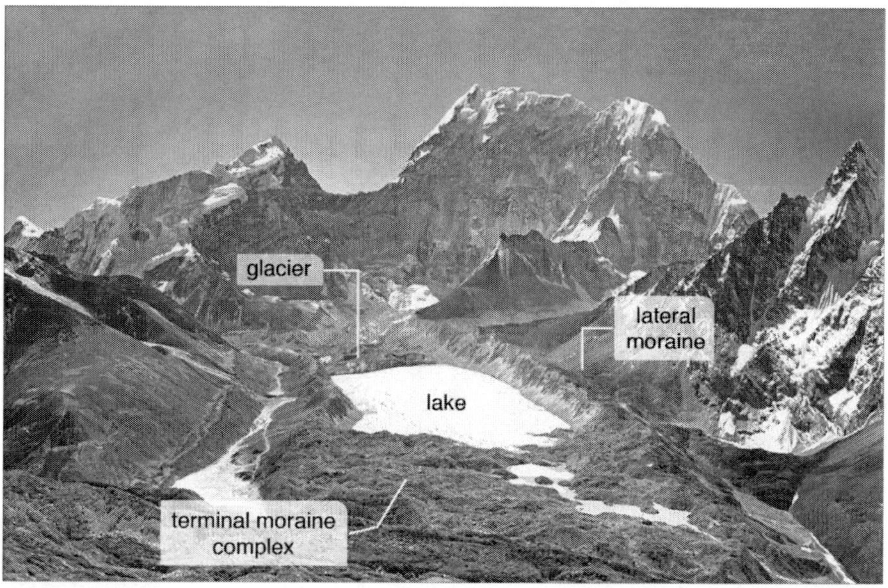

Source: http://www.antarcticglaciers.org/glacier-processes/glacial-lakes/glacial-lake-outburst-floods/

Figure 6. Imja Tsho (or Imja Lake) in eastern Nepal, dammed by a terminal moraine complex. The lake has been growing rapidly since the 1960s as the Imja Glacier has retreated. Photo: Sharad Joshi, Wikimedia Commons, Edited by J. Bendle.

Glacier Runoff and Ocean Salinity

One of the most significant and rapidly changing effects of present day warming is the depletion of sea ice in the Arctic Ocean (Mueller et al, 2018). The loss of sea ice in the arctic has profound implications on local and global scales, ranging from accelerated coastal erosion on Alaska's arctic coast to alterations of the planetary energy balance as the reflectivity of the Arctic Basin drops with increasing open water. The formation and maintenance of sea ice depends on the surface energy balance of the arctic and also on the salinity of Arctic Ocean water (McPhee et al, 1998). Water entering the Arctic Basin via Bering Strait (between Alaska and Siberia) is one of the primary sources of low-salinity sea water in the Arctic Ocean. The salinity of this Pacific sea water is influenced to a significant but poorly constrained degree by the Alaskan Current (Woodgate and Aagaard, 2005), which in turn carries fresh water draining into the Gulf of Alaska from the glaciers of Alaska's south coastal mountains (Chugach & St. Elias Ranges) and interior mountains (Alaska Range, Wrangell Mountains) northward and through Unimak Pass into the Bering Sea (see Figure). The retreat of Alaska's glaciers thus has an effect – probably significant but at this point not well known – and arctic sea ice. Glacier losses in the Canadian Arctic may have a similar influence (Dimitrenko et al, 2017). The influence of Alaska's glaciers on conditions in the Arctic is not well established in part because of the absence of any comprehensive program of observations of freshwater runoff to the Gulf of Alaska. This is one of many examples of the significance of Alaska's glaciers both locally and globally, and the need to invest in research in this area.

- Glacier retreat and gravitational fingerprinting. Like a magnet drawing metal filings around its edges, the large mass of glaciers and ice sheets on land (e.g., the Greenland and Antarctic ice sheets and the glaciers in Alaska) exerts a gravitational pull that draws ocean water toward them. This creates a non-uniform sea level surface, with sea level slightly elevated adjacent to any large mass concentration and slightly lowered elsewhere. This distortion on the mean sea level surface is unique to each ice mass given its size and location, and is informally referred to as the gravitational "fingerprint" of that ice mass. When glaciers shrink, however, the magnitude of their gravitational pull decreases, and the combined result of the melting of any given ice mass is to raise the total amount of water in the ocean (raising global average sea level) and to reduce that particular "fingerprint" distortion of the sea level surface. The combined effect of the gravitational fingerprints from Alaska and, to a lesser extent, Greenland, causes relative sea level to fall all along the west coast of the United States, whereas melting from Antarctica causes a relative sea level rise. The net effect of losses from Alaska, Greenland, and Antarctica on the US west coast is reduce the local sea level rise relative to the global mean value by values ranging from ca. 40% in Washington State to ca. 15% in southern California (National Research Council, 2012). Again, Alaska's glaciers have significant effects both globally and locally.

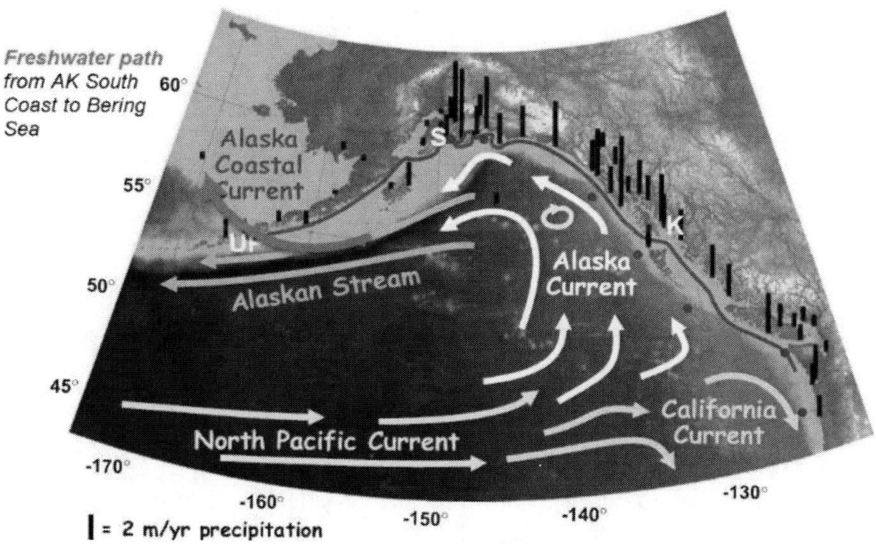

Figure 7. Alaskan coastal transport carrying glacier runoff from coastal mountains in the Bering Sea. Adapted from Weingartner et al. (2005)

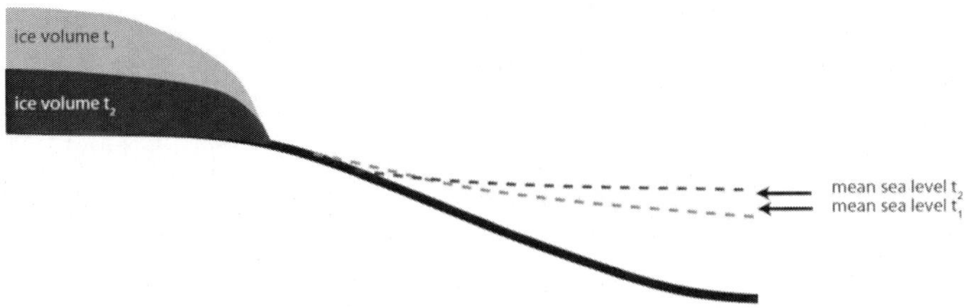

Source: National Research Council, 2012

Figure 8. Gravitational "Fingerprint" of a terrestrially-based ice mass: Globally averaged sea level increases as any terrestrially-based glacier shrinks, but *relative* sea level change is negative adjacent to the declining ice mass and positive away from it.

- Rising sea level from the earth's 200,000-plus glaciers. The total potential sea level rise from these glaciers is very small: only a bit more than 1 foot (Farinotti et al, 2018). However, the present-day *rate* of loss from the glaciers is as great as that coming from the ice sheets, and will in all likelihood continue to match the ice sheet losses for at least the next few decades, when near-term decision making requires the highest level of confidence in projections.

  Since the beginning of comprehensive global observations, virtually all glaciers on Earth have been in a state of mass loss, contributing 0.71 ± 0.08 mm yr^{-1} over the period 2003-2009 (Gardner et al., 2013) corresponding to 29±13% of the observed sea-level rise during that period. The most recent assessment of glacier losses (Zemp et al, 2019) finds a global total loss rate for the period 2006-2016 to be 0.92 ± 0.39 mm yr^{-1}. For context, the most recent ice sheet loss rate assessments show Antarctic contributing 50 ± 26 mm yr^{-1} (2008-2015) and Greenland contributing 0.77 ± 0.005 mm yr^{-1} (2007-12) and 0.53 ± 0.05 mm yr^{-1} (2012-2017).

  Because of their large number and small size, assessments of all 200,000+ glaciers on earth has been difficult, and the calculated aggregate loss rate has varied significantly over time, partly due to limitations in observational methods and partly due to the fact that the rates change over time. Recent research programs have benefitted from rapid developments in remote sensing, including NASA's ICESat satellite (2003 – 2009), the NASA-GFZ GRACE gravity twin satellite mission (2002-2017). Further missions, including the GRACE Follow-On (GRACE-FO), launched in May of 2018, and ICESat-2, launched in September of 2018. These mission investments have aided global glacier assessments enormously and testify to NASA's commitment to earth science generally and glacier monitoring in particular. However, remote sensing methods cannot work alone to continue accurate and validated observations of glacier change, nor can they be used in isolation to solve the numerous outstanding problems faced by modelers seeking to project future glacier behavior. Integration of field and remote sensing observations with model simulations is necessary to accurately project future trends in glacier contribution to sea level. Conventional field observations of mass balance at "benchmark" glaciers, especially those in Alaska, should remain a high priority to ensure the continuity of long-term records, some of which extend back to the 1957-58 International Geophysical Year. Ground-truthing programs are particularly important for large glacierized regions with steep gradients in environmental conditions, where the distant view of an orbiting satellite becomes a liability. Field studies at these and other sites should be expanded to include detailed observations of surface and dynamic processes. Improved parameterization of surface albedo, which controls the dominant term in the surface radiation budget, can be achieved through studies of snow and ice crystal grain sizes (Painter et al., 2009) and parameterization of the impacts of dust/black carbon (Flanner and Zender, 2006) and debris cover (Reznichenko et al., 2010) on surface melt rates. The conversion of volume to mass change in geodetic remote sensing assessments remains a large source of uncertainty (Huss, 2013) and can be informed through field measurements of near-surface densification rates. Glaciers that terminate in lakes or the ocean have the potential for rapid changes through poorly-understood calving mechanisms (Moholdt et al., 2012; Willis et al., 2018), requiring expanded observations of ice thickness, grounding line locations and lake/fiord conditions. Finally, field programs should include observations of stream discharge where possible since this provides valuable information on the integrated water balance of glacierized watersheds.

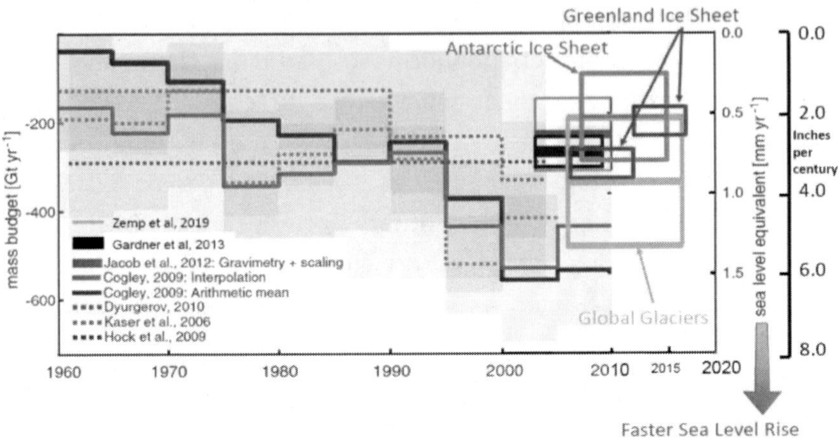

Adapted from Gardner et al, Science, 2013

Figure 9. Global glacier loss rate assessments, 1960 to the present. The most recent loss rates from the Greenland and Antarctic Ice Sheets are included for comparison.

References

Dmitrenko, I. A., Kirillov, S. A., Rudels, B., Babb, D. G., Pedersen, L. T., Rysgaard, S. & Barber, D. G. (2017). Arctic Ocean outflow and glacier-ocean interactions modify water over the Wandel Sea shelf (northeastern Greenland). *Ocean Science*, *13*(6), 1045-1060.

Farinotti, D., Huss. M., Furst, J. J., Landmann, J., Machguth, H., Maussion, F. & Pandit, A. (2019). A consensus estimate for the ice thickness distribution of all glaciers on Earth. *Nature Geoscience*, *12*(3), 168.

Flanner, M. G. & Zender, C. S. (2006). Linking snowpack microphysics and albedo evolution. *Journal of Geophysical Research: Atmospheres*, *111*(D12). Gardiner et al, 2013

Harrison, S., Kargel, J. S., Huggel, C., Reynolds, J., Shugar, D. H., Betts, R. A. & Schaub, Y. (2018). Climate change and the global pattern of moraine-dammed glacial lake outburst floods. *The Cryosphere*, *12*. Huss, 2013

Maurer et al., (2019). Maurer, J. M., Rupper, S. B. & Schaefer, J. M. (2016). Quantifying ice loss in the eastern Himalayas since 1974 using declassified spy satellite imagery. *The Cryosphere*, *10*(5), 2203-2215.

McPhee, M. G., Stanton, T. P., Morison, J. H. & Martinson, D. G. (1998). Freshening of the upper ocean in the Arctic: Is perennial sea ice disappearing?. *Geophysical Research Letters*, *25*(10), 1729-1732.

Moholdt, G., Heid, T., Benham, T. & Dowdeswell, J. A. (2012). Dynamic instability of marine terminating glacier basins of Academy of Sciences Ice Cap, Russian High Arctic. *Annals of Glaciology*, *53*(60), 193-201.

Mueller, B. L, Gillett, N. P., Monahan, A. H. & Zwiers, F. W. (2018). Attribution of Arctic Sea Ice Decline from 1953 to 2012 to Influences from Natural, Greenhouse Gas, and Anthropogenic Aerosol Forcing. *Journal of Climate*, *31*(19), 7771-7787.

Painter et al., 2009.

Richardson, S. D. & Reynolds, J. M. (2000). An overview of glacial hazards in the Himalayas. *Quaternary International*, *65*, 31-47.

Reznichenko, N., Davies, T., Shulmeister, J. & McSaveney, M. (2010). Effects of debris on ice surface melting rates: an experimental study. *Journal of Glaciology*, *56*(197), 384-394.

Weingartner, T., Aagaard, K., Woodgate, R., Danielson, S., Sasaki, Y. & Cavalieri, D. (2005). Circulation on the north central Chukchi Sea shelf. *Deep Sea Research Part II: Topical Studies in Oceanography*, *52*(24-26), 3150-3174.

Willis, M. J., Melkonian, A. K., Pritchard, M. E. & Ramage, J. M. (2012). Ice loss rates at the Northern Patagonian Icefield derived using a decade of satellite remote sensing. *Remote Sensing of Environment*, *117*, 184-198.

Zemp, M., Huss, M., Thibert, E., Eckert, N., McNabb, R., Huber, J. & Thomson, L. (2019). Global glacier mass changes and their contributions to sea-level rise from 1961 to 2016. *Nature*, *568*(7752), 382.

Biography

My name is William Tad Pfeffer. I am a glaciologist employed by the University of Colorado at Boulder, where I am a Professor of Civil, Environmental, and Architectural Engineering, and a Fellow of the University's Institute of Arctic and Alpine Research (INSTAAR). I have been at UC Boulder for 31 years, and have been an active glaciological researcher for 40 years. My particular sphere of expertise is in the study of the world's "small" glaciers – meaning all of the world's ca. 200,000 glaciers exclusive of the two ice sheets covering Greenland and Antarctica. I have worked extensively in glaciological laboratory experiments, numerical modeling, and theoretical analysis, and have conducted hundreds of field expeditions over 35 years in the Continental USA, Alaska, Canadian Arctic, Svalbard, Greenland, Antarctica, the Himalayas, and Africa. I have published over 60 papers in the refereed scientific literature, including several seminal and highly-cited studies of glacier physics and of global glacier contributions to sea level rise. I served as a co-author of the 2012 National Research Council Report "Sea Level Rise for the Coasts of California, Washington, and Oregon: Past, Present, and Future." I was also a Lead Author for Chapter 13 (Sea Level Change) of the IPCC Fifth Assessment (AR5), Working Group 1, in 2013. Most recently, I have shifted my focus to science planning and policy and to the historical development of glaciological and sea level research. Starting in 2013, I was a founding editor of the Oxford University Press Handbook Series on Planning for Climate Change Hazards. I also served in 2015-16 as a National Academy of Sciences Jefferson Fellow; in this capacity, I worked at USAID in Washington DC as a senior science advisor in the Office of Energy and Infrastructure, Europe and Eurasia.

Chairwoman Johnson. Thank you very much. That completes the testimony of our witnesses. We'll now begin our first round of questions. And I'll yield myself 5 minutes.

While a lot of progress has been made in understanding current trends in glacier and ice sheet melt rates and predicting future changes, uncertainties still exist and the potential for tipping points. Whether major parts of the Arctic or Greenland sheets will collapse and when and how much more sea levels will rise is yet still out there.

So for all of the witnesses, as top experts in the field, what are the most pressing needs in glacier research in order to address these uncertainties? And the second question, given the differing impacts of glacier and ice sheet melts on global sea-level rise, ocean temperature and salinity, nutrient cycling and ocean currents, fisheries, and even geopolitical tensions from diminishing drinking water supplies, how can multidisciplinary research approaches help address some of the outstanding questions? And we can start and just go down the panel.

Dr. Alley. OK. Thank you. You raise huge and important issues. You will hear several things as we go along here which are important. I'd like to highlight the people. The students who can really solve these problems have an amazing number of calls on them. They can go to business, they can go to finance, they can do all sorts of things. They have skills that are hugely in demand. We hope that a lot of them go to business and go to finance and do useful things out there, but we would love to have a few of the best students come to us. If those students look at our world and say there isn't funding, there isn't a reliable idea that you can make a career in telling the public what's going on, they all will go elsewhere. And we don't want all of them, but we would really like a few of them. And that means funding for studentships and that means some level of telling the student if you commit to 4 years as an undergrad and maybe 7 years as a graduate student and a bit of postdoc to become a world expert on this, we will support you in doing that. And it is people. And we need a few of them to help us do this.

Chairwoman JOHNSON. Thank you.

Dr. Bell. Well, I think there are three things. As I said before, one is fostering more research, and it's research across the agencies because the U.S. has really been leading in understanding how the ice sheets are changing. And it's becoming increasingly important that research, whether it's supported by NSF, NOAA, USGS (United States Geological Survey), NASA—DOE (Department of Energy) runs a lot of ice sheet models. We have to recognize this incredible resource we have to be at the cutting-edge of who's going to know and be able to provide the answers to communities around the globe.

I echo Richard's workforce question. We really do need to broaden the number of people working on this and not just glaciologists. We need engineers. We need computer scientists. We need to, you know, recognize that this is a significant national security, a national economic issue that requires all hands on deck.

And then the third one is really to foster what NSF is now calling convergent science, science where you really bring together people from different disciplines focused on a problem so that we can address the problem. NSF just released a new priority—they're navigating the new Arctic where that was really the focus of how do we go in a changing arctic. Now getting back to Ranking Member Lucas' question, how do we have science that brings together the people who are going to look at those problems? And NSF is really trying to foster that collaborative problem solving. The word currently is convergent where you actually bring people from different disciplines who are looking to solve a problem, and that's what we absolutely must do, both within the U.S. and globally. This is a problem we cannot solve by ourselves.

Chairwoman Johnson. Thank you. Dr. Moon.

Dr. Moon. Yes. I'd like to emphasize first better understanding the glaciers and ice sheets themselves, the physical processes that we don't yet know. We've never been able to watch ice sheets collapse, so we can't look back into the record. And that requires going there, observing the systems, and doing process studies and then integrating that information with computer simulations.

The second thing is that any projects within this need to be coordinated with other countries. Sea-level rise, ice sheet, glacier melt are international issues, and we need international teams working on them. And anything you can do to help facilitate international collaboration I think is excellent.

And finally, your question regarding connecting different disciplines, this is a very difficult thing to do. Disciplines have been separated for decades. It's the way many elements of our academic system and our research system are built, and it requires long-term investment and an understanding that we have to create those relationships because we're taking our information about glaciers and ice sheets and we're recognizing that they're part of a connected Earth system that includes people, as well as plants, animals, and other physical components.

Chairwoman Johnson. Thank you.

Dr. Wolken. I will echo the comments of my colleagues in that we do need people. It's critical that we have people on these issues. I'll share with you just an example from one of my other projects looking at snow distribution on glaciers in alpine areas. We've gone through three people now in a critical data science position because they can make more money elsewhere. And so it's really hard to retain the people once we have them, and it's really challenging to actually recruit these people.

The other issue is, again, I'll emphasize Alaska is one-fifth the size of the rest of the United States. We have a lot of area, and we are data-poor. We have only a handful of long-term observations in the State, and we have very few long-term records. And when I mean long-term I mean beyond 12 years. So it's very challenging to work in that environment. And so what we really need is more observational information to go off of.

And, you know, we can do a tremendously better job with the science if we have data. The only way to make the models do better is to actually have data that drive the models. And so we don't have that right now. So we're doing the best we can. And one of the most important observation technologies that we had on Operation Ice Bridge in the State that measures the height of glaciers is now gone. It's been discontinued as a program, and so we can't do that anymore.

And so the other thing I'll say is that I'll echo the comments of having a mixed bag of individuals to do critical tasks, and it's important for us to have a diversity of individuals working on these really important issues.

Chairwoman Johnson. Thank you very much. Dr. Pfeffer.

Dr. Pfeffer. Well, after writing down my list of responses to your question, I realize that all I really have to say is I'm with them. We seem to be pretty much on the same page here, and I promise you, we didn't rehearse this in advance. It really boils down to people and support to train those people. We've got these critical questions, the research questions. One of them is the tipping points. What's actually going on in glaciers and ice sheets that causes this occasionally very anomalous behavior? We can't model that problem out of the way with computers or with our knowledge of mathematics and mechanics. We actually have to get in—literally in and under the glaciers to see what's happening.

Decades ago when I was—we don't have to run away and leave the building, do we? OK. Decades ago when I was a graduate student—and Richard will remember this, too—we had programs all through the United States with opportunities for graduate students to work on small glaciers in Alaska, in the Pacific Northwest, and the Scandinavia and Arctic, as well as in Greenland and Antarctica where we could go back a number of times while we were students, really learn what is happening on glaciers. I'd already made 12 trips to Alaska before I finished my Ph.D. That's really not happening anymore. Most departments—and we have—we've got an abundance of programs. We've got a lot of expertise out there searching for students trying to bring them in, but we don't have the support to really go places to train them. And so we are producing a lot of computer modelers, very good, and they're doing very important work, but they're waiting for this knowledge to come in for them to put into these models. And we're really falling behind on that.

Also, as Dr. Wolken mentioned, we're missing what's happening in Alaska. Operation Ice Bridge, which was our best way of tracking the loss of glaciers in Alaska, is—that's vanished for the time being. And monitoring of fresh water flowing into the ocean, we don't know what that is. So we need the answers to those questions, but to get the answers, we need people.

Chairwoman Johnson. Thank you very much. I'm way over. Mr. Lucas.

Mr. Lucas. Thank you, Madam Chairwoman. Dr. Moon, let me turn to you first, and let's discuss for a moment Federal Government spends a lot of money on research, we have a lot of research dollars coming out of National Science Foundation, the various department agencies, the DOD (Department of Defense). We're doing a lot of things. Let's talk for a moment about from your experiences what kind of suggestions you have about, in a more direct way, how do we make sure that the various Federal research activities are better coordinated, integrated, as you noted?

Dr. Moon. Well, I can speak to one example that I think is a nice example for how to work on this. This is a program called IARPC, the Interagency Arctic Research Policy Committee in the Arctic. And that program brings together people who are funded and doing work supported by different agencies. They have regular webinars, they have regular meetings, and it helps people communicate on what's going on. I think having those sorts of tools and having communication between agencies at the program manager level also encourages us on the research scientist level to be able to get information that tells us about what different agencies are interested in so that we have a sense of the interest, the potential funding, and how they might be connected together.

Mr. Lucas. Thank you. And, Dr. Bell, as I mentioned in my opening statement, I have a deep interest in weather patterns. I represent the northwest half of the great State of Oklahoma. I'm a product of my experiences, but I'm also impacted by the experiences of the people who came before me in my district. And where we are on the east side of the Rockies and the Southern Plains, my area was one of those that suffered dramatically in the Dust Bowl of the 1930s, the Great Depression.

But in addition to that horrendous drought, which had a lot of government policy and farming practices that enhanced the misery, we went through a drought in the 1890s, drought in the 1930s, drought in the 1950s, the horrendous drought at the beginning of this decade. So you understand as a farmer in the real world, where I come from, when I asked my questions, how much does the scientific community understand about what will happen in the lower latitudes weather-pattern-wise by what's going on in the glaciers? Do you see where I'm coming from here? Because glacial water changes the chemistry of the oceans, changes the temperature. I did pay some attention in my science portfolio at Oklahoma State.

Dr. Bell. Right. And one of the predictions, you know, one of the very clear predictions from the climate models is we're going to see a lot more of those extremes. We're both going to see a lot more droughts, and we're likely to see a lot more floods—many of your neighbors saw a lot more floods this year because we all know we're here in Washington in the summer and it is hot and muggy. That is because hot air holds more moisture. And in the long run, that's going to make more floods. So, you know, in terms of the direct linkage between the warming climate and the weather, that's the easiest one to think of. We are going to expect more extremes in precipitation, and we're going to expect more extremes in weather. The direct link between the changing land ice and changing climate is something we're still working on is what—we heard it from Tad Pfeffer is what will that water go into the ocean due to the ocean circulation? Where are certainly hypotheses out there that the changing sea ice have contributed to some of the extreme weather we're seeing now. You know, certainly, that's on the table. But again, it's showing how we have not decoded the weather system and the climate system on our planet, but we can see the impacts already.

Mr. Lucas. Please, Dr. Alley.

Dr. Alley. Yes, Representative Lucas, I'm sorry to interrupt. As you know, the great State of Oklahoma is fantastic in educating meteorologists. And you probably also know that there really is scholarship that shows that the Federal investment in meteorology pays handsome dividends for the well-being of farmers, for the well-being of fisherpersons and others and at a level that is a huge payoff on investment. And I can give you chapter and verse if you need it.

There is great optimism now in the community that does weather forecasting that we will be able to move into that area which would give more warning to the farmers of Oklahoma, the fisherpersons of Oregon and Maine as to what's coming. We can't guarantee that, but the optimism is real, it's palpable, and it's exciting.

Dr. BELL. And I'm going to just add one thing—I just want to follow up with one thought following on Richard's is that we have invested a lot in weather forecasting. That's why we now—I'm not a farmer, I'm a sailor, so I think about hurricanes more than droughts. Sorry. But——

Mr. Lucas. You have waterspouts, we have tornadoes.

Dr. Bell. Yes. Yes, yes, I do, and I worry about them, too. They give me goosebumps. But we've been able to narrow our understanding of where those hurricanes are. I can plan much better when I hear there's a hurricane coming than I used to, and that's because we've invested in weather research, everything from the process-based work to the numeric, and that's what we don't have for the ice sheets yet.

Mr. Lucas. Indulge me, Chairwoman. Dr. Wolken, what are you telling the State of Alaska about how to handle the circumstances in the next decade or so?

Dr. Wolken. Well, that's a good question. I mean, we have some of the best climate modelers in the world at the University of Alaska Fairbanks, and they are doing a tremendous job in producing downscale climate models for Alaska. Their products are only as good as the data that they can use to train those, and they've done a tremendous job in predicting out to, say, 2100 what the climate is going to be like. From that, we can make some estimates of how glaciers are going to respond, how the cryosphere in general is going to respond. And, you know, the best tools that we can produce are available, but we need to improve those tools tremendously in order to make better predictions so people can plan.

Mr. Lucas. Thank you, Doctor. Thank you, Chairwoman.

Chairwoman Johnson. Thank you very much. Ms. Bonamici.

Ms. Bonamici. Thank you, Chairwoman Johnson and Ranking Member Lucas. But really thank you to our witness experts who are here today. We do appreciate your expertise.

I'm fortunate because I'm from the Pacific Northwest where we still have glaciers and the Cascade Range and the beautiful Wallowa mountains in eastern Oregon. And over the years the snow and ice masses have really helped delicately balance our water temperature and our ecosystems. The nutrient content, glacial melt water has provided drinking water, and the runoff helps power our communities. Tourism and outdoor recreation are really important in our State. People travel to see our streams, rivers, and lakes, which the glacial sediment makes this iconic teal color. It's a beautiful place. You should all come and visit.

But today, the glaciers that have once filled a lot of the hanging valleys and the moraines and the mountaintops across some of the most pristine regions are rapidly melting and in large part because of anthropogenic emissions.

On Mount Hood, which we can see from Portland in my home State, the Sandy Glacier Caves were once the largest glacier cave system in the lower 48 States, but now, the glacier is melting at an alarming rate. And further north of Oregon at the Columbia River basin at Glacier National Park, they're losing the geologic features that provided its namesake. In fact, when it was founded in 1910 the park had about 150 glaciers. And according to a study from Portland State University and the USGS the park is on track to lose its remaining 26 glaciers in the next few decades.

Dr. Moon, thank you for your testimony. You mentioned the role of glaciers in sustaining ecosystems, and in northwest Oregon the expedited rate of melting of glaciers could have significant consequences for our salmon and steelhead populations and threaten recreational and commercial fishing, tribe species that benefit from healthy salmon runs. And as the glaciers melt and the water flows change and the water temperatures warm in the Columbia basin, the tributaries, the fisheries are threatened. So how quickly are these larger ecosystem changes taking place? And are there potential adaptation and mitigation strategies that we in Congress can support to help at this point in time?

Dr. Moon. I would say—and you might find, too, some—that you have many problems also related to those that are being seen in Alaska and receive comments there. Certainly, we are losing those glaciers very rapidly. Just as you cited, we are seeing retreat and ice loss at rates that have never been seen in these areas. And so those fundamental changes that are happening rapidly and quickly are changing the ecosystem just as quickly. One——

Ms. Bonamici. Right.

Dr. Moon [continuing]. Thing to consider is that in many of these places we initially see a bump in the amount of water because we're getting warmer air temperatures. We still have the glaciers there at the moment, so you actually get a bump in water availability, and we see communities also in other places in the world where they depend even more strongly on glaciers for drinking or irrigation water adjusting to an added level of water input, which then of course is eventually going to decline substantially to levels below what it was——

Ms. Bonamici. Right.

Dr. Moon [continuing]. Previously. So they are rapid changes, and I think that there are many places where the research is not keeping up with the speed of these changes. That's true for us understanding the glaciers and ice sheets themselves and also certainly true for understanding the ecosystems that depend on it. So I think it may be a case where we are changing things that we are not even able to keep up with or see the true level of those changes.

Ms. Bonamici. Thank you. Thank you. I appreciate that very much. And a good place for the Science Committee to get some more research funded.

Dr. Pfeffer, some of your colleagues at the University of Colorado Boulder published a study in 2017 about the effects of dissolved black carbon on glacial melting, that sooty black material that's emitted from gas and diesel engines, coal-fired power plants, and wildfires is a significant portion of particulate matter and contributes to climate change, as we know. The study found that the black carbon from the combustion of biomass and fossil fuels can enhance glacial melting as black carbon deposits on snow and ice surfaces, then the particles decrease the Earth's ability to reflect rays from the sun, so then that results in the absorption of heat and faster melting. But it's also worth noting from the testimony here today that even if anthropogenic emissions were halted immediately, we'd still see the reciprocal effects on glaciers.

So, Dr. Pfeffer, what are the most apparent gaps in the current modeling of glacial recession for various emission scenarios? And assuming that the U.S. achieved a net-zero carbon emission policy by midcentury, where should we invest more Federal resources in responding to the consequences of glacier melting?

Dr. Pfeffer. So the process that you bring up, black carbon, it's hard to see when you're actually out there. It's quite a subtle effect but very small particulate matter, which is carried into the air, and this was particularly a problem prior to the collapse of the Soviet Union because there was a lot of coal burning industry in Siberia. It's far enough north for their emissions to get trapped in this atmospheric gyre in the Arctic. That's reduced a little bit but not by a large degree. And emissions from further south still get into the Arctic basin and also elsewhere. Not all of Greenland is in the Arctic basin, for example. Southern Greenland is exposed to air masses that come off of Europe and North America, so there's a lot of mixing. And this material continues to be deposited.

I think that understanding the surface energy balance, things like if you make the surface of an ice sheet just a tiny bit darker, how much effect will that have, that understanding is pretty well in hand but we need observations. Simply knowing that it happens isn't enough. I really do think, though, that the basic needs go beyond that to simply making the observations. There are so many parts of the world that were, until recently, really in the dark. A lot of high mountain Asia, Himalayas and other ranges, that's been partially addressed by remote sensing, but again, not all of it. Some of this work just has to be done on the ground.

Ms. Bonamici. Thank you. And I see I'm out of time, but, Chairwoman Johnson, I request unanimous consent to enter into the record this study from the University of Colorado.

Chairwoman Johnson. Without objection.

Ms. Bonamici. I yield back. Thank you.

Chairwoman Johnson. Thank you. Mr. Brooks.

Mr. Brooks. Thank you, Madam Chairwoman.

Is anyone on the panel not familiar with the Earth's last glacial maximum, roughly 20,000 years ago? OK. Everybody is? Good. For those in the audience who are not, by way of background, during the last glacial maximum, northern Europe was under ice, roughly 90 percent of Canada and almost all of the Continental United States of America north of Missouri and the Ohio Rivers and east of New York City under ice. According to the United States Geologic Survey, during the last glacial maximum, again, 20,000 years ago, sea levels were roughly 410 feet lower than today. Stated differently, for 20,000 years, sea levels have risen on average 2 feet per century versus the much less roughly 1 foot per century rising rate since 1993 that is reflected in Dr. Alley's written testimony.

Finally, per Zurich University of Applied Science, Earth's average temperature 20,000 years ago was 48° F versus 59° F today. That's an 11-degree increase in global temperature average over the last 20,000 year period.

So my question to each of you is, and we'll start over here with Dr. Pfeffer and move from my right to left, did human beings cause the global warming that started 20,000 years ago and continues through today or, if not, what did?

Dr. Pfeffer. So the examples from 20,000 years ago that Mr. Brooks gave us, they are excellent examples of the kind of natural variability that the Earth experiences. And there's no question that in the past, there have been changes in temperature and sea-level rise and weather patterns and climate generally as dramatic or more dramatic than what we may be experiencing in the future. And of course they weren't human-caused 20,000 years ago or in the last million years. All of these variable events have been occurring throughout the Earth's modern history.

Mr. Brooks. Well, my first question, in your judgment, did human beings cause the global warming that began 20,000 years ago during the last glacial maximum?

Dr. Pfeffer. No. No, absolutely not. It's an example of spontaneous natural variability, one of the many ways that this whole system, whether you look at it in terms of sea-level rise or temperature, storms, can be varied.

Mr. Brooks. Are you familiar——

Dr. Pfeffer. Natural——

Mr. Brooks [continuing]. With the phrase snowball Earth or slush ball Earth——

Dr. Pfeffer. Oh, yes. Yes.

Mr. Brooks [continuing]. Roughly 600 million years ago——

Dr. Pfeffer. Yes.

Mr. Brooks [continuing]. When we were almost entirely ice or slush?

Dr. Pfeffer. Entirely natural variation.

Mr. Brooks. Versus the Paleocene and Eocene thermal maximum of about 55 to 56 million years ago when the average temperature was roughly 73°F, which is 14 degrees warmer than what we are experiencing now.

Dr. Pfeffer. Yes.

Mr. Brooks. If you don't mind, Dr. Wolken, let's go to you. Did human beings cause the global warming that began 20,000 years ago?

Dr. Wolken. No, absolutely not. That was just a product of natural variability in the climate system. Yes.

Mr. Brooks. Dr. Moon?

Dr. Moon. Humans weren't around in nearly the numbers we are today, so we certainly weren't available to be combusting fossil fuels at the rate we are today or putting emissions into the atmosphere. You can consider we've built America in the last 243 years, and we're changing things at a much more rapid rate.

Mr. Brooks. So you also agree then that the global warming that has occurred over the last 20,000 years, that 11°F increase in temperature was not human caused at least when it began 20,000 years ago?

Dr. Moon. So I would agree that when it began 20,000 years ago when we were coming out of the last glacial, that was not caused by humans.

Mr. Brooks. All right.

Dr. Moon. The warming of the last hundred years most certainly was.

Mr. Brooks. Out of curiosity, how do you explain that the sea- level rise average over the last 20,000 years has been 2 feet per century, yet we're down to 1 foot per century?

Dr. Moon. So much of our rise in sea levels that you're talking about came earlier in that 20,000 years.

Mr. Brooks. For 6,000 or 7,000 years.

Dr. Moon. Over this last 10,000 years, we've been sitting with very stable sea levels. And those stable sea levels have allowed us to develop the coasts of the world.

Mr. Brooks. All right. Thank you, Dr. Moon. And I only have about 30 seconds left for Dr. Bell. Dr. Bell, in your judgment, 20,000 years ago, global warming when it began, was that caused by humans?

Dr. Bell. In my judgment, the variation that we were seeing 20,000 years ago was part of the pulse of the planet. It pulses at 100,000 year glacial, interglacial. When I started graduate school, we were expecting to go into the next glacial period——

Mr. Brooks. Yes.

Dr. Bell [continuing]. Except that we as human beings in the last hundred years—and you can see the kick up—since we invented the steam engine, you can see the temperature moving up.

Mr. Brooks. All right. I'm out of time. Madam Chairwoman, I appreciate your indulgence. I just wish I had sufficient time to actually get into what the cause of the global warming that began 20,000 years ago was if not humans. Thank you.

Chairwoman Johnson. Excuse me. Go ahead, Doctor.

Dr. Pfeffer. I just wanted to respond a bit further to your question. The changes in the past have—there are two significant differences between those events and the events today. One of them is that they were triggered by natural variations, not by human agency.

And let me just give you an analogy of your house. Your house might burn down, and it might burn down for entirely natural reasons. It might be struck by lightning. But it could also burn down if you're careless and you, you know, drop a cigarette in the crack in the sofa. Both of those are triggers that result in your house burning down. The presence of one of them doesn't really say much about the other except that they both lead to the same endpoint.

The other thing is that while there were these very dramatic temperature changes and sea-level rises in the past, which were entirely natural, we weren't there to deal with them. The problem here is with people. How do we respond to environmental change?

The Earth will take care of itself. It doesn't really care what happens. It's what people do. And if this had happened, you know, a long time ago when the population of the Earth was a few hundred million, it probably wouldn't have mattered either because we could have just gotten out of the way. But as it is today with the numbers of people that we have and the infrastructure, we're very sensitive to changes of this kind. We don't handle change very well.

For example, suppose that the conditions for growing crops that exist today in California picked up and moved to North Dakota for a couple of hundred years. There are variations like that in the fairly recent geologic past that occurred. How would we deal with that? It's an entirely different world than what we were not here to experience but we know about 20,000 years ago. We're much more sensitive. We don't deal well with change, and to deal with it, we need to know a lot about it.

Mr. Brooks. Dr. Pfeffer, thank you for your additional insight.

Chairwoman Johnson. Thank you very much. Mr. McNerney.

Mr. McNerney. Well, I thank the Chair for calling this hearing, and I thank the witnesses. I appreciate all your testimony this morning.

While the planet is continuing to warm up and I believe we are going to blow past the 2° centigrade marker that people say is the limit of tolerability, we need to be looking at all the potential tools in the climate solutions toolbox, especially if we're to take action to prevent the collapse of the West Antarctic and Greenland ice sheets. That's why I introduced the *Geoengineering Research Evaluation Act* last Congress. It didn't pass, but just introducing that caused the National Academies of Science to explore the state of research in climate intervention strategies, as well as the need to implement a governance structure of those technologies.

Dr. Bell, given the complexity of the climate system and the risks associated with further human interference, how do you think the U.S. should approach the field of research on climate intervention?

Dr. Bell. Both the National Academy and AGU, the American Geophysical Union, have statements that say this is an issue that we must research. If done wrong, it could be terrifying. But, again, it is the same problem that we have been saying before. We don't have the sufficient workforce looking at the issue, evaluating it, and building the body of knowledge to evaluate whether or not it is a good idea.

To me, I come back to the very, very few examples of geoengineering of the ice sheets that are out there. And to give you the idea of how many groups have done it, I think two groups have put it on the table. You know, one is basically for one approach— you get a bunch of snow blowers and put more snow back on the ice sheet. The problem is it turns out if you put snow blowers on the ice sheet, it gets steeper and it flows back into the ocean. It didn't work. The other idea is to build bigger than the Panama Canal many times walls to keep the ice sheets from being attacked by the warming ocean.

These are ideas being put on the table by a small cadre of glaciologists. What this illustrates is that we need, as a species, to research this, and we need not just glaciologists, not just atmosphere scientists, but we need to bring the full suite of talent to the table to think about this because, as we address climate change, we're going to probably need to look at every tool that we have available. That's what we found when we did the building down the street.

Mr. McNerney. Thank you.

Dr. BELL. We couldn't reach our goal by doing just one thing.

Mr. McNerney. Dr. Alley, do the risks of abrupt change in the Arctic and Antarctica indicate that we should be serious about technological interventions such as sunlight reflection to maintain stability?

Dr. Alley. So I would echo what the National Academy and what the American Geophysical Union have said, which is that we need the knowledge base that will allow you, all of you in this learned body, to actually make wise decisions. We don't yet have that knowledge base. There are real issues with international governance, as you raised, and thank you. There are real issues with reception by people. I can tell you stories of—geoengineering cloud seeding that led ultimately to a professor from Penn State having a hole shot in his car door because the local farmers were very unhappy with the idea of cloud seeding. Sort of how this plays out into the broader populace is sometimes not as obvious and as simple as you might imagine. So I think gaining this knowledge base so that you would then have the capability of making wise decisions is wise.

Mr. McNerney. Thank you. Again, Dr. Alley, the West Antarctic ice sheet has been noted to have the greatest amount of uncertainty in the melting and breaking rates. How much of the uncertainty related to West Antarctic ice sheet can be addressed by additional research, and how much is dependent on the future rates of warming?

Dr. Alley. Right. Certainly, the uncertainty can be reduced by the research, but it is already very clear that the faster and the more we warm, the more likely a failure will be. So in our world mitigation, trying to slow down the warming, buys you time. It buys you time to learn. There is always some danger with a tipping point that you pass it before you see it, and it's too late to slam on the brakes. It's too late to turn and avoid the iceberg. And very rapid warming, that becomes more likely for West Antarctica as we run at the future.

Mr. McNerney. Well, what are some of the—Dr. Bell perhaps— what are some of the major concerns about the collapse of the West Antarctic ice sheet?

Dr. Bell. The major concerns are that it could go fast, and we don't actually know how fast. It's back to the ice sheet we know there used to be an ice sheet in New York and many of the States here. We didn't see that one collapse or the residents of New York then who didn't record what was happening. So we as a species don't have the record of how an ice sheet collapses, so we worry about how it collapsed—what happens to the ocean, how the ocean chews at the bottom of it as the ocean warms. We worry about what happens when the surface melt, where does that water go? Does it fall into cracks and act like a jackhammer to open it up, or does it run off like a river? There are some major fundamental understandings about how warming air, warming ocean impacts ice. And in that sentence alone you see how we have to have different disciplines talking to each other.

Dr. Alley. So Dr. Moon is working on this problem in Greenland, and Dr. Pfeffer is working on this problem in Alaska, as is Dr. Wolken, so the truth is the—what we learn spreads broadly.

Mr. McNerney. Well, I hope the other three panelists don't feel neglected, but I only have 5 minutes, so I'll yield back. Thank you.

Chairwoman Johnson. Thank you very much. Mr. Babin.

Mr. Babin. Yes, ma'am.

Chairwoman Johnson. Changing of the guard.

Mr. Babin. All right, musical chairs. Sorry about that.

Dr. Wolken, in addition to serving here on the Science Committee, I also serve on Transportation and Infrastructure. And I represent southeast Texas and have four ports in my district. I recognize the importance of our navigational ship channels.

With that being said, one of the things I find very interesting on this topic that's relevant to my Committees and my district is the possibility of two trans-Arctic commercial shipping routes that are opening up. This isn't to say that I want to see all the glaciers melt and the sea levels rise uncontrollably, but if there are inevitable changes, I want to make sure that the United States is positioned to be economically fortified. And I know that the Russians are certainly exploiting newly opened up shipping lanes, ice-free zones, and even claiming certain areas that were considered in international waters are no longer that but belongs to Russia.

So how do you see the Department of Transportation or even the U.S. Coast Guard interacting with coordinated multi-agency collaboration that you say is needed?

Dr. Wolken. Yes, thanks for the question. I'll answer in two ways. The first is that what you speak of is really an incredibly important issue and, you know, economics and national security really do come to mind. And that's a sea-ice issue in the north, principally. And reduced sea ice of course is offering opportunities to enter into the Arctic and explore and ship, and that comes with fantastic opportunities of course and a lot of perilous conditions that could cause lots of environmental damage if not done right.

Having a multi-agency approach is incredibly important a little bit farther south. And you mentioned the Coast Guard. We have changes in Alaska that are impacting many of the fjords and the transportation routes in the South, and some of the changes in the cryosphere or changes in the snow, the ice, and the permafrost in the mountains are unpredictable to us right now. We don't have enough information. And so the Coast Guard communicating with various universities and agencies about how stable the slopes are, about how fast conditions are changing in certain areas could really be an asset to the Coast Guard as they respond to emergencies or possible disasters from cruise ships or fishing boats in different areas.

I will point out an example in 2015, there was one of the world's largest snow/rock avalanches into the Tyndall fjord, and in the process of that collapse, the tsunami that resulted from the rock falling into the fjord was enormous. It caused a trimline like the bathtub ring that was around 600 feet high. And any fishing boat caught in that or Coast Guard vessel or tourist ship would have been destroyed.

Mr. Babin. Right.

Dr. Wolken. So communication about the data that we have to the individuals who will be working in these different areas, Federal agencies such as the Coast Guard, it's critical that we have this conversation.

Mr. Babin. Absolutely. Thank you very much. And one other question. Some experts have predicted that our currently available mapping and navigation and ship capabilities are going to limit just how frequently and successfully we use these potential routes. And, Dr. Wolken, and to all of our witnesses, when conducting research on ice depth and volume, is there also efforts to improve commercial shipping potential such as data needed for mapping? Dr. Wolken, I'll ask you first, and then I'll go to Dr. Pfeffer.

Dr. Wolken. Yes, so a lot of the work that's being done in the fjords in Alaska are specifically focusing on the nearshore environment, and so the exchange of dynamics of interactions between the glaciers and the water in that environment. And so in the process of doing that, wonderful maps of the fjord are being generated; lots of different surveys of the coastlines are being generated in the process. And so the really great part about this is that we can have overlapping interests being served with good research in the right areas. And I think that's where this idea of having these interagency collaborations, these multiple perspectives, this team approach is really important.

Mr. Babin. Great. And, Dr. Pfeffer, I think you wanted to say——

Dr. Pfeffer. Yes, I wanted to respond because your description of the situation in Texas reminds me a little bit of an experience that I had about 5 or 6 years ago where I was employed as a consultant for the Prince William Sound Citizens Advisory Committee, which is an environmental group that was set up in the State of Alaska following the Valdez oil spill to provide environmental oversight in Prince William Sound, which includes the town of Valdez and southern terminus of the Alaska Pipeline. And their specific concern was icebergs.

The Columbia glacier in Alaska, which is one of the glaciers that I've worked on for many years, was a major iceberg producer. And those icebergs came out into the shipping lanes. And the Alyeska, which is the operating company for the Pipeline, and the Coast Guard were both concerned about what future iceberg hazards were going to look like. Specifically, they had an ice detection radar system that had come to the end of its useful lifetime, and they had to replace it. And what their specific question was, you know, do we have to be worrying about icebergs for the next hundred years or the next 5 years?

And so I worked with them for about a 2-year period developing some simple models based on how much of the glacier was left and our best prediction of what the retreat would look like to give them some sense of what the iceberg discharge would look like. It was a good opportunity to collaborate with a State-level agency and also with the Coast Guard. We have a limited amount of bathymetry for that region. It would be good to have more, and NOAA has done some surveying in there. But that kind of interagency cooperation could be a lot more frequent than it is, and when it does happen, it's extremely beneficial. It certainly was a great help to us in Alaska.

Mr. Babin. All right. Thank you. Thank you very much. I appreciate it.

Dr. Bell. I have a quick addition to that in that just last summer we saw one of the first groundings—I actually saw the vessel before it grounded of a Russian icebreaker that ended up grounding in the Northwest Passage, you know, exactly the places we're hoping or were thinking may be opportunities for more connections across the high Arctic. So it is a critical issue because it ran aground on an uncharted rock——

Mr. Babin. Right.

Dr. Bell [continuing]. In essence. And the other piece is that the Coast Guard provides critical infrastructure to support the work we do in Antarctica. Without the U.S. Coast Guard and the heavy ice-breakers, we could not, the U.S. could not run the flagship programs they do. And we are seeing the Asian countries invest deeply in icebreakers. The Chinese Government has invested in two. The Koreans have a beautiful new icebreaker. We need strong ice-breaking capability both for ability to engage in the Arctic and continue to be leaders in Antarctica.

Mr. Babin. And we have a shortage of icebreakers, do we not?

Dr. Bell. Yes.

Mr. Babin. Yes.

Dr. Bell. That's why I thought this was a moment to remind you that——

Mr. Babin. Yes. Thank you.

Dr. Bell [continuing]. With science, it's really clear—boats run aground, and we need icebreakers.

Mr. Babin. All right. Thank you very much. My time has long expired. Thank you.

Chairwoman Johnson. Thank you very much. Mr. Casten.

Mr. Casten. Thank you, Madam Chair. Thank you all so much for being here.

Earlier this Congress at the Environment Subcommittee hearing on the impacts of climate change on our oceans and coasts, our experts were talking about what we need to do to stay below 1.5 degrees of warming. And I asked them if we got rid of all CO_2 emissions tomorrow, how much sea-level rise is already baked in? And the answer was an unequivocal 2 feet. I think that's consistent with your testimony, Dr. Alley. That is frightening, but in some ways, I have a bigger fear that's the deficiencies of our little Homo sapien brand.

And I want to demonstrate this and I want all of you out in the audience, you get to participate now. We're going to do a little experiment. So what we're going to do—this is real easy. I'm going to say two things. You give me the next in order. A, B? A little louder, come on. You got this. This isn't hard. Thank you. All right. Second one, 2, 4? You're all wrong. I was looking for 8.

This is the problem, right? We have all of these nonlinear trends, and our little brain says 2, 4, 6 and we see all these things that are going on. And, Dr. Alley, I think you alluded to this in your testimony. And so if 2 feet is baked in and if the likely skew of that data is not a bell curve but on the more frightening end of the spectrum, what sea-level rise should we be planning for within the zone of possibility?

Dr. Alley. I surely wish I knew. This is a frustration for us at a level that is deep and I wake up at 2 in the morning and I look at the ceiling and I say what do I tell somebody? I can remember coming back from Old Ironsides on the water taxi while doing the Freedom Trail in Boston and sitting in the water taxi and putting West Antarctica into Boston Harbor and not knowing what to do, which is—I mean, I'm sorry, it is very self-serving for me to sit here and tell you that funding research is good because it might go to me or my students, but we want to know.

Mr. Casten. So I'm not asking you for certainty, and I appreciate—look, I started my career doing—I got a master's of science in chemical engineering. I get the caution. But we've got to sit on this side of the dais and make decisions, so I'm just asking if you were in our seats with uncertainty of information, what is the range that we should be thinking about in our zone of possibility?

Dr. Alley. Yes. Don't go below the IPCC and start thinking about flexibility. Think about adaptive capability, the——

Mr. Casten. I'm just asking for like a number of feet.

Dr. Alley. Yes, I can't give you a number.

Mr. Casten. How about a——

Dr. Alley. I wish I could.

Mr. Casten. How about a timing? How long do we have before 2 feet is locked in?

Dr. Alley. Yes, very soon if not and, so for the 2 feet, you're getting close. But the big numbers, it really is, you know, I mean, a good businessperson looks for the black swan, but they don't know when a whole flock of black swans is coming, and so they really do look to their best people to be ready, which is here. You know, that's you.

Mr. Casten. I want to get to a couple other things in my time, but the reason I ask this question is in part because the same day that we had that hearing—I sit on Financial Services. We had Federal Reserve Chairman Powell in, and I said to him we just had this hearing. You are responsible for helping us write 30-year mortgages. Do you factor in whether or not those mortgages are going to be paid off in low-lying coastal areas? And the answer was that he thought we probably should start thinking about that but we haven't yet.

We have a whole host of issues here that go just beyond whether the sea level's a little higher, right? We got housing. I live in Illinois where we've got, you know, polar vortex because—and polar bombs or whatever the term is of this year because, as that ice melts, we're destabilizing global weather flows and shifting that cold air down temporarily until we all get a lot hotter.

Dr. Wolken, I had a little fun doing a little Googling on the weather report on Moose Mountain where you live. I understand you got a huge unseasonable amount of rain a few hours ago. I understand that is pretty positive because you've got some concerns up there. Can you just help explain to me what's happening on Moose Mountain that makes that rain good and how that is related to the falloff in sea ice?

Dr. Wolken. Wow, that's a really good question. I will preface this with some history about the winter. It was a very low snow year. We didn't have near the snow that we would normally have. And this is a trend especially across the Arctic. And this year it's been unseasonably hot. In fact, this week in Alaska many records have been broken. And this is common as well in recent years.

I left Alaska the other day to evacuation notices, so before I came here, we were planning to evacuate our house because fires were raging just near our house. And so the rain coming is a great idea. The whole State is suffering from smoke right now because there are so many fires really resulting from a chronic low-snow issue and having warmer temperatures that are really fueling the fires. And so this is a major issue for us, and it's become quite personal for me.

Mr. Casten. So just last question for the whole panel, has anybody estimated how many people's homes are at risk because of this combination of sea-level rise, spreading wildfires, flooding in the Northwest? How many people do we need to be thinking about dealing with right now? Do we have any estimates of that answer, Dr. Moon?

Dr. Moon. So I'm going to give you an estimate that's just a fraction of those things that you just asked about. This is just an estimate on homes. It doesn't include power plants, airports, military bases, anything else, just homes. If we're looking at 1 foot by 2035, that would be about 140,000 homes. If we're looking at 4 feet of sea-level rise, that's about 1.2 million homes. If we're looking at 2 feet, that's about 300,000 homes. So it's in the hundreds of thousands, and if we look at levels where we're reaching 6.5 feet of sea-level rise by 2100, we're looking into the trillion-dollar kind of mark just for homes. That's not other roads, other infrastructure, et cetera.

Mr. Casten. And I would presume that's just coastal. That doesn't include Dr. Wolken's house that may be at risk——

Dr. Moon. And it doesn't include wildfires or any of those other things that you mentioned that will be also addressed by addressing climate change.

Mr. Casten. Thank you. I yield back.

Chairwoman Johnson. Thank you very much. Mr. Baird.

Mr. Baird. Thank you, Madam Chair and Ranking Member. I guess my question deals across the board. We got all doctors here as witnesses, so the question I have simply, what Federal programs are most critical to gaining a greater certainty on the future change in ice sheets and those effects on sea level? So you can go in any order you want. Dr. Alley, you want to start?

Dr. Alley. Right. It is interagency. The National Science Foundation provides so many of the people support, and they do the lead agency in Antarctica and in some other things. NOAA, we have to have what they are doing. NASA has been keynote not only on Operation IceBridge, which we have been talking about, but the satellite monitoring. The DOE has a role in modeling, and so I've hit a lot of the high ones, but it really is the interagency, the U.S. Geological Survey. When I gave the number on how rapidly the icebergs were breaking off when John Muir was watching, that number came from the United States Geological Survey. So it is having these wonderful centers of excellence that you have built that live in the U.S. Government and give us leadership, they are not localized in one place. They are in several agencies, and they work together, they know each other, and they can do this with support.

Mr. Baird. Thank you. I'll remind you that Madam Chairwoman gave me 5 minutes, so we can spread that out.

Dr. Bell. Well, I will echo the NSF for understanding why, NASA for monitoring how it's changing, the USGS for incredibly important measurements of the glaciers, DOE for modeling, and NOAA for lots of information about how the ocean is changing and what the fundamental tide gauges are doing.

Mr. Baird. Thank you.

Dr. Moon. You asked about narrowing our range of what's going to happen into the future. On the science side, integrating better observations and understanding of the physical system into our models, our models can't make up that information on their own. But I also want to reiterate that it is our mission's pathway that is going to make a tremendous difference in what that future number of sea-level rise and our future number for ice loss is. That's not the science part.

Dr. Wolken. Yes, I just want to echo the comments of my colleagues here and really just add that we're doing this in Alaska already. We're getting as many people together as often as possible to try to solve some of these issues. The only way to really do this is through an interagency perspective. And there's really no other way to address such a large issue. And all of the Federal Government programs are critical to what we do.

Dr. Pfeffer. OK. Well, again, I'm echoing what all of my colleagues have said, but I want to add to this. The problem of collaboration and communication between these agencies is not an easy task. One example,

NSF operates on a principle that could be summarized as turn the brightest people loose on the most interesting questions. The fundamental function of NSF is to support these investigator-based science where each one is evaluated on its own scientific merits. It's not a mission-driven agency in the way that, say, NASA is. That has produced—it's been extraordinarily successful by letting scientists decide on what's the best thing for them to study.

But in a situation like sea-level rise, I think that more—well, it's for climate change generally, not just sea-level rise, I think that a more coordinated approach is necessary. Back in the early 1970s the National Science Foundation had a brief program called RANN, Research Applied to National Needs, where basically a management structure was experimentally imposed on research programs. And it was a notable failure. Almost everybody that you talk to that knows about RANN say, oh, boy, yes, that was a bad time at NSF but not everybody.

It's a little bit like the Manhattan Project. If the Manhattan Project had started out with the, you know, advisor saying, OK, we need to understand about atomic energy, all of you pick an interesting problem and go work on it and come back in 5 years, you know, that's not the way the Manhattan Project worked. And I don't think we're going to solve this problem that way either. I'm not talking about the magnitude of the project or how much money should go into it, but I am talking about coordination and the need for some really innovative thinking about how those agencies should interact because it's hard to steer scientists and, you know, it really is a herding-cats problem.

But particularly with all these agencies, there needs to be some really imaginative way of figuring out what gets done first and how long do we have to solve it. And I don't have any answers to that, but I think that's a really strong need.

Mr. Baird. Thank you, and I'm out of time and I yield back.

Chairwoman Johnson. Thank you very much. Ms. Wexton.

Ms. Wexton. Thank you, Madam Chairwoman, for yielding. And I thank the witnesses for appearing today.

It has been quite alarming to say the least to read your testimonies and also listen to the responses to some of my colleagues' questions here today. It is absolutely clear to me and it should be clear to everyone that we are at a tipping point of our Earth changing dramatically and irreversibly due to human-caused climate change. It's even more alarming that we're locked into 2-feet of sea-level rise—everybody seems to agree about that—and that, given the melting of the Greenland and Antarctic ice sheets, we could be looking at 11-feet of sea-level rise.

And how you prepare for that is something that is really important to all of us and certainly to me in my home State of Virginia. We have a lot to contend with over that. We are home to Naval Station Norfolk, which is the largest naval base in the world, and Langley Air Force Base. They are already having to deal with the effects of sea-level rise and the effects it has on our national security. And I'm also the mom of two kids, and I worry about what kind of a planet we're leaving for them and their kids.

I know that we had some questions about climate ice cover and sea levels and how they routinely change from season to season and over time. Some claim that this natural variability means we shouldn't be concerned with humans changing the climate.

Dr. Alley, I know that Mr. Brooks asked a little bit about this and Dr. Pfeffer did give some explanation of what is actually happening. But, Dr. Alley, can you explain what the science tells us and why we should be concerned with the changes in ice and sea level and climate that we're seeing right now? What makes it different from what happened, you know, over the past 20,000 years?

Dr. Alley. Yes. So thank you very much. It's wonderful that people take interest in what we do, you know? So, as you know, on a dry, hot summer day, you know, the hills of Virginia have always burned when there was a lightning storm. And because you know that, if you see kids headed out on a dry day with illegal fireworks, you were very worried about it. We know that people have always died, so we have metal detectors at the front of your building here. We know that climate has always changed, and that proves that climate is changeable. And you've never met the person who said the hills have always burned, so we won't worry about arson. But you have met the person who said the climate has always changed, so we won't worry about humans changing the climate.

The climate has always changed proves that the climate is changeable. The climate change has always affected living things, which proves it's important. Climate has changed for a lot of reasons, but CO_2 has been especially important. And that points a finger at us. Now, if you were an arson investigator, you better know natural fires. You do CSI (crime scene investigation) fire. If you're a homicide investigator, you do CSI homicide. We do CSI ice. We do CSI climate. And we actually have very high confidence that what is going on now is human, not natural. If anything, over the last small number of decades nature has tried to cool it off a little, so how much of the warming has been us is a little bit more than all of it is the central estimate.

But the fact that nature has done these huge things in the past, that when nature warmed a little bit, sea-level rose a lot. And then you say, well, we could cause a whole Ice Age of warming with our CO2 in the future. And the last end of an Ice Age gave us 400-feet of sea level. There's 200 more left.

So I believe that climate has always changed is a very, very strong argument to be concerned about what we're doing for climate in the same way that burnable hills make you nervous about arson. And when——

Ms. Wexton. And related to that, Dr. Alley, in your testimony you discussed several studies that suggest that the IPCC report is overly conservative and underestimates the rate at which ice sheets are and will continue to melt.

Dr. Alley. I have great difficulty finding any evidence that they are overly alarmist, and there certainly are things that point to the possibility that they have been low in the past. And, yes, that's fairly clear. When you look at the history of——

Ms. Wexton. Can you discuss this current scientific research on estimates for tipping points for the Greenland ice sheet, Arctic ice, and Antarctica ice? What are the tipping points or what does the science tell us?

Dr. Alley. So Greenland, as it gets thinner, it gets warmer. As it gets warmer, it melts faster and gets thinner, and at some point it will be committed to loss. It probably will melt fairly slowly. West Antarctica, if it starts doing what the glaciers in Alaska have done, the coastal glaciers have done, it could go very, very rapidly. We're cautiously optimistic that the sea ice in the Arctic will act like a dial rather than a switch, but we're not entirely sure of that. We are worried a little bit about circulation in the Atlantic and other places that act more like a switch or a tipping point. The National Academy of Sciences looked at tipping points in 2013. They especially pointed to tipping points in ecosystems and in human systems. So at what point when people are stressed and they're having to move their houses or change what they do, at what point do the people become very mad and then tip into some other level of behavior. And so when you look, there are some physical tipping points, there are more ecological tipping points, and there may be a whole lot of people tipping points.

Ms. Wexton. Thank you very much. I see my time——

Dr. Alley. Thank you.

Ms. Wexton [continuing]. Is expired. I yield back.

Mrs. Fletcher [presiding]. Thank you. I'll now recognize Mr. Gonzalez for 5 minutes.

Mr. Gonzalez. Thank you, Madam Chair, and thank you, everybody, for being here for this important hearing.

I want to focus at least the beginning of my time on adaptation and resiliency. I think it unfortunately seems like there's a lot that's sort of locked in that we're going to be dealing with over the next however many years. And I'll start with Dr. Pfeffer. What are you seeing or what guidance can you give us with respect to making sure that we can adapt as sea levels rise and that we're building more resilient infrastructure?

Dr. Pfeffer. So I mentioned in my early comments I really am concerned with—in the work that I've done in the near term, the next 30, 40, 50 years where this constellation of factors has to be considered. One of the very interesting and extended conversations that I had was with a man named David Behar, who works for the San Francisco—the city of San Francisco as a coastal engineer. And one of the problems that they have to deal with are—it's a very large dike system that basically surrounds San Francisco Bay. And they need to know how far do they have to raise this dike system, which is extremely expensive? It's in the billions of dollars for a very small rise. And so it was not adequate to simply say, well, let's just be safe and figure on 10-feet of sea-level rise and then, you know—and you only get 1 foot and you've spent an awful lot of money.

Mr. Gonzalez. Yes.

Dr. Pfeffer. In the same sense, one of the questions—and this goes back to an earlier question about how many people may be displaced by sea-level rise. If you take an overly conservative number meaning let's take worse-case scenario and you draw a line on the coast saying, OK, this is going to be inundated by such-and-such a date, what happens to the value of those homes on the basis of that line that you've drawn? And the nearer in time you get, the more important that becomes. So you really have to have a tight bound on sea-level rise and a tighter bound to the nearer to the present that you get. We don't really have that yet. In some places we do, and it's—very often is a group of scientists that live in a particular region like Hudson River, for example.

Mr. Gonzalez. Yes.

Dr. Pfeffer. San Francisco Bay is another example where you can look at all of the causes of sea-level rise, including things like isostatic depression or rebound in an area as—partly as a result of large-scale things like ice sheets disappearing 20,000 years ago and partly local things like putting buildings on that land.

Mr. Gonzalez. Yes.

Dr. Pfeffer. There are a lot of different factors that have to be considered and different time scales you deal with different factors. And I think it's another thing that points to this interagency collaboration.

Mr. Gonzalez. Got it.

Dr. Pfeffer. But one of the things that I've tried to emphasize in the past is there's certainly a cost to neglecting sea-level rise, but there's also a cost to overestimating.

Mr. Gonzalez. Yes. And I think that's actually a really important point is, you know, when we talk about resiliency and adaptation, there is a cost to all of this, right?

Dr. Pfeffer. Yes.

Mr. Gonzalez. And we can't completely ignore that. We can't be too conservative or too aggressive or——

Dr. Pfeffer. That's right.

Mr. Gonzalez [continuing]. You know, we're going to be wasting a lot of money.

Dr. Wolken, if I could shift to you quickly, in your testimony you mentioned that in Alaska there are only three long-term continuous records of glacier mass for the entire State. Considering remote sensing and computer modeling are used to predict future scenarios due to the lack of ground-based observational data, how reliable and accurate are remote sensors and computer modeling in measuring glacial melt and predicting future changes?

Dr. Wolken. Yes, we're doing really well with these different tools, and I think, you know, one of the things that you can envision is if you go to the hardware store and you get a laser range-finder, for instance, from the shelf and, you know, you do some home renovations at your house, well, that laser is actually quite accurate. It's a laser, and it's very precise and accurate. And we use tools like that to really gauge how the ice is responding. We use other remote-sensing tools to do similar things, to see how much it's changing in this direction. And those are incredibly useful, and that's how we do things. We do those with both airborne and satellite-based assets.

There is a need in places like Alaska where the topography is so extreme and where the changes are so great to actually have ground observations. And so when you're using these different remote-sensing tools, the resolution isn't quite there some of the times, and so having ground observations to validate in some way or to correct in other ways is really the way to go. And so more ground observations truly do help us. With a lack of that, we have no option but to use the tools that are in front of us, and really, remote-sensing-based opportunities are where it's at for us.

Mr. Gonzalez. Great. Thank you, and I yield back.

Mrs. Fletcher. Thank you. I'll now recognize Dr. Foster for 5 minutes.

Mr. Foster. Thank you, Madam Chair. And I'd like to thank really the Ranking Member and all Members of this Committee and the witnesses about the tremendous increase in the level of serious discussion that we're having on issues like this over the last 2 years.

I think if you Google my name along with Greenland, you're led to a video of a previous witness who was a lawyer trying to convince this Committee that it was a matter of scientific debate whether or not it was a good thing that the Greenland ice sheet melted, OK? And so we're having a long-overdue and very high- quality discussion here.

Now, my next question, how many of you knew Charlie Bentley? Wow.

Dr. Alley. He was my Ph.D. advisor.

Mr. Foster. Well, oh, wow. I grew up next door to Charlie Bentley on Lake Mendota in Madison, and, you know, and I remember sitting on his porch discussing what he did. You know, he would disappear every couple of years and study the ice sheet in Antarctica, which did seem goofy. And I think it's a lesson on curiosity-driven research, that this thing, over the course of his career, went from something that was done by, you know, sort of an eccentric professor to something that is now going to be an absolutely crucial thing in deciding how we deploy trillions of dollars of capital to try to mitigate the damage of this.

Charlie passed away I think a couple years ago, and I understand there is a mountain named for him in Antarctica. Anyway, I was pleased to see the recognition among the Committee here.

Now my next question I had is, what is known about the speed of response of the ice sheet system to changes in temperature? You know, there are natural experiments when you get volcanoes going off with a couple degrees swing for a few years, is that long enough to actually be seen in the response of the ice sheet?

And the reason I'm asking this question is I think it's likely that we'll be able to decarbonize the U.S. economy. I think it is much less likely, you know, since we're 5 percent of the world population that we're going to be able to convince the rest of the world to decarbonize as quickly as necessary. And if that happens, then I think it's likely we'll be looking at things like albedo modification which has the potential of very rapidly changing the temperature. There's an article in Nature earlier this year that used state-of-the- art climate models to say, OK, you know, will it work or are we going to get cyclones and so on? And the first look was that it might be feasible.

But they didn't, to my remembrance, model anything having to do with the ice sheet. And so I was worried that maybe there was sea-level rise locked in just due to the thermal time constants, that even if you rapidly bring down the temperature of the atmosphere, that it will take a while. And so what is known in modeling or in data about that issue?

Dr. Bell. The ice sheets respond slowly—they're slow. I mean, when Richard and I started studying ice, we couldn't imagine they'd change as fast as they are today. I mean, Charlie actually— one of my first papers I wrote told me I couldn't write that they were going to change fast because even in the 1980s we couldn't imagine the speed at which we're seeing now. And now you can actually occasionally hear fear in scientists' voice because they are changing faster than Charlie thought they could when you grew up next to him, that we just couldn't imagine these thick pieces of ice changing, and he couldn't either.

But now we know they're changing due to the ocean warming and the atmosphere. The atmosphere is a faster driver than the ocean. So it will—there's—we don't have a good handle on how fast it's going to respond——

Mr. Foster. Do you have models that even make a decent approximation? Can you see, for example, in response to volcanic eruptions and the swing there, can you see changes in the rate of ice accumulation or de-accumulation?

Dr. Alley. We can do a lot of it right and a lot of it not yet, and so I could brag on the progress that we've made and some of it with Charlie's help. And I could bore you or scare you with where we're missing, especially the couplings into the ocean. So if you start blocking the sun, what it does to the atmosphere is fairly straight- forward. What that does to the ocean, which is interacting with the ice, is not at all straightforward, and that really needs work. And there is——

Mr. Foster. Are these computing-limited problems or knowledge-limited problems?

Dr. Alley. Yes, especially knowledge-limited. The computing is coming. We could use a little more, but it's primarily knowledge- limited.

Mr. Foster. Yes.

Dr. Bell. It also has to do with those measurements, our lack of knowledge of even what the ocean temperature is around Antarctica. We can look from space—we can measure the top of the ocean, but we're still so limited in understanding what's going on at depth, and that's what matters because the critical parts of the ice sheet that are really—the sensitive switches, those are down low, and we don't have the good measurements.

Mr. Foster. Let's see. No, I'll abuse another 20 seconds. Yes, for the last——

Dr. Pfeffer. Yes, I wanted to add, and it's already been mentioned that the IPCC's fifth assessment, their discussion of sea-level rise is very conservative. I was one of the lead authors on chapter 13, which is the sea-level chapter, and that discussion that we had about what number are we going to put in for our upper limit, and I remember that very vividly. And essentially what we did is we said we just do not know yet enough about the rapid tipping point mechanics to be able to attach a number to this rapid response.

Mr. Foster. Thank you. And to the extent these are compute-limited modeling problems, you're very welcome to use the super-computers at Argonne National Lab in my district—I can't think of a better use for them.

Dr. Pfeffer. Yes.

Mr. Foster. Thank you. Yield back.

Mrs. Fletcher. Thank you. I'll now recognize Ms. Stevens for 5 minutes.

Ms. Stevens. Thank you, Madam Chair.

We got to talk about the psychology here. I just went from another hearing on heat effects in the workplace and the overheating in the workplace from warehouses to fields and how that's impacting human health. We're willing the have the dialog on climate change.

So, Dr. Bell, you had a chart that kind of showed the sea levels and made that point about your father's life. How long have we been able to actually talk about rising sea levels and their impacts on us? How has modern science been able to influence this discussion and the question of what we can actually do to combat this?

Dr. Bell. The answer is, you know, people have been living with changing ice for a while, but the real understanding of the linkage between the changing ice—because people who live up near the mountains like the people who live in Alaska are very aware of the changing ice and the people who live at the beach. It's really I'd say in the last two decades that it's gotten very strong, that connection. In fact, back to your psychology point, it's really in the last decade that we're starting to see the conversation about the psychology of how we handle it.

And it was only this year that—actually last month that the Earth Institute handled the—convened the first-ever conference on could we ever talk about managed retreat? What would that mean? How do we have that conversation? So in fact we are just opening this door of connecting the work that we all do on frozen stuff and beautiful places far away with what happens to our assets at the coast, to our beaches, to the naval ports, to the airports. We happen to like to build airports. You know, of the 10 impacted airports on the planet, five of them are ours because that's a good place. So we're just starting——

Ms. Stevens. Well, and we're coming up with new terms, I mean, with these extreme weather events and what it means. And, you know, we can study the free-rider principle. We can study, you know, you start to think of like nuclear warfare or weapons of mass destruction and when faced with that threat and what do you as an individual do. What do you as a society do? What do we as a body of Congress do? What is it going to take for us to take this seriously?

I'm in Michigan, and I don't have a lot of sea around me but I got a lot of lake. And this is going to impact us. You know, I stumbled across a video. What is a world without ice? Is it going to take a modern society to see a full city go underwater for us to take climate change seriously, for us to take rising sea levels seriously and the grand challenge that we actually can do something about it, that it's a uniquely positioned challenge for us as the great America to take on?

So I don't know who else, with all your great expertise and your phenomenal science and all your great background can provide some guidance here, some common sense for us to not just talk about it but to do. And I don't know who can chime in here because we do this on recycling, with the plastics crisis, and what the individual can do, what the body can do, the body that we're currently in, and then on. Thank you.

Dr. Alley. Right. So you raised very important questions. I wish we had good answers. But you know the Nobel Prize in economics, corecipient last year, William Nordhaus from Yale developed tools which allow decisionmaking or inform decisionmaking. And he showed that efficient response on climate change helps the economy, right? If you want a bigger economy with more jobs, you take actions that honor the science on this. Many of our medical professionals, through their organizations, have said this is a serious health issue, that actions that would reduce the warming will have health benefits. Our military leaders have been very clear on the national security issues of not dealing with this. So environment and ethics are actually in the direction of economy and employment, as well as national security and health. And we can see futures in which very expensive sea-level rise happens and large other changes happen, and we can see very bright futures where we use our knowledge.

The Yale climate communications people have surveyed what America thinks about climate and climate science. Most Americans tend to accept climate science but not all. But many, many, many Americans are very excited by the solution space. And if you ask them should we solve it, even if maybe they're not sure there's a problem, they're happy to go look for solutions.

Ms. Stevens. Yes.

Dr. Bell. Representative Crist, I used to always show pictures of drowning Florida in my presentations. You know, I'd show how much I'd drown Florida. I drown Florida a lot. I decided I couldn't do that because it was depressing people, turning them off, and they were not listening. I see change—we have to move—before we drown Florida or New York where I'm from, we have to actually start thinking about how we as individuals—I very much worry about what I as an individual—I worry about my community I live in. I worry about my professional society, what we can do. And I'm very happy that you are asking—I worry about our local government, and I'm very happy that you are engaged—and it's essential that you take leadership on this, too, because we can lead if we step forward before we drown a city or a State.

Mrs. Fletcher. Thank you. I'll now recognize Mr. Crist for 5 minutes.

Mr. Crist. Thank you, Madam Chair. I appreciate the opportunity. I have some very well-prepared questions that my staff has put together for me, but you've inspired me to kind of go off script. I am from Florida, and my colleague asked I thought an excellent question about, you know, what is it going to take before, you know, Congress takes concrete action, each of us as individuals do so.

I live in St. Petersburg, Florida—it's on the west coast—and have lived there since 1960. And I live downtown. My parents live in the northeastern part of the city. We both live on the coast. So when I drive to visit my parents, who, thank God, are still alive, I go along the coast and I can see the difference in the sea level in the bayou that they live on. And I have noticed it significantly greater in the past 5 years. And I don't think that's just an anecdotal thing. I think it's a real thing.

You know, I previously served as Governor of my State, so I would travel the whole State quite a bit. And whenever I would go down to Miami, on Miami Beach in particular, there's a road called Alton Road. And Alton Road will flood when it's not raining. And I remember President Obama visiting south Florida and would talk about that example of, you know, the climate changing, the rising sea level. And so we Floridians get it because we've seen people drown a lot. And we're witnessing that occurring, you know, so it kind of freaks us out.

And, you know, it seems to me that we need to figure out a way to sort of get off the dime. And I'm sure, given your illustrious professions and dedication to what you study that it's got to be frustrating for you as academics to not see a whole lot of action in this area.

And I'm going to ask almost the same question but maybe in a different way. What kind of advice can you give us—as hopefully decent communicators to Americans—to motivate action?

Dr. Moon. So I want to reiterate Dr. Alley's point in emphasizing solutions. I had an opportunity to give a TEDx talk to my community in January, and I emphasized the solution space of it. And I had a friend then a month or two later sent me an article about the U.N. report, which told you about all the horrible things that are coming down the Pipeline for us. And she said is this true? This seems really radical. And I said, yes, all the information in there is true. And she said, you know, this hasn't motivated me at all, but your talk did.

So I want to emphasize that we need to be talking about solutions. That's motivation to people who don't even necessarily think about climate change because they wanted to be getting renewable energy, becoming energy independent, which is something that we can do with that sort of thing. The solution space is very inspiring. As Americans, we have led, we have innovated, we have created new paths for the world, and I think that we can convince people that we can do that in this space as well because, in fact, we can do that in this space.

And then the one other thing that I want to say in this area, too, is that it's about encouraging people to talk about this and come together with each other, too. We simply don't talk about this enough. And if we talk about solutions, we can also think about how we're directly helping people. I mean, in the last couple weeks we've heard about hundreds of people being laid off from coal mining jobs because of bankruptcies or other problems the—in decline in coal. But if we're thinking aggressively about moving forward, we can think about how are we going to give these people other jobs? How are we going to support them as we're losing this industry instead of just putting our head in the sand as we lose this industry, which is hurting people on both sides.

Mr. Crist. Well, if I could follow up, I have a little time left. In speaking about the solutions, what are the most obvious ones to you that you would be willing to share with us?

Dr. Moon. Well, I'll tell you, I'm a scientist, so in my personal——

Mr. Crist. Thank God.

Dr. Moon. In my personal solution space, a lot of it is in communication. I don't envy you as policymakers and having the much more difficult job in discussing all of the elements, not just science, that go into your policy decisions. And unfortunately, many of the questions on those solutions lie on your desks. And I really would love to see us depoliticizing climate change so that all of you can spend your time discussing which of these solutions we're going to implement and how.

Mr. Crist. Thank you very much. I yield back. Thank you, Madam Chair.

Mrs. Fletcher. Thank you, Mr. Crist. I'm going to recognize myself for 5 minutes, and then we'll continue with the hearing. This has been a really great panel, so I want to thank all of you for the time that you've taken with us this morning. And I want to thank Ranking Member Lucas for holding the hearing. We've heard great questions and great answers.

And there's a lot that I'd love to cover, but, Dr. Moon, I want to go back to something that you talked about in your opening. You were talking about sea-level rise and sort of potential possible rise levels. And you mentioned that it was going to be 2.5- to 3-feet potentially in the next 80 years, and that number would be higher on the Gulf Coast. As a Representative of Texas' 7th congressional District in Houston right off the Gulf Coast, that of course perked up my ears. And I would love to learn more about why this is, what is the best estimate for the Gulf Coast region? Certainly, all of our eyes are on New Orleans right now. All of us are focused on the impacts of hurricanes and overall sea-level rise on our coastline and the Gulf Coast. And so part of the question is why is it and also what can we do about it?

Dr. Moon. There are a variety of things that determine sea-level rise in your local spot. So where we're losing ice on the Earth makes a difference. You are going to be influenced differently by losing ice in Antarctica than Greenland. There's also the ways that ocean currents and atmosphere currents move, pushes oceans one way or another, and also what's happening in your local region as far as your land naturally rising or falling already. And that's a— land subsidence is something that we see broadly across the Gulf Coast.

So there are these multiple different elements that all stack up to make what you in your individual city are going to see as far as sea-level rise. And it's quite consistent that in the Gulf Coast region we will be seeing substantially more than the global average over the—since roughly 1960, many areas along the Gulf Coast have already seen 8 inches or more, which is much more than the global average during those periods.

Mrs. Fletcher. Thank you. That's helpful, helpful information. And something that we do talk about a lot and we talk about resilience and rebuilding a resilient infrastructure, there are a lot of issues, and I think it is top of mind in a way that it might not be for some other folks in terms of sea-level rise. But I think one consistent theme I've heard from every witness today is that we need more people doing the research, helping us get the information that we want to know so that we can make smart policy decisions and that we can know what we're dealing with.

So I really want to put this out to the entire panel to talk about how we are recruiting and training the next generation of glaciologists and where there's room for us to help. What kind of policy can we implement here, what kinds of things can we do in addition to funding that would be helpful for you, and for anyone who wants to take that on and talk about what we can do to increase that number from 1,400 to—and maybe what number you think would be good overall.

Dr. Bell. Well, there are 13,000 people who are members AMS just to give you an idea of what—who—and that's the American Meteorological Society, so who's working and worrying about the weather in the U.S. We have 13,000 people doing that, and we have 1,400 around the globe doing ice. So numbers should be higher than 1,400. Let me give you an order of magnitude.

What can we do? I think it's partially making it so everybody can talk about the science and back to Twila's point about it not being politicized and also making it so—I think we're driving some of the young talent from the field because it seems like it's a hard place to be, not because it's hard to go to the field and see beautiful places but because it's hard because you're under attack.

I think embracing science so we have within our communities science-based, evidence-based planning for the future I think will attract more people because young people want to make a dif- ference in the world. And if they see there's science, even if you're studying how ice deforms and flows, is going to matter to what happens in your district, that's one way we can help attract it, by working on—even by holding this hearing is huge, but by working to ensure we have scientists intimately involved with developing the policy on how we're going to lead in the future.

Mrs. Fletcher. Thank you. Would anyone else like to weigh in on that?

Dr. Pfeffer. Yes, if I could add——

Mrs. Fletcher. Dr. Pfeffer.

Dr. Pfeffer [continuing]. A couple of comments to this. I mean, I think we're suffering to a certain extent from sort of, you know, the boiling frog syndrome of things changing around us at the moment at a rate which is, you know, gradual enough that we can say, oh, you know, this is just sort of natural variability or I remember something like this happening 20 years ago. I think making climate change generally a reality for people involved, somehow bringing it out of sphere of scientists. You know, a news report will say, OK, here's a scientist in Antarctica who has done such-and-such and thinks this, and then they show the picture of an icebreak or something, which to the ordinary, you know, person on the street, it looks like these scientists are on a different planet. It's all kind of removed from them and—in the hypothetical.

And somehow this link—and I think things like this hearing are creating this link that's not just scientists in this hypothetical space discovering this thing which can only be detected through sophisticated measurements but that it's actually happening in a way that everybody is feeling, and it's happening now. We're no longer waiting for the evidence that climate is changing. We've got it. We've got buckets of it. And that boils down to communication.

And I've done a lot of public presentations. As was mentioned earlier, I was involved in the movie "Chasing Ice" and have done a lot of that kind of public communication both before and after, and very often I get questions from people about, you know, what can we do? And it can be very hard to answer that question, especially if they're asking what can I do personally about climate change because it just seems like such a big problem? And one of the things that I do say to them is, you know, things like installing fluorescent light bulbs and, you know, buying a more fuel-efficient car doesn't seem like much, but we did create the problem one airplane seat at a time, one car at a time, one truck at a time. And the individual action does matter if everybody does it. And so recruiting people to understand and accept that this is a reality is sort of the first step.

Mrs. Fletcher. Thank you. I yield back my time, and I'm going to recognize Mr. Tonko for 5 minutes, who will then close the hearing. Thank you all very much for your time today.

Mr. Tonko [presiding]. Thank you, Chair Fletcher, for what I think is a very important hearing. Thank you to the panelists for setting such a respectful tone for science, refreshing.

I represent New York's 20th congressional District, upstate New York, and it's home to much innovative pioneering work, the topic before us. At Union College in Schenectady, for example, Professor Rodbell has been working for more than 30 years to document glacier fluctuations in the Peruvian and Ecuadorian Andes. Professor Rodbell and his students are conducting ongoing research on glaciation in the Andes with a specific focus on determining rates of current ice retreat compared to natural rates of ice retreat in the geologic past.

At the University of Albany, Dr. Mathias Vuille, a professor in the Department of Atmospheric and Environmental Sciences, is researching climate impacts and glacier retreat in the tropical Andes. In February of this year, Dr. Vuille testified in a public hearing held by the New York

State's Senate Standing Committee on Environmental Conservation. He noted that sea-level rise is resulting from warming of the ocean and added water mass due to ice melting glaciers and ice sheets in Greenland and West Antarctica. He noted in particular that sea-level rise is not equal everywhere and sea-level rise in the mid-Atlantic and New England coasts are much larger than the global average. He also emphasized that since we have no glaciers in New York State, impacts can seem far away and irrelevant, but glacial melt affects us nonetheless.

So, Dr. Bell, can you describe the indirect impacts of glacial and ice sheet melt in States like New York that I represent that do not have glaciers?

Dr. Bell. Well, thank you very much for that question. I'm also from New York, so—and the ones I'm going to speak of are—actually the nice examples are in New York because—because of sea- level rise, the number of people impacted by Sandy was significantly larger because—because of that in New York it's about 9 inches in the last hundred years the sea-level has—you can see the record right from the Battery. And you can see how many more homes were flooded, how many more people were impacted, and today, we're seeing that those are the homes that are actually being built up along the edge of the Hudson. It now looks like you're in New Orleans. The homes are being elevated right there in Haverstraw. You have homes that you could see in New Orleans.

So that's the kind of impacts we're seeing. You're seeing that we've had Sandy. We impacted far more people, tens of thousands more people than we would have, and now we're responding to it.

Mr. Tonko. Thank you for that. And what more can you tell us about the uneven distribution of sea-level rise across our country? What will sea-level rise look like, for example, on the East Coast versus the West Coast or in New York City versus Washington, D.C.? What are the wide-ranging impacts of sea-level rise?

Dr. Bell. The National Climate Assessment did a beautiful job of laying out those variations and going through the different parts of the U.S. and really explaining the difference. But briefly, each community has to worry about which ice sheet you're close to. If you're close to an ice sheet, it turns out it doesn't matter as much, so for New England, Greenland—Antarctica matters way more than Greenland, whereas the Representative from Florida is going to see both Greenland and Antarctica full on. So there's the proximity.

Then there's how close you are to ocean currents. That's some of the change we've seen in New England is the warming ocean has impacted New England. And then the Representative from Virginia is seeing the tremendous impacts of local subsidence around Norfolk because you've withdrawn water, so the land is going down at 4 millimeters a year. You add that onto the sea level going up, suddenly, you have a problem.

Mr. Tonko. And to anyone on the panel, what mountainous regions around the world are most at risk, and what adaptation measures can be taken to avoid large flows of environmental refugees?

Dr. Pfeffer. If I could——

Mr. Tonko. Yes, Dr. Pfeffer.

Dr. Pfeffer [continuing]. Address that, there are potential for environmental refugees at sort of both ends of the hydrologic cycle. Let's discuss the Himalayas, for example. Earlier, I mentioned the various geologic hazards that people in the immediate vicinity of glaciers, these high valleys, high density of people in those valleys. As we go downstream, there are people who are very dependent upon runoff from those mountains for crop irrigation, so this goes out of Nepal and into India. And then the people on the coast—and Bangladesh is very often used as the example—that are at risk from sea-level rise.

So everything from geologic hazards to changes in water supply to sea-level rise, each one of those has a population which is put at risk. And as far as mountainous regions where this really matters, certainly the Himalayas, also portions of South America, Alaska is subject to certain risks, but the primary influences there I think are going to be environmental on the changes in water and immediately coastal effects.

But the people I think really in the Indian subcontinent, they're at very high risk, and that is a global problem. It's not just a problem for them, and I think that's probably very clear.

Mr. Tonko. Thank you. Thank you very much.

Look, that concludes, I believe, all who have chosen to ask the witnesses any questions. Before I bring the hearing to a close, I do want to thank this panel. Thank you so much as witnesses for testifying here before the Committee. And I want to thank both our Chair and our Ranking Member for hosting this hearing.

The record will remain open for 2 weeks for additional statements from the Members and for any additional questions the Committee may ask of the witnesses. And we ask that you respond as efficiently as possible.

And then finally, I will say the witnesses are excused, and the hearing is now adjourned.

[Whereupon, at 12:24 p.m., the Committee was adjourned.]

APPENDIX I: ANSWERS TO POST-HEARING QUESTIONS

Answers to Post-Hearing Questions. Responses by Dr. Richard B. Alley

Answers for the Record to Member Questions
House Committee on Science, Space and Technology

From the hearing
Earth's Thermometers: Glacial and Ice Sheet Melt in a Changing Climate
United States House of Representatives
Room 2318 of the Rayburn House Office Building
10:00 a.m., July 11, 2019

Dr Richard B. Alley*
Pennsylvania State University

*Any opinions, findings, conclusions, or recommendations expressed in this publication are those of the author and do not necessarily reflect those of the Pennsylvania State University, the Intergovernmental Panel on Climate Change, the National Academy of Sciences, or other organizations.

Submitted by Chairwoman Eddie Bernice Johnson

1. What is the biggest knowledge gap that is essential to fill to improve our projections of future sea level rise due to glacial melt?
   - The short answer, in my opinion, is that the biggest knowledge gap is the lack of understanding needed to predict possible "tipping point" behavior of Thwaites Glacier, West Antarctica and other such outlets that could rapidly and greatly raise sea level.
   - The longer answer is that actionable results are most likely to be supplied from sustained, integrated research that addresses the complex range of issues required to project future sea-level rise due to glacial melt, including efforts to maintain and expand participation by young scientists.

My written testimony briefly outlined the possible tipping-point behavior that could greatly and rapidly raise sea level. An international effort, especially funded by the US National Science Foundation (NSF) and the UK National Environment Research Council (NERC), with contributions from other nations and agencies, is addressing this issue, but in my opinion is small compared to the scale of the challenge and the potential cost of the sea-level rise if Thwaites Glacier collapse proceeds rapidly. I am happy to provide extensive information on this topic if desired, with pointers to broad-based assessments of key unknowns and ways to address them, with as much technical detail as needed. The Scambos et al. (2017) paper cited below is a useful starting point. Simply put, the research community has considerable understanding of the known and unknown unknowns that stand between us and the ability to project whether sea-level rise under strong warming will be two feet, or three feet or four, or more than ten feet by the end of the century, the community is working hard with the available resources, but resource limitations are still reducing our ability to move forward. (I know this is self-serving-I am a member of that community, and various colleagues and I personally have benefitted from the resources already made available-but I believe my assessment is accurate.)

More broadly, though, the key question of ice-sheet tipping points is inextricably coupled to a larger research framework. Allow me to make a few points, which I believe reflect the insights of the community, although I am presenting them as my opinions as an individual:

1. The solutions involve helping young people enter and remain in our field. I am privileged to teach, and so I get to speak to many young people who are interested in helping their country and the world prepare for the future. These bright students can go in any direction they want-they can conduct seismic surveys of great value to oil companies, sift through "big data" in ways that interest tech firms and banking companies, write computer code that could be used almost anywhere in the economy, and more, while communicating accurately and engagingly. Most of these young people will go to industry, finance, and other parts of the economy, but a few of them want to work in the climate-energy-economy nexus to extend our knowledge of climate change and use that knowledge to help people, even though these students expect to take a pay cut compared to their other options. But, these students express grave doubts about the wisdom of studying climate; they are not confident that the US government and other funding sources are sufficiently committed to this area to support a career. (These reports are anecdotal, and collected off-the-record so I cannot quote names and dates for you, but I have heard these worries repeatedly.)

   Our family just toured the amazing "Destination Moon: The Apollo II Mission" exhibit from the Smithsonian Institution in partnership with The Museum of Flight in Seattle, a wonderful testament to problem-solving and exploration. Many of the displays highlighted the unsung heroes at home who were essential to the success of the moon landing, including those in Houston: "When Apollo missions began flying in 1968, Mission Control had become a magnet for the best and brightest young engineers in the country." Solving the problems of climate, energy and economy requires getting a few more oftoday's best and brightest scientists and engineers to help us now, not scaring them away.

2. The solutions are broad-based. NSF is the US lead on studies of Thwaites Glacier now, but the observational and scientific contributions of NASA are essential, ice-sheet changes are tightly coupled to atmospheric and oceanic climate in ways that must involve NOAA, the modeling capability of DOE is increasingly important, the US Army Corps of Engineers has long contributed through their Cold Regions Research and Engineering Lab, the Coast Guard icebreakers have been essential while the 109th Airlift Wing of the New York Air National Guard provides heavy-lift aircraft for deep-field access on ice sheets, and contributions of USGS, the National Park Service, and others are important and can grow. Philanthropic funding has helped us in many ways, and we are deeply grateful for that, but does not in any way replace the broad-based, publicly accountable role of the US Government.
3. The known-knowns are important as well as the less-known or unknown unknowns of tipping-point behavior. If tipping points are not crossed, important uncertainties still remain about mountain glaciers, surface melting in Greenland, and more. The question, below, from Congresswoman Lofgren highlights possible linkages between ice melt and changes in ocean circulation, but those in tum will affect the pattern of sea-level rise by changing the size and location of the places that winds and currents pile up water against the shore or move water away. Coastal planners surely want to know whether they face more than ten feet of sea-level rise or "just" three feet, but they also are very interested in knowing whether that three feet could be two feet or four feet.
4. Research addressing the known-unknowns contributes to understanding the tipping points of the big ice sheets. For example, those of us who started out working on the big ice sheets have relied on work by many people including Professor Pfeffer on the

behavior of Alaskan tidewater glaciers; knowledge of the history and mechanisms of rapid ice retreat in coastal locations informs our understanding of possible rapid retreat in Antarctica. For various reasons, iceberg calving is often best studied on Greenland's ice now. Better ways to combine data and models, and continuing improvement in data collected through and beneath the ice, are frontiers where much essential progress can be made.

Again expressing my opinion, I believe that if you pressed most members of the community to pick a single research target, they would choose further research on ice-sheet tipping points. But, I also believe that they would make that recommendation without enthusiasm, because they understand the broad-based nature of the problem and the integrated approach needed to find solutions, surely involving helping students as well as satellites and other research tools.

References

Scarnbos, T., Bell, R. E., Alley, R. B., Anandakrishnan, S., Bromwich, D. H., Brunt, K., Christianson, K., Creyts, T., Das, S., DeConto, R., Dutrieux, P., Fricker, H. A., Holland, D., MacGregor, J., Medley, B., Nicholas, J. P., Pollard, D., Siegfried, M. R., Smith, A. M., Steig, E. J., Trusel, L. D., Vaughan, D. G. & Yager, P. L. (2017). How Much, How Fast?: A Review and Science Plan for Research on the Instability of Antarctica's Thwaites Glacier in the 21st century. *Global and Planetary Change, 153*, 16-34.

Submitted by Congresswoman Zoe Lofgren (AC-19)

1. Studies suggest that slower Atlantic Ocean circulation would cause colder winters and hotter summers in Europe, change rainfall patterns in the tropics, and pool warmer water along the U.S. East Coast which would contribute to sea level rise and more extreme storms. We've already seen how warmer water off the East Coast

fueled extremely destructive hurricanes like Hurricane Maria and Superstorm Sandy.

(a) How do melting glaciers affect ocean circulation, and to what extent is research exploring this feedback loop?

The biggest role of melting glaciers is to raise sea level, flooding coasts and also potentially increasing storm surges by putting deeper water closer to buildings along the shore. But, this question does correctly identify a possible role for meltwater in ocean circulation. The most important sites for such influence are in or near the polar regions, in the north Atlantic and in the Southern Ocean, although changes are possible elsewhere.

Ocean circulation is driven in part by small differences in the saltiness of the water affecting the water density, so adding meltwater can indeed affect circulation. The best-known concerns point at the likelihood that freshwater added to the North Atlantic is slowing the rate at which water sinks to the depths there, affecting weather, climate, ecosystems and more. We are moderately confident that there will not be large, rapid and catastrophic changes, but important uncertainties remain despite rapid progress. Key monitoring and research are in place moving forward, but the ocean and the ice are changing rapidly and may get ahead of our learning, so I believe that additional research could be justified strongly, involving data collection today, model-data integration, and further work on the history of this complex and critical system.

Recent attention has focused on related changes in the Southern Ocean. Observational advances have shown us how much we don't know, and more clearly illustrated how tightly coupled the ice, ocean, air and ecosystems are in that vast ocean around Antarctica. Accelerating flow and melting of the Antarctic ice sheet dumping water and icebergs into the Southern Ocean could actually slow warming or cause cooling at the surface far to the south, but allow warmer waters beneath to reach the ice sheet and drive faster melting and ice loss, perhaps crossing the threshold for a tipping point. Deep uncertainty is attached to this possibility, with much research needed to project how this will evolve in a warming world. Again, sustained observations of ongoing changes, data-model integration, and studies of climate history can move understanding forward.

(b) Dr. Alley, in your testimony, you discuss how ice sheet melt is causing Earth's rotation to slow down very slightly and likens it to a spinning ice skater sticking out her arms. What are the implications of a slower rotation of the Earth?

The direct implications of changing Earth rotation are relatively small, but the science is important. Experts maintaining precise time must relate our noon-midnight timekeeping on a world with slowing rotation to the exact passage of time, and so must account for changes in the length of the day, but the technical means of doing so are well-known and functional. The detailed pattern of sea-level rise is actually affected a little by the changes in Earth's rotation, and that effect must be included in projecting future sea-level rise; this is broadly functional, although ongoing research can improve the results.

The point I hoped to make by introducing this topic in my testimony is related to just how much effort the scientific community puts into getting the right answers, how strongly interwoven the threads of knowledge are, and thus how reliable the results are that the community brings forward through assessments to policymakers such as yourself and your Committee. Measurements of growth or shrinkage of ice sheets are made by repeatedly weighing the ice sheets using satellites, by watching the surface elevation of the ice sheets change, and by comparing the snow added and the ice lost by melting and iceberg calving. These measurements agree within the known uncertainties of the measurements, and together give very high confidence that the ice sheets are losing mass. The water from this ice loss goes into the ocean, and the rise of sea level is what is expected from loss of mountain glaciers and ice sheets together with expansion of the ocean as it warms, providing a further check on the different measurement techniques. But, these also make predictions about changes in Earth's rotation, and when scientists check, these predictions are borne out-the different measurement techniques agree with each other, with the measured sea-level rise, and with the changes in Earth's rotation. We have even looked back in time at the record of eclipses as observed by past civilizations because the changing length of day affects those observations, confirming our understanding of the history of sea-level rise and the recent acceleration as human-caused warming melts ice.

You may occasionally hear some critic dismissing scientific knowledge as being uncertain and unreliable. Earth's rotational change is one of many examples showing that the scientific results are confident and reliable. When the scientific community brings forward to policymakers assessed results through the National Academies of Sciences, Engineering and Medicine, or the Intergovernmental Panel on Climate Change, those results rest on an interwoven fabric of evidence, such that even if a weakness were found in one thread, the fabric would remain intact. The occasional caricature of science as resting on one study or one investigator is simply wrong. If Einstein were erased from history our understanding of relativity would not change because so much knowledge has accumulated; we celebrate the pioneers, but rely on their insights only after extensive testing followed by assessment. The reliability of climate science never rests on one study or one scientist but instead on multiple strong lines of evidence.

Responses by Dr. Robin E. Bell

Response to Supplementary Questions: Glacier Melt Hearing Prof. Robin Elizabeth Bell August 8, 2019

Chairwoman Johnson Question 1. You have described how the work of understanding changing ice and glacier melt is spread across multiple agencies. Is this effective and if not what changes would you recommend?

The federal work on changing ice and glacier melt and its impacts is spread across multiple agencies. The scope of this impressive work ranges from fundamental research on how ice sheets work (NSF and NASA), to earth observations from space of how the ice is changing (NASA), to building models that will forecast future change (NSF, DOE, NASA). But the changing ice does not stay at the poles as documented by measurements spanning over a century (NOAA) and decades (NASA). The impacts of glacial melt and changing ice are being observed and quantified along our changing coastlines by the USGS. Simultaneously states, counties, municipalities and many branches of the Federal government including the Army Corps of Engineers, National Park Service, USDA, and Department of Defense, are dealing with the consequences. All these stakeholders are looking to the glacial scientific community for information and projections to help them plan. The National Climate Assessment provides a snapshot of sea level projections and some of what has been done in communities and is a very important resource, but there is no comprehensive plan on how our nation will respond to glacial melt and its impacts on communities.

Now is the time to act. We must develop a strategy to link the rich glacial melt research underway with the needs of the nation especially the agencies and communities now facing accelerating change. A plan on how to holistically bridge this gap is essential. The scientific community knows improving our projections of future ice will melt is essential but a gap still remains between the changing ice community and the changing coastline community. This gap is between global projections and what local communities need to build effective responses. Different communities have different needs. We must develop a plan to respond to changing sea level in the coming century even as we work to constrain the actual numbers. Preparing for sea level rise is like preparing for an earthquake. Our communities benefit greatly from planning before an earthquake even if we do not know exactly when the earthquake will come or how large it will be. Melting glaciers, changing ice and changing coastline require the same planning. I believe it is absolutely essential to create a better framework to prepare the change we know is coming. A major study by the

National Academies of Sciences, Engineering, and Medicine linking effective preparations for changing sea level with the science of the changing ice, warming oceans and changing coastlines could help identify gaps in our knowledge and build synergies that will better prepare communities at risk. A study would encompass an inventory of the existing programs, strategies for response, and develop frameworks for communities, agencies and individuals to respond. This study would strengthen the connectivity among the experts working all the facets of the issue to those responsible for planning and mitigation efforts. Such a study would build on the recent National Climate Assessment and strengthen the next. A congressional recommendation for an Academy study to develop a robust response to changing ice and changing coastlines would be a major service to all.

Chairwoman Johnson Question 2. Does the Notional Climate Assessment play a role in preparing our coastal communities for the impacts of glacial melt we can anticipate in the coming decades and if so what is that role?

Every four years, the National Climate Assessment presents the state of the art of the science behind sea level rise in the climate science section and evaluates the impacts and adaptation both across sectors and for all the regions of the U.S. It is an illustration of how the science and our communities can work together to quantify the risks to changing climate and share impacts and responses. This structured national process is as essential to planning for our future as is the census- the Founding Father's just has not considered climate change yet. With this structured review of the science agencies, communities, and individuals have an opportunity to understand what is happening, how it is impacting communities and what sort of solutions are being developed. A remarkable resource, the National Climate Assessment plays an important role in making it identifying the impacts of climate with detail and clarity.

Chairwoman Johnson Question 3. What is the biggest knowledge gap that is essential to improve our projections of future sea level rise due to glacial melt?

The biggest gap that is essential to improve our predictions of future sea level rise is our knowledge of how Antarctica will change in the future. We know where it is changing most quickly, and we see recently accelerating hot spots. What we do not know is how the warming ocean is reaching the ice -the trigger that has started the speedup - and we do not know what will happen when the air warms up and the surface of Antarctica begins to look like Greenland. Antarctica impacts the most people in the U.S. and globally, and it the biggest wild card in our sea level rise projections. Addressing these questions will require major focused fieldwork and probably large scale international collaborations. Antarctica is simply hard to work in. It is big, cold and remote. Even though it is warming quickly it is still inhospitable and a challenging place to work.

Chairwoman Johnson Question 4. Given the understanding that sea levels will be different for each community due to gravitation fingerprinting and other local factors, how do you think we can make personal for communities the connections between global changes in ice with the changes we will see here in the U.S. and around the globe?

When I began to realize how complex sea level rise it struck me that it was essential to move past making isolated statements of sea level rise from individual ice sheets or even a global mean. Who knew how complex and beautiful our planet could be? Communities need to know everything from where the ice is melting, to whether their region is sinking or rising and how the ocean is warming. An effective approach can frame how sea level has changed in the past, considering what range of change a community will see in the future and will look at the assets close to sea level. It is essential to convey that we can develop a plan, work to minimize the change and mitigate the uncertainties and surprises around sea level rise.

Chairwoman Johnson Question 5. In your research field, what is the importance of convergent research? How will it help the U.S. address glacial melt and foster resilience in our coastal communities?

Convergent science emerged from the biomedical science where advances on complex problems are only possible with a broad team of experts working closely together. The ROSETTA-Ice team working on

Antarctica's Ross Ice Shelf, which I lead, illustrates the power of the convergent approach. The ROSETTA-Ice team of geologists, geophysicists, oceanographers, atmospheric scientists, engineers and glaciologists produced novel insights because it brought such diverse expertise to the table. The team discovered that the geologic framework of the Ross Sea is preventing warm ocean water from reaching the grounding line of the ice shelf but discovered the sensitivity to change is at the ice shelf front where local winds can bring in heat. The next 100 years must focus on convergent collaborations to advance our understanding on how ice sheets change. This type of science is hard. It is easier to look at a problem from the safe base of your knowledge but reaching across disciplines can be rewarding and provide key insights. Our earth and the ice our planet does not care about disciplines. More broadly convergent changing ice research will link the changing ice in the poles to our impacted coastlines and communities. Connecting our understanding of changing ice to changing coastlines is beginning but must expand in the coming decade. Each community will need an individual projection that includes ocean warming, melting ice, changing land levels and coastal dynamics. It is essential that we educate next generation convergent changing ice and changing coastlines scientists. This convergent workforce will actively engage with the entire breadth of scientists, engineers, social scientists and policymakers who are essential to providing communities with the information they need to plan their infrastructure investments and land use in the coming century

Congresswoman Hill Question 1a. Dr. Bell can you expand on your role as Chair of the National Academies review of the National Climate Assessment?

Peer-review of science is one of the gold standards of the modern scientific process. I served as the chair of a panel of 16 experts who, through the National Academies of Sciences, peer reviewed the National Climate Assessment. I lead the committee through the process of reviewing all the chapters and identifying issues within the document. As we reviewed the report we considered whether the information was broadly accurate and represented the current state of understanding at a level that would inform decision-making and be accessible to a broad, non technical audience.

The National Climate Assessment is a key document that communicates how climate change is occurring in our backyards and, that together with strategies informed by research, our communities can be resilient. In addition to sea level rise, the report covers human health and community well-being, the built environment, businesses and economies, ecosystems and natural resources. Hundreds of experts from federal, state, and local governments, academia, non-government organizations, and the private sector assembled the report. The Assessment process also gathers from community engagement events and public comment. By bringing together such a deep array of perspectives on climate change impacts and resilience, the effort represents a tremendous large-scale collaboration. Because the climate change report highlights adaptation and mitigation efforts, it serves as a valuable resource for many. We delivered our report (that itself was peer reviewed through the National Academy process) to the National Climate Assessment team and they addressed the issues we raised. We did ask they increase discussion of adaptation and mitigation efforts emerging as these are critical to communities as they address the change.

Reading the 1500 pages on the impact of the climate on our country, from the Virgin Islands to the Pacific territories, gave me and the committee a profound understanding of the scope of climate change that our nation is already experiencing. Melting glaciers and changing sea level is one of the strongest signals of the change across the U.S. but the change is right in our backyards.

Congresswoman Hill Question 1b. What is the importance of this document to understanding the consensus in this field?

The National Climate Assessment provides both the state-of-the art science across the U.S. and its territories and a detailed tracing of the source of the information. The National Climate Assessment presentation of the data (such as temperature change) is at such a detailed level it is possible to see how much temperature has changed in individual communities. Some news outlets have developed useful ways to look up change in their community by entering a zip code. Here we are seeing the sharing of the NCAA again and again. The document not only demonstrates the consensus of scientific knowledge but also allows citizens to see where they and their communities fit into the patchwork of change and impacts

Congresswoman Hill Question 1c. What is its importance to our understanding of climate science and its effects on society more broadly?

The National Climate Assessment is a very powerful process that brings together climate experts, the state of the art data and projections to communities who are both witnessing and planning for this change. Breaking out the impacts and adaptations by region is a powerful tool for communities and citizens to understand how climate change is evolving around them. Often individuals have a difficult time visualizing the change greenhouse gases (invisible) and a relatively small temperature rise can mean. However, the evidence of the longer growing seasons-up to 40 days longer in California-is a powerful way to advance the understanding of climate impacts. People's daffodils and crocuses are coming up sooner in the spring and their tomatoes and basil are lasting longer in the fall. Seed companies have remapped hardiness zones in their catalogues. Climate change is effecting what fish people pull from the ocean, because cold loving fish shift north in the Atlantic while in the Gulf of Mexico cold loving fish are being replaced by warmer (less tasty) species. The National Climate Assessment documents how climate change is impacting local communities from the plants they grow to the take at the end of their fishing lines. No other efforts document local effects comprehensively. This process enables communities to understand the change and begin to develop resilience plans.

Responses by Dr. Twila A. Moon

Questions for the record for the House Committee on Science, Space, and Technology for the hearing on Earth's Thermometers: Glacial and Ice Sheet Melt in a Changing Climate provided on August 2, 2019 by Twila

Moon, PhD Research Scientist, National Snow and Ice Data Center, Cooperative Institute for Research in Environmental Sciences, University of Colorado Boulder.

Question 1) Given the importance of glacial meltwater as a freshwater supply for certain animals and plants, such as salmon in the Pacific Northwest, what are the implications of increased glacial melt rates to these species and ecosystems?

Response: As the climate warms, glacier melt is increasing. In a glacierized basin, glacier melt makes up some percentage of the stream and river flow, with that contribution changing through the seasons and depending on weather conditions. As the glaciers begins to melt more, there can initially be an increase in glacier melt input (potentially lasting across many years). However, as the glacier area continues to decline, there is less ice remaining to melt and glacier runoff will decrease. The result is that ecosystems must respond to complex variability that can include both increases and decreases in glacier meltwater input, influencing for example temperature, turbidity, and total stream flow amount.

The implications of increased glacial melt for salmon, other species, and the ecosystems they depend on varies by location and in some cases is not known at all. For example, in the Pacific coastal temperate rainforest *of* Alaska, scientists foresee possible positive and negative consequences. Streams in this area with particularly high glacier meltwater content can be too cold for salmon rearing and overwintering. So reduction in glacier melt in some streams may make them better salmon habitat. However, water temperatures that are too warm in the summer because glacier melt input has dropped too low may hurt salmon productivity. It may be that different salmon species are helped or hurt by glacier loss in specific locations. Further discussion of the Pacific coastal temperate rainforest ecosystem and connections between ecosystem and glacier loss can be found in O'Neel et al. (2015). For many other locations, the impacts of losing glacier ice are more clearly negative. For example, the meltwater stonefly *Lednia tumana,* which is endemic to Waterton-Giacier International Peace Park in Canada and the U.S. and has been petitioned for listing under the

U.S. Endangered Species Act, lives in a restricted range near glaciers, snowfield, and springs, and changes in habitat from glacier loss-including increasing water temperatures, changes in the timing and amount of glacial meltwater input, changes in dissolved oxygen levels, and alteration of stream flow character are expected to negatively impact this species (e.g., Muhlfeld et al. 2011). Reductions in ecosystem health in the base of the food chain is likely to negatively impact other dependent species.

Overall, however, it is my assessment that research into links between glacier loss and dependent plants, animals, and ecosystems is limited. This is particularly worrisome given that glaciers are changing rapidly in parallel with other potentially large changes, such as increased temperature and changing precipitation. It is likely that multiple climate impacts will be at play in any given ecosystem, and it will be important for research to consider connections across human, biological, and physical systems.

References

Muhlfeld, C. C., Giersch, J. J., Hauer, F. R., Pederson, G. T., Luikart, G., Peterson, D. P., Downs, C. C. & Fagre, D. B. (2011). Climate change links fate of glaciers and an endemic alpine invertebrate, *Climatic Change*, *106*(2), 337-345,doi:10.1007/s10584-0l1-0057-l.

O'Neel, S., et al. (2015). lcefield-to-Ocean Linkages across the Northern Pacific Coastal Temperate Rainforest Ecosystem, *BioScience*, *65*(5), 499-512,doi:10.1093/biosci/biv027.

Question 2) Local communities that rely on glacial streams for water for drinking, industry, or agriculture may be impacted by drought in the future as glacial runoff diminishes. What are the impacts of melting glaciers on local and global drinking water supplies in the U.S. and around the world?

Response:

There is no doubt that worldwide loss of glacier ice will impact water supplies around the globe. First, as mentioned above, it is important to recognize that increasing glacier melt can produce both an initial bump in water availability and a subsequent, long-lasting decline. Glacier meltwater is also most available during summer, when air temperatures are above freezing, and is therefore particularly important as a water source from spring through fall.

Depending on the local or regional environment, glacier melt can be vital as a source for drinking water, irrigation and agricultural water, and hydropower and other industry. For example, in south Asia (including Pakistan, Afghanistan, Tajikistan, Turkmenistan, Uzbekistan, and Kyrgyzstan), 800 million people depend on water resources served by Himalayan glaciers. The role of glacier melt in this area is particularly strong during drought years, when glacier melt offsets the low water input from rain. Projected future declines in glacier area across the Himalaya will impact water availability, adding another stressor to a region that already grapples with geopolitical tensions (for a full discussion of this case study see Pritchard (2019)). These challenges are not unique to south Asia (e.g., Carrey et al. 2017). In South America, glaciers provide valuable water resources for many countries. In Bolivia, for example, the capital city of La Paz is growing rapidly, but is also dependent on water resources that include substantial glacier meltwater. During 1963-2006, glaciers contributed 14% of La Paz's water resources in the wet season and 27% in the dry season. While increased glacier melt during the latter period of this 1963-2006 record had so far offset the loss of 50% of the glacier area, eventual complete loss of the region's glaciers will cause a 9% reduction in total runoff during wet seasons and 24% reduction during the dry season (for further discussion on this case study see Soruco et al. (2015)). Even in the United States, glacier meltwater can be an important local source of runoff, for example in the Pacific Northwest (Nolin et al. 2010). It is my personal opinion, however, that the wide-ranging impacts of glacier loss within the United States are understudied.

References

Carey, M., Molden, O. C., Rasmussen, M. B., Jackson, M., Nolin, A. W. & Mark, B. G. (2017). Impacts of Glacier Recession and Declining

Meltwater on Mountain Societies, *Annals of the American Association of Geographers*, *107*(2), 350-359, doi:10.1080/ 24694452.2016. 1243039.

Nolin, A. W., Phillippe, J., Jefferson, A. & Lewis, S. L. (2010). Present-day and future contributions of glacier runoff to summertime flows in a Pacific Northwest watershed: Implications for water resources, *Water Resour. Res.*, *46*, W12509, doi:10.1029/2009WR008968.

Pritchard, H. D. (2019). Asia's shrinking glaciers protect large populations from drought stress, *Nature*, 1-20, doi:10.1038/s41586-019-1240-1.

Soruco, A., Vincent, C., Rabatel, A., Francou, B., Thibert, E., Sicart, J. & Condom, T. (2015). Contribution of glacier runoff to water resources of La Paz city, Bolivia (16° S), *Annals of Glaciology*, *56*(70), 147-154, doi:10.3189/2015AoG70A001.

Responses by Dr. Gabriel J. Wolken

The Impacts of Glacier Change in Alaska, Testimony of Gabriel J. Wolken, Alaska Division of Geological & Geophysical Surveys and University of Alaska Fairbanks For the hearing entitled Earth's Thermometers:Glacial and Ice Sheet Melt in a Changing Climate Before the Committee on Science, Space, and Technology, United States House of Representatives. 11 July 2019

Questions for the Record

1. Given that Alaska has the highest number of glaciers out of any state in the United States and those glaciers are melting at substantial rates, what are the projected economic impacts of glacial melt to Alaska's economy, and does the state of Alaska have any adaptation or mitigation strategies for addressing loss of its glaciers?

Response: The rapid and sustained melting of glaciers in Alaska will negatively impact Alaska's economy.

As glaciers in Alaska continue to lose mass, a rise in sea level will lead to more lowland flooding and coastal erosion that damages or destroys infrastructure, threatens public safety, security and livelihoods, and increases the financial pressure on economies at all levels.

By the end of this century, glaciers in Alaska are projected to lose about 30-50% of their current total mass. This will have a large negative impact on runoff volume, modifying the chemical composition and decreasing the temperature of many of the streams that support key fisheries, including salmon, and the billion dollar fishing industry on which many Alaskans depend. Changes in runoff will also affect hydropower production. As glacier mass loss continues, the storage capacity of snow and ice will eventually diminish, which is projected to have significant impacts on future water resource availability, even in basins with low glacier coverage. This will lead to an economic burden on the communities in Alaska that currently rely on hydropower generation from glacierized catchments.

Alaska's tourism industry may also be negatively impacted by continued glacier mass loss. Tourism in Alaska is a key economic sector that supports one out of every ten jobs in the state, delivers 1.5 billion dollars in labor income, and has an economic impact of 4.5 billion dollars. Recent glacier retreat has already been significant in Alaska, making it more difficult for tourists to access glaciers using Alaska's road system, and tourism operators are being forced to modify the manner in which their clients experience glaciers.

Few adaptation and mitigation strategies have been implemented, or developed for addressing the impacts of glacier melt in Alaska. The State of Alaska, and some municipalities, have updated the language in their hazard mitigation plans to include glacier-related hazards associated with continued glacier mass loss, and exposure to extreme runoff events. Some communities are experiencing the impacts in a very direct way.

For instance, the City of Valdez has spent nearly 6 million dollars in adaptation responses to glacier lake outburst floods resulting from the rapid retreat of Valdez Glacier. In other parts of Alaska, lowland flooding and coastal erosion, resulting in part from rising sea level, are forcing communities and governments to evaluate their adaptive capacities to these rapidly changing conditions; in some locations, relocation is the only option.

Alaska will continue to experience the effects of climate change more than most states in the U.S. because of its high northern latitude location and abundance of snow, ice and permafrost. Alaska is warming at twice the rate as the rest of the U.S., and rapid changes in the snow, ice, and permafrost will increase the occurrence, and likely the magnitude, of hazards throughout the state (e.g., avalanches, floods, erosion, slope instability, glacier collapses, and glacier lake outburst floods). If these hazards are not properly assessed and monitored, and if the appropriate adaptive measures are not considered, these cryosphere hazards will have a damaging effect on Alaska's communities and infrastructure, as well as on the security and livelihoods of Alaskans.

Responses by Dr. W. Tad Pfeffer

W.T. Pfeffer Testimony to US House of Representatives Committee on Science, Space, and Technology 11 July 2019.

Response to question from Rep. J. McNerney (CA-09): "You mentioned in your testimony the fact that the Arctic gyre is known to trap air. Can you describe how this phenomenon would impact climate interventions, such as the injection of cloud aerosols, in the Arctic, and vice versa?"

My expertise does not extend to arctic climate dynamics and especially not to questions of geoengineering, but I can provide some background to your question and suggest other researchers who are expert in these areas.

The presence of a circumpolar closed low-pressure system has been known since the 19th century, but the presence and significance of an upper-level (upper tropospheric- stratospheric) cyclonic circulation was not detected until the mid-1950s when high-altitude weather reconnaissance missions were flown from Alaska into the high Arctic (Shaw, 1995). Later analysis of data acquired during these flights help to identify the mechanism responsible for "Arctic Haze," episodes of reduced visibility in the arctic first described early in the mid-18th century, at the outset of the industrial revolution. This haze is caused by scattering and attenuation of light by small aerosol particles, and from the work of Shaw and others has been shown to originate predominantly from high-latitude coal burning industries in Eurasia. The 1995 Shaw paper (PDF copies of this and other papers accompanies this response) gives a good summary overview of the Arctic Haze phenomenon and what it reveals about arctic atmospheric circulation.

Regarding the climate-change implications of arctic atmospheric circulation, probably the most important aspect is the changing strength, position, and mobility of the upper level arctic jet and its effects on mid latitude weather and climate.

These looping waves tend to wander about fixed average positions on various time scales ranging from synoptic (i.e., weather systems) to seasonal (winter vs. summer), but also appear to be moving and weakening in response to climate change forcings. Seasonal changes in the position of the polar jet cause familiar regular changes in seasonal climate (e.g., in New England, the transition from hot/humid weather in July to cool/dry weather in August), while snoptic-scale changes cause other familiar phenomena like cold incursions into the Midwest in the winter (the "Alberta Clipper"). Recent anomalous events like the January 2019 cold outbreak in central USA are manifestations of a polar jet changing in new and unexpected patterns. Climate change appears to be altering the dynamics of the arctic jet (and arctic circulation generally), and more anomalous weather and seasonal patterns may be expected in the future. I have included some papers covering this issue (Chen et al, 2914; Zhang et al; 2016), and recommend Dr. Jennifer Francis (Woods Hole Institute) as an expert in this subject.

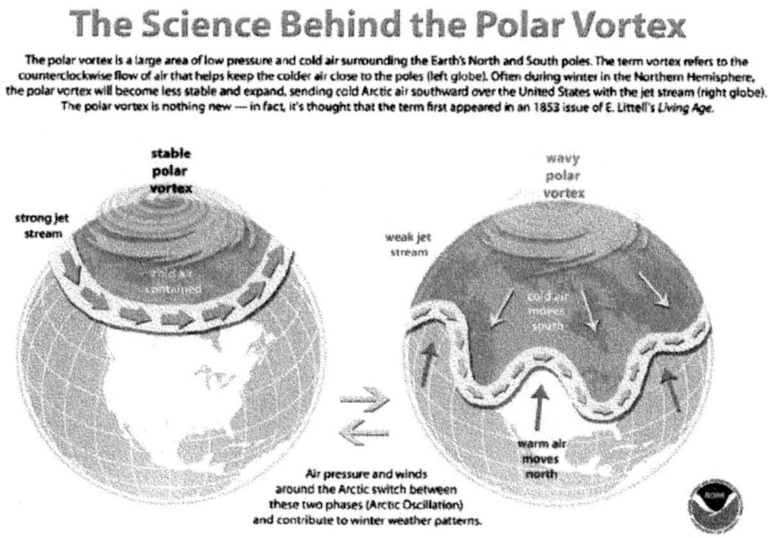

The arctic jet is an upper level band of strong westerly wind that separates the cold arctic air mass to the north from the warmer mid-latitude air to the south (see the accompanying graphic). The jet does not move in a simple circular pattern from west to east at a fixed latitude (as shown in the "stable polar vortex" in the graphic), but wanders in a looping pattern (as shown in the "wavy polar vortex" in the graphic).

I can offer no general information or advice regarding geoengineering and arctic climate beyond noting that geoengineering approaches are being discussed (see Caldeira and Wood, 2008; and Dykema et al, 2014). I recommend Prof. Alan Robock (Rutgers University) as an expert in this subject.

Further Reading

Arctic Haze entry in Wikipedia (https:ljen.wikipedia.org/wiki/Arctic haze).

References

Caldeira, K. & Wood, L. (2008). Global and Arctic climate engineering: numerical model studies. *Philosophical Transactions of the Royal Society A: Mathematical, Physical and Engineering Sciences*, *366*(1882), 4039-4056.

Cohen, J., Screen, J. A., Furtado, J. C., Barlow, M., Whittleston, D., Coumou, D. & Jones, J. (2014). Recent Arctic amplification and extreme mid-latitude weather. *Nature geoscience*, *7*(9), 627.

Dykema, J. A., Keith, D. W., Anderson, J. G. & Weisenstein, D. (2014). Stratospheric controlled perturbation experiment: a small-scale experiment to improve understanding of the risks of solar geoengineering. *Philosophical Transactions of the Royal Society A: Mathematical, Physical and Engineering Sciences*, *372*(2031), 20140059.

Francis, J. A. & Vavrus, S. J. (2012). Evidence linking Arctic amplification to extreme weather in mid latitudes. *Geophysical Research Letters*, *39*(6).

Shaw, G. E. (1995). The Arctic haze phenomenon. *Bulletin of the Amencan Meteorological Society*, *76*(12), 2403-2414.

Zhang, J., Tian, W., Chipperfield, M. P., Xie, F. & Huang, J. (2016). Persistent shift of the Arctic polar vortex towards the Eurasian continent in recent decades. *Nature Climate Change*, *6*(12), 1094.

APPENDIX II: ADDITIONAL MATERIAL FOR THE RECORD

Letter Submitted by Representative Suzanne Bonamici.

Geophysical Research Letters

Dissolved Black Carbon in the Global cryosphere: Concentrations and Chemical Signatures

Alia L. Khan[1,2], Sasha Wagner[3,4], Rudolf Jaffe[3], Peng Xian[5], Mark Williams[6], Richard Armstrong[1], and Diane McKnight[2]

[1]Department of Civil and Environmental Engineering and the Institute of Arctic and Alpine Research, University of Colorado Boulder, Boulder, Colorado, USA, [2]National Snow and Ice Data Center and Cooperative Institute for Research in Environmental Sciences, University of Colorado Boulder, Boulder, Colorado, USA, [3]Southeast Environmental Research Center and Department of Chemistry and Biochemistry, Florida International University, Miami, Florida, USA, [4]Skidaway Institute of Oceanography, Marine Sciences Department, University of Georgia, Savannah, Georgia, USA, [5]Aerosol and Radiation Section of the Marine Meteorology Division, Naval Research Laboratory, Monterey, California, USA, [6]Department of Geography and the Institute of Arctic and Alpine Research, University of Colorado Boulder, Boulder, Colorado, USA

Abstract

Black carbon (BC) is derived from the incomplete combustion of biomass and fossil fuels and can enhance glacial recession when deposited on snow and ice surfaces. Here we explore the influence of environmental conditions and the proximity to anthropogenic sources on the concentration and composition of dissolved black carbon (DBC), as measured by benzenepolycaroxylic acid (BPCA) markers, across snow, fakes, and streams from the global cryosphere. Data are presented from Antarctica, the Arctic, and high alpine regions of the Himalayas, Rockies, Andes, and Alps. DBC concentrations spanned from 0.62 μg/L to 170 μg/L. The median and (2.5, 97.5) quantities in the pristine samples were 1.8 μg/L (0.62,12), and nonpristine samples were 21 μg/L (1.6, 170). DBC is susceptible to photodegradation when exposed to solar radiation. This process leads to a less condensed BPCA signature. In general, DBC across the data set was composed of less polycondensed DBC. However, DBC from the Greenland Ice Sheet (GRJS) had a highly condensed BPCA molecular signature. This could be due to recent deposition of BC from Canadian wildfires. Variation in DBC appears to be driven by a combination of photochemical processing and the source combustion conditions under which the DBC was formed. Overall, DBC was found to persist across the global cryosphere in both pristine and nonpristine snow and surface waters. The high concentration of DBC measured in supraglacial melt on the GRIS suggests that DBC can be mobilized across ice surfaces. This is significant because these processes may jointly exacerbate surface albedo reduction in the cryosphere.

Plain Language Summary

Here we present dissolved black carbon {OBC) results for snow and glacial melt systems in Antarctica, the Arctic, and high alpine regions of the Himalayas, Rockies, Andes, and Alps. Across the global cryosphere, DBC composition appears to be a result of photochemical processes occurring en route in the atmosphere or in situ on the snow or ice surface, as well as the combustion conditions under which the DBC was formed. We show that samples from the Greenland Ice Sheet (GRIS) have a distinct molecular chemical signature, consistent with deposition of BC from Canadian wildfires occurring the week before sampling, The concentration range observed in this global cryosphere study indicates significant amounts of DBC persist in both pristine and human impacted snow and glacial meltwater. Our results are significant for understanding the controls on meltwater production from glaciers worldwide and the feedbacks between combustion sources, wildfires, and the global cryosphere. Wildfires are predicted to increase due to climate change, and albedo cannibalism is already influencing meltwater generation on the GRIS. Anticipated longer summer melt seasons as a result of climate change may result in longer durations between snowfalls, enhancing exposure of recalcitrant DBC on snow/ice surfaces, which could further exacerbate surface albedo reduction in the cryosphere.

Introduction

Many inherent challenges remain in quantifying and predicting melt of polar ice sheets and glaciers. In particular, the complex influence of light-absorbing aerosols is not well understood (Bond et al., 2013). Biomass burning and fossil fuel combustion are both sources of black carbon (BC) to the cryosphere (Fellman et al., 2015). Organic matter from these sources can be stored in glaciers for millennia (Hood et al., 2009, 2015; Stubbins et al., 2012a) and mobilized during glacial melting (Hodson, 2014), A study on the Greenland Ice Sheet (GRIS) attributed two widespread melting events (1889 and 2012) to BC produced from Northern Hemisphere wildfires (Keegan et al., 2014).

This attribution was based on higher BC and ammonia concentrations in ice lenses in firm cores, along with air mass back trajectory analysis; however, direct chemical characterization of the BC was not obtained. Once deposited, BC aerosols can be solubilized and transported in the aqueous phase as dissolved black carbon (DBQ (Dittmar, 2008). In Alaskan glacier rivers, DBC was interpreted as being sourced from atmospheric deposition of anthropogenic combustion products (Ding et al., 2014a). In Antarctica, DBC from wildfires in preindustrial eras has accumulated over millennia in the saline bottom waters of closed-basin lakes; the DBC in the bottom waters has a more polycondensed chemical signature in comparison to DBC in surface waters susceptible to influence from current local anthropogenic point sources of BC [Khan et al., 2016]. Therefore, atmospheric deposition of both natural and anthropogenic BC aerosols represents an important source of DBC to remote areas of the cryosphere.

DBC is quantified and characterized using the benzenepolycarboxylic add (BPCA) method, which produces individual BPCA molecular markers upon the oxidation of condensed aromatic structures (Dittmar, 2008). The condensed aromaticity of DBC can be inferred from the relative proportion of produced BPCAs (Schneider et al., 2010; Abiven et al., 2011). The degree of aromatic condensation for particulate BC is primarily a function of pyrolysis temperature, and links between BPCA composition and pyrogenic source are not always dear (e.g., Schneider et al., 2010; Mcbeath eta/., 2013a, 2013b; Wiedenmeyer et al., 2015).1n addition, DBC molecular signatures are altered via photodegradation, either en route or in situ, changing BPCA com position and obscuring potential links to its pyrogenic source (Schneider et al., 2010; Ziolkowski and Druffel, 2010; Stubbins et al., 2012b; Ward et al., 2014). Although BPCA composition cannot be used to directly infer the pyrogenic source of OBC, we can use them to investigate potential biogeochemical drivers of the quality of DBC which persists in cryospheric environments.

In this study, we hypothesize that a more condensed DBC signature will be characteristic of samples for sites with freshly deposited BC and limited time for photodegradation, while OBC enrichment in less condensed aromatics will be characteristic for sites where the DBC may have been exposed to solar radiation for long durations of time either in the atmosphere or on the snow/ice surface. In terms of sites with freshly deposited BC, we investigated sites on the GRIS where collection occurred shortly after Canadian wildfires. We also investigated sites where continuous inputs of local anthropogenic sources of BC may lead to a fresher signature. In addition to the GRIS, an area known to be impacted by wildfire-derived pyrogenic carbon, we included other polar regions with high solar exposure, which are exposed to continuous sunlight almost 4 months per year, as well as high-altitude mountain sites. Sampling locations in this study are largely remote, experience extended lengths of sunlight exposure, and receive diverse inputs of atmospheric BC from both natural and anthropogenic combustion sources. As such, our overall hypothesis is that in the cryosphere, degradative processing during DBC transit in the environment, as well as the combustion conditions under which the DBC was formed, are important drivers of the DBC BPCA profile across the cryosphere.

Here we assess variation in DBC content and composition in samples that represent the typical global cryosphere, as well as sites with known local sources of DBC (Figure 1: map). Data are presented from glacial melt systems in Antarctica, the Arctic (GRIS and Svalbard), and high alpine regions of the Himalayas, Rocky Mountains, Andes, and Alps. Snow samples are from the Rocky and Andes Mountains, as well as the Arctic. Samples from the Norwegian Arctic were collected on Svalbard, including around an active coalmine and coal burning power plant, fueling the largest settlement, Longyearbyen. Sites are categorized as pristine" (P; >5 km from a fuel combustion source) and "nonprlstlne" (NP; <5 km from a fuel combustion source).

To support the interpretation of the GRIS results, we determined biomass burning smoke aerosol optical thickness (AOT) and air mass transport from the Navy Aerosol Analysis and Prediction System (NAAPS) over several days in the week before the samples were collected.

Methods

Water samples were obtained that represent the diverse aquatic environments of the cryosphere. The lakes and streams sampled include supragladal systems found on the surface of glaciers, systems fed directly by glacial melt and/or snow melt, as well as proglacial lakes, which form at the tongue of glaciers. There are also surface water ponds included from the McMurdo Dry Valleys (MDVs), including DL-Hoare Pond (DLH Pond), latitude: –77.622901916504, longitude: 162.902999877930, which is adjacent to Lake Hoare: http://www.mcmlter.org/content/dirty*little-hoare pond. Lake Hoare was named for physicist Ray Hoare, of the eighth Victoria University Expedition (1963-1964). Additional samples were collected at Cape Royds in Antarctica. Regionally, the samples are divided into Arctic, Antarctic, and Alpine. The Arctic data set is mostly composed of samples from the island of Svalbard in the Norwegian Arctic, as well as two samples from the GRIS, one fresh snow and one supraglacial melt. Antarctic samples were collected from the MDV and Cape Royds. Alpine samples are from snow and snowmelt in the Colorado Rocky Mountains, supraglacial melt from the Ngozumba Glacier, the longest glacier in Nepal located in the Gokyo Valley of the Nepalese Himalayas, snow and glacier fed streams in the central Chilean Andes, and a glacier-fed stream from the Mer de Glace in the French Alps.

Water and snow samples were collected in acid-rinsed and precombusted amber glass bottles or acid-rinsed high-density polyethylene (HDPE) bottles. Snow was melted at room temperature in precteaned add-rinsed Nalgene buckets. All samples were filtered on precombusted glass fiber filters (Whatman GF/F, pore size: 0.7 m), acidified to pH = 2 with concentrated HCI (analytical grade), and stored at 4°C until analysis.

For all samples, a 60 mL aliquot was poured into add-rinsed and precombusted amber glass bottles and analyzed for dissolved organic carbon on a Shimadzu TOC-V-CSN with a detection limit of 0.07 mg C/L.

DBC (<0.7 μm) was measured using the BPCA method (Dittmar, 2008; Dittmar et al., 2008) and optimization for freshwater DBC (Ding et al., 2014b). This method is based on the oxidation of polycondensed DBC structures to BPCAs with three to six carboxylic add groups (B3CA to B6CA). The BPCA method was applied to the dissolved organic matter (DOM) isolated by solid-phase extraction (SPE). First, a SPE technique (Dittmar, 2008; Dittmar et al., 2008) was applied to all samples using Bond Elut PPL cartridges, composed of a styrene divinylbenzene polymer solid phase. SPE efficiency for marine samples is typically -45% and -60% in fresh water (Dittmar et al., 2008). In this study we measured SPE efficiency on eight samples (five Antarctic surface waters (37%, 34%, 46%, 50%, and 67%), a glacier-fed stream from the Andes (34%), and two snow sample (47% and 73%).The average efficiency of these eight samples was 48% ± 15%, with a range from 34% to 73%.

The PPL cartridges were transported frozen to Boulder, CO, and stored in a freezer. The PPL cartridges were transported to Miami, FL, and completely dried under ultrahigh-purity nitrogen gas. The PPL cartridges were then eluted with 10-20 mL of MeOH until the eluent was colorless. Aliquots of MeOH were then quantitatively transferred to 2 ml glass ampules and evaporated to dryness under a stream of ultrahigh- purity nitrogen. The dried DOM extract was redissolved in concentrated nitric acid (65%) before ampules were sealed and oxidized in sealed glass ampules in a programmable oven for 6 h at 160°C in order to produce BPCAs (Ding et al., 2014b). They were then dried in a sand bath under ultrahigh purity nitrogen gas. BPCAs were then redissolved in the mobile phase buffer, separated, and quantified on a Sunfire C18 reversed phase column (3.5 μm, 2.1 x 150 mm; Waters Corporation) by an HPLC system coupled with a photo diode array detector (Dittmar, 2008; Ding et al., 2012).

DBC concentrations were calculated from B3CA to B6CA concentrations based on a previously reported algorithm (Dittmar, 2008). B6CA was excluded from BPCA proportion analysis due to its low signal and low resolution in most samples. Analytical replicates were measured in triplicate with <10% standard deviation. Sample duplicates were collected for 5% of the samples and fell within this <10% standard deviation.

The degree of condensed aromaticity of DBC in each sample was estimated based upon the BPCA distribution (Glaser et al., 1998; Ding et al., 2014b) using the ratio of (B3CA + B4CA)/B5CA. Although a complete understanding of all biogeochemical factors that influence BPCA ratios remains elusive, the ratios can be used to provide preliminary assessments of the environmental dynamics of DBC.

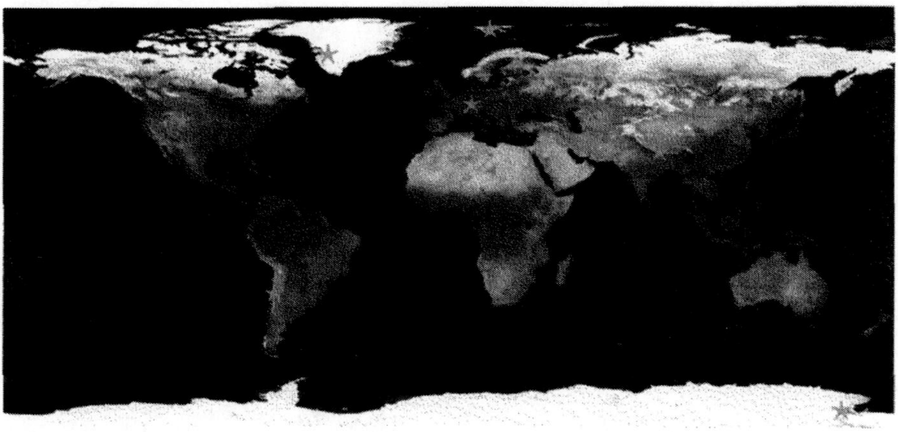

Figure 10. Map of global distribution of sample sites overlaid on a NASA Visible Earth image. Data are presented from glacial melt systems in Antarctica, the Arctic, and the Himalayas, Snow samples are from the Rocky and Andes Mountains, as well as the Arctic.

For instance, photodegradation can significantly alter BPCA composition, reducing the overall condensed aromaticity of the DBC pool once it reaches surface waters (Stubbins et al., 2012b; Ward et al., 2014).

As such, the DBC signature of samples exposed to solar radiation over long durations may indicate extensive photodegradation, resulting in low signals of B6CA. BPCA ratios can also vary significantly with pyrogenic source material and environmental conditions. For example, higher abundance of BSCA + B6CA, indicating more condensed aromatic DBC, has been associated with DBC in wildfire-impacted watersheds (Wagner et al., 2015), whereas DBC solubilized from urban dust has been shown to be more enriched in B3CA + B4CA (i.e., less polycondensed BC) (Ding et al., 2014a, 2014b). Although we understand that the source of pyrolized carbon (i.e., fossil fuels or biomass burning) can influence DBC quality, we expect BPCA compositions for the current sample set to be influenced by the atmospheric residence time of the original aerosol BC source and the degree of exposure to solar radiation (Table 1). Thus, we expected to observe lower (B3CA + B4CA)/BSCA ratios when atmospheric transport time was long and/or opportunities for photodegradation were high.

The biomass burning smoke aerosol optical thickness (AOT) data used in this study were obtained from the Navy Aerosol Analysis and Prediction System (NAAPS) reanalysis (Lynch et al., 2016). NAAPS reanalysis is a decade-long global 1 x 1° and 6-hourly 550 nm AOT reanalysis product, which was recently developed and validated at the Naval Research Laboratory. This reanalysis utilizes a modified version of the NAAPS as its core and assimilates quality-controlled retrievals of AOT from Moderate Resolution Imaging Spectroradiometer (MODIS) on Terra and Aqua and the Multiangle Imaging 5 pectroradiometer (MISR) on Terra (Zhang and Reid, 2006; Hyer et al., 2011; Shi et al., 2014). NAAPS characterizes anthropogenic and bio genic fine aerosol species (including sulfate and primary and secondary organic aerosols), dust, biomass burning smoke, and sea-salt aerosols. Smoke from biomass burning is derived from near-real-time satellite based thermal anomaly data used to construct smoke source functions (Reid et al., 2009), with additional orbital corrections on MODIS-based emissions and regional tunings. The reanalyzed fine- and coarse-mode AOT at 550 nm is shown to have good agreement with the ground-based global-scale Sun photometer network Aerosol Robotic Network ACTs (Holben et al., 1998). The figure below is a series of snapshots at 18Z showing wildfire-related smoke AOT at 550 nm between 18 and 21 June 2014 at 550 nm.

Table 1. Samples Grouped by Pristine and Nonpristine in Ascending Order of (B3CA + B4CA)/B5CARatio[a]

	(B3CA + B4CA)/B5CA	DBC (μg/L)	DOC (mg/L)	%DOC of DBC	Region	Medium	Geography
			Pristine Snow Sample				
GRIS snow/surface hoar	0.82	1.76	0.67	0.003	GRIS	Snow <1, hoar >5 days	Arctic
Chilean Andes Snow-ET	3.0	1.80	0.83	0.002	Andes	Snow <2 days	Alpine
Niwot Ridge B	3.0	8.93	nd	nd	Rockies	Snow <1 day	Alpine
San Francisco Glacier Snow	3.4	1.44	0.36	0.004	Andes	Snow <2 days	Alpine
WoodFjorden, Svalbard	4.2	1.26	0.54	0.002	Svalbard	Snow <2 days	Arctic
Larsbren	6.3	1.63	0.39	0.004	Svalbard	Snow >5 days	Arctic
Storm Peak Lab	17	9.65	nd	nd	Rockies	Snow >3 days	Alpine
Niwot Ridge A	27	9.27	nd	nd	Rockies	Snow >3 days	Alpine
			Pristine Meltwater Samples				
GRIS supraglacial melt	0.32	12.4	0.14	0.088	GRIS	Supraglacial melt	Arctic
Mer de Glace, Chamonix	1.8	1.51	4.13	0.000	Alps	Glacial stream	Alpine
Bano Morales Stream	1.9	6.12	0.50	0.012	Andes	Glacial stream	Alpine
Lake Joyce 7 m[b]	2.5	2.67	0.50	0.005	Antarctic	Lake	Antarctic
Alatna Pond	3.2	3.60	1.00	0.004	Antarctic	Surface water	Antarctic
Lake Bonney East Lobe 5 m	3.6	1.44	0.40	0.004	Antarctic	Lake	Antarctic
Von Guerrard Stream[b]	3.6	0.89	0.63	0.001	Antarctic	Glacial stream	Antarctic
Ngozumba Glacier, Nepal	3.7	1.05	0.10	0.010	Himalayas	Supraglacial lake	Alpine
Ngozumba Glacier, Nepal	3.8	0.62	0.12	0.005	Himalayas	Supraglacial lake	Alpine
Alaskan Glacial Rivers[c]	4.3	0.53 ± 0.09	0.010 ± 0.004	1.9 ± 0.6	Alaska	Glacial river	Arctic
Lake Bonney West Lobe 5 m[b]	5.4	1.22	0.40	0.003	Antarctic	Lake	Antarctic
Green Lake 4, Colorado	5.4	4.40	nd	nd	Rockies	Lake	Alpine
Ngozumba Glacier, Nepal	6.3	1.54	0.12	0.012	Himalayas	Supraglacial lake	Alpine
Ngozumba Glacier, Nepal	6.7	1.94	0.40	0.005	Himalayas	Supraglacial lake	Alpine
Ngozumba Glacier, Nepal	6.8	1.39	0.83	0.002	Himalayas	Supraglacial Lake	Alpine
			Nonpristine Snow				
Mine 7	3.3	71.4	1.03	0.070	Svalbard	Snow >5 days	Arctic
Upwind of the Mine	7.7	2.36	5.43	0.000	Svalbard	Snow >5 days	Arctic
Next to Coal Power Plant	17	6.84	0.32	0.021	Svalbard	Snow >5 days	Arctic
Upper Snowmobile Track	43	44.4	4.68	0.009	Svalbard	Snow >5 days	Arctic
Lower Snowmobile Track	59	21.6	0.54	0.040	Svalbard	Snow >5 days	Arctic
			Nonpristine Meltwater				
High Park Fire PNF0714[d]	1.3	177	2.97	0.06	Rockies	Terrestrial stream	Alpine
Marble Point	1.5	33.6	4.23	0.008	Antarctic	Surface water	Antarctic
Clear Lake	2.2	47.3	4.30	0.011	Antarctic	Surface water	Antarctic
Lake Fryxell 5 m	2.2	8.28	1.80	0.005	Antarctic	Lake	Antarctic
DLH Pond	3.1	14.0	2.30	0.006	Antarctic	Surface water	Antarctic
Pony Lake	3.4	170	19.0	0.009	Antarctic	Surface water	Antarctic
Longyear River	3.5	4.68	5.83	0.001	Svalbard	Glacial stream	Arctic
Longyearbyen Fjord Water	4.3	20.5	0.33	0.062	Svalbard	Ocean	Arctic

[a]A lower ratio is more indicative of wildfire-derived DBC, and a higher ratio is more indicative of fossil fuel combustion and/or photodegradation.
[b]From *Khan et al.* [2016].
[c]From *Ding et al.* [2014a, 2014b].
[d]From *Wagner et al.* [2015].

Results and Discussion

As expected, pristine sites had lower median and (2.5, 97.5) quantiles of DBC concentrations (1.8 μg/L (0.62, 12)) than nonpristine sites (21 μg/L (1.6, 170)). In particular, pristine snow samples exhibited low DBC concentrations (1.8 μg/L (1.3, 10)). The median (B3CA + B4CA)/B5CA ratio for pristine samples from snow was 3 (1, 27) (Table 1 and Figure 2). The remote pristine snow/surface hoar sample from the GRIS had a distinct composition compared to the rest of the data set, containing high levels of BSCA and a ratio of 0.3, below the 2.5% quantile. Similar to the glaciers of Antarctica (Khan et al., 2016), the only potential source of BC to the GRIS is long-range atmospheric transport. The GRIS sample was collected in late June 2014, as wildfires blazed across the Canadian Arctic. As shown by the reanalysis from the NAAPS model, continuous smoke activities occurred over the northwest of Canada between 18 and 21 June 2014 and were transported eastward to the western region of the GRIS. Fresh aerosols from these fires, which appear to have been transported to the western region of the GRIS, would have remained in the surface hoar with limited time for photodegradation prior to sample collection and provide an explanation for the highly condensed aromatic DBC signal observed for this sample. The other most remote pristine snow sample was from Svalbard, which had a DBC concentration of 1.1 μg/L and BPCA ratio of 4. This sample was collected in Woodfjorden, which is several hundred kilometers from the closest town, thus only impacted by atmospheric long-range transport.

Pristine snow samples from the Rocky Mountains had similar DBC concentrations, from 8.9 to 9.7 μg/L, but the (B3CA + B4CA)/B5CA ratios ranged from 3 to 27. Two of the pristine snow samples yielded (B3CA + B4CA)/B5CA ratios higher than the Woodfjorden sample, 4.2. The (B3CA + B4CA)/B5CA ratio was 17 at Storm Peak Lab in Northern Colorado and 27 at Niwot Ridge in central Colorado (Table 1), suggesting that the BC in these remote Colorado samples is less condensed than the remote sample from Svalbard, which may be indicative of more photodegradation of DBC at the Colorado sites prior to sample collection, or a different pyrolized source. These two samples were aged snow samples and exposed to solar radiation longer than the Woodfjorden sample (Table 1). In contrast, a fresh snow sample from a similar location in the Rocky Mountains only yielded a (B3CA + B4CA)/B5CA ratio of 3. The DBC in this fresh snow sample had limited time for photodegradation, which may result in the less condensed BPCA ratio, relative to the aged snow samples from the Rocky Mountains. Again, these sites are >5 km away from known point sources so the enrichment in less condensed DBC may be related to accumulated solar exposure from the time since snow deposition to the time of sample collection and/or the long-range transport of fossil fuel derived soot particles, which have been found on remote glaciers (Stubbins et al., 2012a; Ding et al., 2014a).

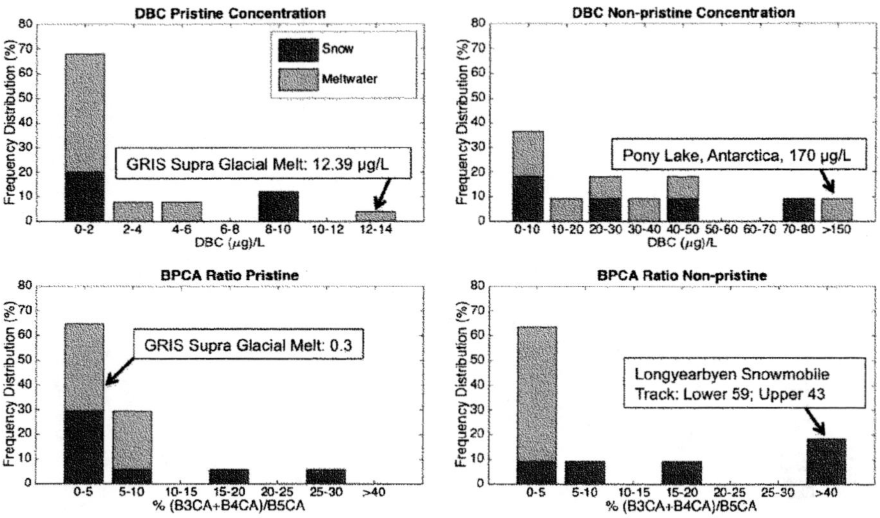

Figure 11. Frequency distribution of pristine and nonpristine snow and meltwater samples from unpublished values in Table 1. Meltwater samples are composed of both snow and ice melt. Note that the maximum *x* axis value for DBC pristine concentration is 14 µg/L versus 150 µg/L for DBC nonpristine concentration.

Median DBC concentrations were higher in nonpristine snow samples (14 µg/L (1.6, 71)) than pristine snow samples (1.8 µg/L (13, 10)). Furthermore, the median (B3CA + B4CA)/BSCA ratio for nonpristine samples from snow was much higher than pristine snow samples, 12 (3, 59) and 3 (1, 27), respectively (Table 1 and the figure above). The highest DBC concentrations in snow were found in samples collected from Longyearbyen, Svalbard, next to an active coalmine (71 µg/L), a coal burning power plant (6.8 µg/L), and along the primary Longyearbren snowmobile track (22 µg/L) upper track; 44 1-19/L lower track), which receives continuous inputs of fresh anthropogenic BC.

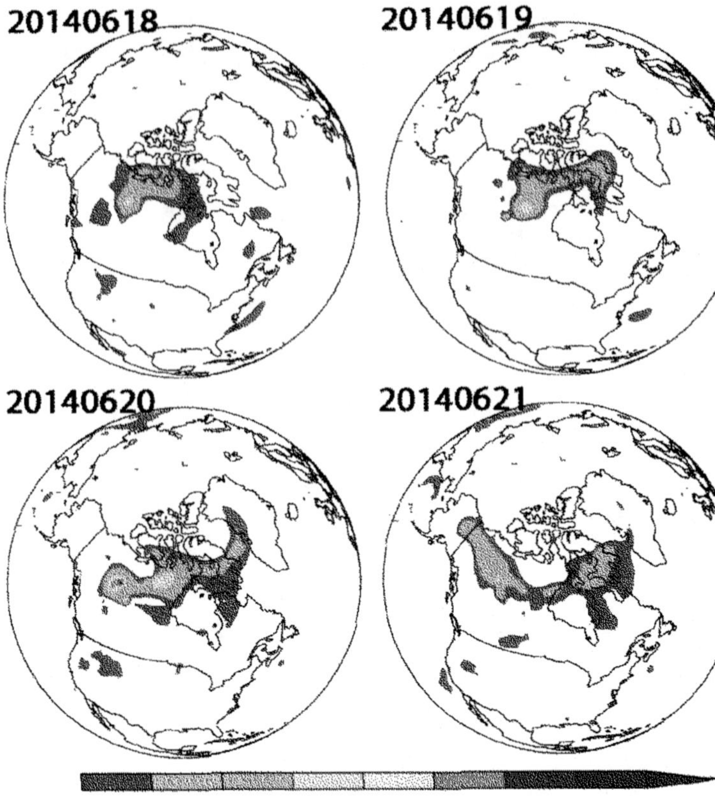

Figure 12. Smoke aerosol optical thickness (AOT) reanalysis at 18Z from the Navy Aerosol Analysis and Prediction System (NAAPS) between 18 and 21 June 2014 shows that wildfire aerosols were transported from the Canadian Arctic to the western region of Greenland Ice Sheet in June 2014. The color bar is atmospheric total column smoke aerosol optical thickness (AOT) at 550 nm, which is unitless.

In contrast to our hypothesis that continuous inputs would result in a signature less indicative to photodegradatlon, these samples also feature the highest (B3CA + B4CA)/B5CA ratios, with the snow next to the coal burning power plant featuring a ratio of 17, and even higher ratios along the snow mobile track (43 upper track; 59 lower track).

Pristine meltwaters had low median DBC concentrations (1.7 μg/L (0.6, 12)), similar to the pristine snow samples (1.8 μg/L (1.3, 10)). The median (B3CA + B4CA)/B5CA ratio for pristine meltwaters, 4 (0.3, 7) (Table 1 and Figure 11), was marginally higher than for pristine snow samples, 3 (1, 27).In contrast to the Nepal meltwaters, the GRIS supraglacial meltwater had a high concentration of DBC (12.4 μg/L) corresponding to the 97.5% quantile. In addition, the GRIS meltwater had a similar low ratio of (B3CA + B4CA)/B5CA to the GRIS snow/surface hoar sample. These BPCA ratios (0.3 and 0.8, respectively) were the lowest in this data set and are lower than the previously reported ratios from terrestrial rivers impacted by wildfires (Wagner et al., 2015). As noted previously, these low ratios likely reflect fresh and recent BC deposition sourced from Canadian Arctic wildfires. The higher DBC concentration compared to the Nepal supraglacial waters could reflect not only the substantial long range wildfire inputs but also the large surface area of the GRIS, which may provide a larger supraglacial catchment for accumulation and transport of DBC to the ablation area where the sample was collected. Further, the high DBC in the GRIS meltwater provides support to the potential positive feedback proposed by Keegan et al. (2014) whereby DBC enriched meltwater is retained as refrozen ice layers in the snowpack and enhances future melting.

Meltwater from supraglacial lakes on the Ngozumba Glacier in Nepal, located at 5000 m, had the lowest DBC concentrations (average of five samples, 1.3 ± 0.49 μg/L) of all meltwater samples. The ratio of these samples was also composed of less condensed DBC, with an average ratio of 5, than supraglacial melt from the GRIS which was composed of highly condensed DBC and had a ratio of 0.3. Although no nearby snow samples were obtained for direct comparison, these low values from the Ngozumba Glacier may be associated with heavy glacial debris coverage in the majority of the ablation region. The debris cover, when thick enough, provides insulation from glacial melt (Pratap et al., 2015) and may reduce supraglacial meltwater generation, along with transport and accumulation of DBC. Although these lakes form an ice cover from about November to March, in the summer they receive large solar radiation inputs, enhanced by the high altitude, which <:auld drive photodegradation of DBC. The less condensed DBC signature could also suggest different combustion sources of DBC between these sampling locations.

Similar to the observations for .snow and ice, nonpristine meltwaters had higher median DBC concentrations, (27 µ/L (4.7, 170)) than pristine meltwaters (1.7 µg/L (0.6, 12)). Closed basin nonpristine Antarctic surface waters, such as DLH Pond, Marble Point Pond, Clear Lake, and Pony Lake, contained high concentrations of DBC ranging from 14 to 170, µg/L These lakes are located near active point sources of fossil fuel combustion, such as helicopter flight paths and fueling stations, and exhibited elevated DBC concentrations relative to snow and meltwater. Because these surface water ponds have no outflowing streams, any DBC deposited in the systems will remain and accumulate for millennia, similar to the MDV lakes (Khan et al., 2016), or may be photodegraded. The BPCA ratio of these four nonpristine Antarctic surface waters ranged from 2 to 3. Similarly, the BPCA ratios for Long year River and Advent fjord ocean water in the Norwegian Arctic, both located near the town of Longyearbyen and the local coal burning power plant, were both 4. Such elevated DBC concentrations and BPCA ratios suggest that site proximity to continuous inputs from these particular anthropogenic sources results in high concentrations of less condensed DBC and that long-term photooxidation may not be the only cause of less condensed DBC in the cryosphere.

Conclusion

This study reveals that within the cryosphere, DBC concentrations and composition are widely variable. Regionally, concentration differences exist between remote-pristine sites and those near combustion point sources. Overall, these data support our hypothesis that DBC in the cryosphere that is exposed to longer durations of solar radiation results in less condensed DBC due to photodegradation. Additionally, sites in the cryosphere receiving fresh BC from wildfires, such as on the GRIS where aerosols were deposited directly from Canadian Arctic wildfires, consist of more condensed DBC This signatuie of the DBC from the GRIS was distinct from the sites which receive continuous inputs from local anthropogenic sources of BC, such as along the heavily used snowmobile track in Svalbard, as well as sites which have prolonged sunlight exposure, thus enhancing effects of photodegradation.

While the influence of wildfire inputs throughout the cryosphere may be important regionally, as observed in the samples from the GRIS, the intense solar exposure in high alpine and polar regions, and the characteristics of some anthropogenic combustion sources subject to solar exposure during long-range atmospherics trans port (Cooke and Wilson, 1996), may cause the similarity in DBC composition observed throughout the rest of the global cryosphere data set. This includes photodegradation of background levels of wildfire-derived BC, which may also contribute to B3 + 4CA enrichment in the cryosphere (Stubbins et al., 2012b; Ward et al., 2014). As such, the influence of photodegradation may sometimes be "overwhelmed" by discrete wildfire sources, which likely account for DBC composition of the GRIS samples. Although the current data set allowed for preliminary assessments of contributions to and persistence of DBC in the cryosphere, the observed variability of DBC composition is likely derived from a combination of biogeochemical processes in the environment and different pyrogenic sources.

Previously reported accumulation of millennia old DBC in Antarctic takes (Khan et al., 2016) suggests that DBC is recalcitrant in the cryosphere. The ranges of concentrations observed in this diverse data set of DBC from the global cryosphere indicate that relatively high DBC concentrations persist in both pristine and nonpristine Arctic and Antarctic snow and surface waters. The relatively high concentration of DBC measured in one supraglacial stream on the GRIS may suggest that DBC can be mobilized across ice surfaces. Wildfires are predicted to increase due to climate change (Flannigan et al., 2009), and albedo cannibalism is already influencing meltwater generation on the GRIS (Tedesco et al., 2015). Anticipated longer summer melt seasons as a result of climate change may result in longer durations between snowfalls, enhancing exposure of recalcitrant DBC on snow/ice surfaces, which could further exacerbate surface albedo reduction in the cryosphere.

References

Abiven, S., P. Hengartner, M. P. W. Schneider, N. Singh, and M. W. I. Schmidt (2011), Pyrogenic carbon soluble fraction is larger and more aromatic in aged charcoal than in fresh charcoal, *Soil Biol. Biochem.*, *43*(7), 1615–1617, doi:10.1016/j.soilbio.2011.03.027.

Bond, T. C., et al. (2013), Bounding the role of black carbon in the climate system: A scientific assessment, *J. Geophys. Res. Atmos.*, *118*, 5380–5552, doi:10.1002/jgrd.50171.

Cooke, W. F., and J. J. N. Wilson (1996), A global black carbon aerosol model, *J. Geophys. Res.*, *101*(D14), 19,395–19,409, doi:10.1029/96JD00671.

Ding, Y., Y. Yamashita, W. K. Dodds, and R. Jaffé (2012), Dissolved black carbon in grassland streams: Is there an effect of recent fire history?, *Chemosphere*, *90*(10), 2557–2562, doi:10.1016/j.chemosphere.2012.10.098.

Ding, Y., Y. Yamashita, J. Jones, and R. Jaffe (2014a), Dissolved black carbon in boreal forest and glacial rivers of central Alaska: Assessment of biomass burning versus anthropogenic sources, *Biogeochemistry*, *123*, 15–25, doi:10.1007/s10533-014-0050-7.

Ding, Y., K. M. Cawley, C. N. da Cunha, and R. Jaffé (2014b), Environmental dynamics of dissolved black carbon in wetlands, *Biogeochemistry*, *119*, 259–273, doi:10.1007/s10533-014-9964-3.

Dittmar, T. (2008), The molecular level determination of black carbon in marine dissolved organic matter, *Org. Geochem.*, *39*(4), 396–407, doi:10.1016/j.orggeochem.2008.01.015.

Dittmar, T., B. Koch, N. Hertkorn, and G. Kattner (2008), A simple and efficient method for the solid-phase extraction of dissolved organic matter (SPE-DOM) from seawater, *Limnol. Oceanogr. Methods*, *6*, 230–235, doi:10.4319/lom.2008.6.230.

Fellman, J. B., E. Hood, P. A. Raymond, A. Stubbins, and R. G. M. Spencer (2015), Spatial variation in the origin of dissolved organic carbon in snow on the Juneau icefield, Southeast Alaska, *Environ. Sci. Technol.*, 150908131727005, doi:10.1021/acs.est.5b02685.

Flannigan, M., B. Stocks, M. Turetsky, and M. Wotton (2009), Impacts of climate change on fire activity and fire management in the circumboreal forest, *Glob. Chang. Biol.*, *15*(3), 549–560, doi:10.1111/j.1365-2486.2008.01660.x.

Glaser, B., L. Haumaier, G. Guggenberger, and W. Zech (1998), Black carbon in soils: The use of benzenecarboxylic acids as specific markers, *Org. Geochem.*, *29*(4), 811–819, doi:10.1016/S0146-6380(98)00194-6.

Hodson, A. J. (2014), Understanding the dynamics of black carbon and associated contaminants in glacial systems, *Wiley Interdiscip. Rev. Water*, *1*(April), 141–149, doi:10.1002/wat2.1016.

Holben, B. N., et al. (1998), AERONET—A federated instrument network and data archive for aerosol characterization, *Remote Sens. Environ.*, *4257*(98).

Hood, E., J. Fellman, R. G. M. Spencer, P. J. Hernes, R. Edwards, D. D'Amore, and D. Scott (2009), Glaciers as a source of ancient and labile organic matter to the marine environment, *Nature*, *462*(7276), 1044–1047, doi:10.1038/nature08580.

Hood, E., T. J. Battin, J. Fellman, S. O'Neel, and R. G. M. Spencer (2015), Storage and release of organic carbon from glaciers and ice sheets, *Nat. Geosci.*, *8*(2), 1–6, doi:10.1038/ngeo2331.

Hyer, E. J., J. S. Reid, and J. Zhang (2011), An over-land aerosol optical depth data set for data assimilation by filtering, correction, and aggregation of MODIS Collection 5 optical depth retrievals, *Atmos. Meas. Tech.*, *4*, 379–408, doi:10.5194/amt-4-379-2011.

Keegan, K. M., M. R. Albert, J. R. Mcconnell, and I. Baker (2014), Climate change and forest fires synergistically drive widespread melt events of the Greenland Ice Sheet, *Proc. Natl. Acad. Sci. U. S. A.*, *111*, 7964–7967, doi:10.1073/pnas.1405397111.

Khan, A. L., R. Jaffé, Y. Ding, and D. M. McKnight (2016), Dissolved black carbon in Antarctic lakes: Chemical signatures of past and present sources, *Geophys. Res. Lett.*, *43*, 5750–5757, doi:10.1002/2016GL068609.

Lynch, P., et al. (2016), An 11-year global gridded aerosol optical thickness reanalysis (v1. 0) for atmospheric and climate sciences, *Geosci. Model Dev.*, *9*, 1489–1522, doi:10.5194/gmd-9-1489-2016.

Mcbeath, A. V., R. J. Smernik, and E. S. Krull (2013a), Organic geochemistry: A demonstration of the high variability of chars produced from wood in bushfires, *Org. Geochem.*, *55*, 38–44, doi:10.1016/j.orggeochem.2012.11.006.

Mcbeath, A. V., R. J. Smernik, E. S. Krull, and J. Lehmann (2013b), ScienceDirect: The influence of feedstock and production temperature on biochar carbon chemistry—A solid-state ^{13}C NMR study, *Biomass Bioenergy*, *60*, 121–129, doi:10.1016/j.biombioe.2013.11.002.

Pratap, B., D. P. Dobhal, M. Mehta, and R. Bhambri (2015), Influence of debris cover and altitude on glacier surface melting: A case study on Dokriani Glacier, central Himalaya, India, *Ann. Glaciol.*, *56*(70), 9–16, doi:10.3189/2015AoG70A971.

Reid, J. S., et al. (2009), Global monitoring and forecasting of biomass-burning smoke: Description of and lessons from the Fire Locating and Modeling of Burning Emissions (FLAMBE) program, *IEEE J. Sel. Top. Appl. Earth Obs. Remote Sens.*, *2*(3), 144–162.

Schneider, M. P. W., M. Hilf, U. F. Vogt, and M. W. I. Schmidt (2010), The benzene polycarboxylic acid (BPCA) pattern of wood pyrolyzed between 200°C and 1000°C, *Org. Geochem.*, *41*(10), 1082–1088, doi:10.1016/j.orggeochem.2010.07.001.

Shi, Y., J. Zhang, J. S. Reid, B. Liu, and E. J. Hyer (2014), Critical evaluation of cloud contamination in the MISR aerosol products using MODIS cloud mask products, *Atmos. Meas. Tech. Discuss.*, 1791–1801, doi:10.5194/amt-7-1791-2014.

Stubbins, A., et al. (2012a), Anthropogenic aerosols as a source of ancient dissolved organic matter in glaciers, *Nat. Geosci.*, *5*(3), 198–201, doi:10.1038/ngeo1403.

Stubbins, A., J. Niggemann, and T. Dittmar (2012b), Photo-lability of deep ocean dissolved black carbon, *Biogeosciences*, *9*(2002), 1661–1670, doi:10.5194/bg-9-1661-2012.

Tedesco, M., S. Doherty, X. Fettweis, P. Alexander, J. Jeyaratnam, E. Noble, and J. Stroeve (2015), The darkening of the Greenland ice sheet: Trends, drivers and projections (1981–2100), *Cryosphere Discuss.*, *9*(5), 5595–5645, doi:10.5194/tcd-9-5595-2015.

Wagner, S., K. M. Cawley, F. L. Rosario-Ortiz, and R. Jaffé (2015), In-stream sources and links between particulate and dissolved black carbon following a wildfire, *Biogeochemistry*, *124*, 145–161, doi:10.1007/s10533-015-0088-1.

Ward, C. P., R. L. Sleighter, P. G. Hatcher, and R. M. Cory (2014), Insights into the complete and partial photooxidation of black carbon in surface waters, *Environ. Sci.: Processes Impacts*, *16*(4), 721–731, doi:10.1039/c3em00597f.

Wiedenmeyer, M., et al. (2015), Aromaticity and degree of aromatic condensation of char, *Org. Geochem.*, *78*, 135–143, doi:10.1016/j.orggeochem.2014.10.002.

Zhang, J., and J. S. Reid (2006), MODIS aerosol product analysis for data assimilation: Assessment of over-ocean level 2 aerosol optical thickness retrievals, *J. Geophys. Res.*, *111*, D22207, doi:10.1029/2005JD006898.

Ziolkowski, L. A., and E. R. M. Druffel (2010), Aged black carbon identified in marine dissolved organic carbon, *Geophys. Res. Lett.*, *37*, L16601, doi:10.1029/2010GL043963.

In: Climate Change
Editor: George De Haas

ISBN: 978-1-53617-841-8

Chapter 2

CHANGES IN THE ARCTIC: BACKGROUND AND ISSUES FOR CONGRESS (UPDATED)*

Ronald O'Rourke, Laura B. Comay, Peter Folger, John Frittelli, Marc Humphries, Jane A. Leggett, Jonathan L. Ramseur, Pervaze A. Sheikh and Harold F. Upton

ABSTRACT

The diminishment of Arctic sea ice has led to increased human activities in the Arctic, and has heightened interest in, and concerns about, the region's future. The United States, by virtue of Alaska, is an Arctic country and has substantial interests in the region. The seven other Arctic states are Canada, Iceland, Norway, Sweden, Finland, Denmark (by virtue of Greenland), and Russia.

The Arctic Research and Policy Act (ARPA) of 1984 (Title I of P.L. 98-373 of July 31, 1984) "provide[s] for a comprehensive national policy dealing with national research needs and objectives in the Arctic." The National Science Foundation (NSF) is the lead federal agency for implementing Arctic research policy. Key U.S. policy documents relating

* This is an edited, reformatted and augmented version of Congressional Research Service, Publication No. R41153, dated January 23, 2020.

to the Arctic include National Security Presidential Directive 66/Homeland Security Presidential Directive 25 (NSPD 66/HSPD 25) of January 9, 2009; the National Strategy for the Arctic Region of May 10, 2013; the January 30, 2014, implementation plan for the 2013 national strategy; and Executive Order 13689 of January 21, 2015, on enhancing coordination of national efforts in the Arctic. The office of the U.S. Special Representative for the Arctic has been vacant since January 20, 2017.

The Arctic Council, created in 1996, is the leading international forum for addressing issues relating to the Arctic. The council's mandate explicitly excludes the issue of military security. The council's membership consists of the eight Arctic states, and all decisions of the council and its subsidiary bodies are by consensus of the eight member states. The council also includes six indigenous permanent participants and 39 observer states and organizations. The United Nations Convention on the Law of the Sea (UNCLOS) sets forth a comprehensive regime of law and order in the world's oceans, including the Arctic Ocean. The United States is not a party to UNCLOS.

Record low extents of Arctic sea ice over the past decade have focused scientific and policy attention on links to global climate change and projected ice-free seasons in the Arctic within decades. These changes have potential consequences for weather in the United States, access to mineral and biological resources in the Arctic, the economies and cultures of peoples in the region, and national security.

The geopolitical environment for the Arctic has been substantially affected by the renewal of great power competition. Although there continues to be significant international cooperation on Arctic issues, the Arctic is increasingly viewed as an arena for geopolitical competition among the United States, Russia, and China. Russia in recent years has enhanced its military presence and operations in the Arctic, and the other Arctic states are now taking steps to enhance their own military presences and operations in the region. China's activities in the Arctic have been growing steadily in recent years.

U.S. military forces, particularly the Navy and Coast Guard, have begun to pay more attention to the Arctic in their planning and operations. The Department of Defense (DOD), the Navy, and the Coast Guard have all issued Arctic strategy documents in recent years. Whether DOD and the military services are adequately resourcing these strategies and taking appropriate and adequate actions for defending U.S. interests in the Arctic has emerged as a topic of debate and congressional oversight. The Coast Guard has two operational polar icebreakers—the heavy polar icebreaker Polar Star, and the medium polar icebreaker Healy. The Coast Guard has received funding for the procurement of the first of three planned new heavy polar icebreakers.

The diminishment of Arctic ice could lead in coming years to increased commercial shipping on two trans-Arctic sea routes—the Northern Sea Route close to Russia, and the Northwest Passage close to Alaska and through the Canadian archipelago—though the rate of increase in the use of these routes might not be as great as sometimes anticipated in press accounts. International guidelines for ships operating in Arctic waters have been recently updated.

Changes to the Arctic brought about by warming temperatures will likely allow more exploration for oil, gas, and minerals. Warming that causes permafrost to melt could pose challenges to onshore exploration activities. Increased oil and gas exploration and tourism (cruise ships) in the Arctic increase the risk of pollution in the region. Cleaning up oil spills in ice-covered waters will be more difficult than in other areas, primarily because effective strategies for cleaning up oil spills in ice-covered waters have yet to be developed.

Large commercial fisheries exist in the Arctic. The United States is currently meeting with other countries regarding the management of Arctic fish stocks. Changes in the Arctic could affect threatened and endangered species, and could result in migration of fish stocks to new waters.

Under the Endangered Species Act, the polar bear was listed as threatened on May 15, 2008. Arctic climate change is also expected to affect the economies, health, and cultures of Arctic indigenous peoples.

INTRODUCTION

The diminishment of Arctic sea ice has led to increased human activities in the Arctic, and has heightened interest in, and concerns about, the region's future. Issues such as Arctic territorial disputes; commercial shipping through the Arctic; Arctic oil, gas, and mineral exploration; endangered Arctic species; and increased military operations in the Arctic could cause the region in coming years to become an arena of international cooperation, tension, or competition.

The United States, by virtue of Alaska, is an Arctic country and has substantial political, economic, energy, environmental, and other interests in the region. Decisions that Congress makes on Arctic-related issues could significantly affect these interests.

This report provides an overview of Arctic-related issues for Congress, and refers readers to more in-depth CRS reports on specific Arctic-related issues. Congressional readers with questions about an issue discussed in this report should contact the author or authors of the section discussing that issue. The authors are identified by footnote at the start of each section.

This report does not track legislation on specific Arctic-related issues. For tracking of legislative activity, see the CRS reports relating to specific Arctic-related issues that are listed at the end of this report, just prior to Appendix A.

BACKGROUND[1]

Definitions of the Arctic

There are multiple definitions of the Arctic that result in differing descriptions of the land and sea areas encompassed by the term. Policy discussions of the Arctic can employ varying definitions of the region, and readers should bear in mind that the definition used in one discussion may differ from that used in another. This CRS report does not rely on any one definition.

Arctic Circle Definition and Resulting Arctic Countries

The most common and basic definition of the Arctic defines the region as the land and sea area north of the Arctic Circle (a circle of latitude at about 66.34° North). For surface locations within this zone, the sun is generally above the horizon for 24 continuous hours at least once per year (at the summer solstice) and below the horizon for 24 continuous hours at least once per year (at the winter solstice). The Arctic Circle definition includes the northernmost third or so of Alaska, as well as the Chukchi

---

[1] Except for the subsection on the Arctic and the U.N. Convention on the Law of the Sea, this section was prepared by Ronald O'Rourke, Specialist in Naval Affairs, Foreign Affairs, Defense, and Trade Division.

Sea, which separates that part of Alaska from Russia, and U.S. territorial and Exclusive Economic Zone (EEZ) waters north of Alaska. It does not include the lower two-thirds or so of Alaska or the Bering Sea, which separates that lower part of the state from Russia.

The area within the Arctic Circle is about 14.5 million square kilometers, or about 5.6 million square miles.[2] This equates to about 2.8%, or about 1/36th, of the world's surface.[3] About 4 million people, or about 0.05% of the world's population, live in the Arctic, of which roughly half (roughly 2 million) live in Russia's part of the Arctic.[4]

Definition in Arctic Research and Policy Act (ARPA) of 1984

Section 112 of the Arctic Research and Policy Act (ARPA) of 1984 (Title I of P.L. 98-373 of July 31, 1984)[5] defines the Arctic as follows:

> As used in this title, the term "Arctic" means all United States and foreign territory north of the Arctic Circle and all United States territory north and west of the boundary formed by the Porcupine, Yukon, and Kuskokwim Rivers [in Alaska]; all contiguous seas, including the Arctic Ocean and the Beaufort, Bering, and Chukchi Seas; and the Aleutian chain.

This definition, which is codified at 15 U.S.C. 4111,[6] includes certain parts of Alaska below the Arctic Circle, including the Aleutian Islands and portions of central and western mainland Alaska, such as the Seward Peninsula and the Yukon Delta. The U.S. Coast Guard states that "The

---

[2] Source: Polar Discovery (Woods Hole Oceanographic Institution), "Arctic: Location and Geography," accessed October 24, 2018, at http://polardiscovery.whoi.edu/arctic/geography.html. This source states that 14.5 million square kilometers is equal to about 5.5 million square miles, but 14.5 million square kilometers equates to about 5.598 million square miles.

[3] The surface area of the world is approximately 510 million square miles. Source: Universe Today, "What is the Surface Area of the Earth?" February 10, 2017, accessed October 24, 2018, at https://www.universetoday.com/25756/ surface-area-of-the-earth/.

[4] Sources: National Snow & Ice Data Center, "Arctic People," accessed October 24, 2018, at https://nsidc.org/cryosphere/arctic-meteorology/arctic-people.html; and United Kingdom, House of Commons, Defence Committee, On Thin Ice: UK Defence in the Arctic, Twelfth Report of Session 2017–19, August 15, 2018 (Ordered by the House of Commons to be printed 19 July 2018), p. 6.

[5] Title II of P.L. 98-373 is the National Critical Materials Act of 1984.

[6] As codified, the definition reads, "As used in this chapter."

U.S. Arctic encompasses some 2,521 miles of shoreline, an international strait adjacent to the Russian Federation, and 647 miles of land border with Canada above the Arctic Circle. The U.S. Exclusive Economic Zone (EEZ) in the Arctic contains approximately 889,000 square miles of ocean."[7] Figure 1 below shows the Arctic area of Alaska as defined by ARPA; Figure 2 shows the entire Arctic area as defined by ARPA.

Other Definitions

Other definitions of the Arctic are based on factors such as average temperature, the northern tree line, the extent of permafrost on land, the extent of sea ice on the ocean, or jurisdictional or administrative boundaries.[8] A definition based on a climate-related factor could circumscribe differing areas over time as a result of climate change.

The 10°C isotherm definition of the Arctic defines the region as the land and sea area in the northern hemisphere where the average temperature for the warmest month (July) is below 10° Celsius, or 50° Fahrenheit. This definition results in an irregularly shaped Arctic region that excludes some land and sea areas north of the Arctic Circle but includes some land and sea areas south of the Arctic Circle. This definition currently excludes all of Finland and Sweden, as well as some of Alaska above the Arctic Circle, while including virtually all of the Bering Sea and Alaska's Aleutian Islands.[9]

---

[7] Coast Guard, *Arctic Strategic Outlook*, April 2019, p. 11.

[8] For discussions and (in some cases) maps, see Susan Joy Hassol, *Impacts of a Warming Arctic* [Executive Summary]. Cambridge, Cambridge University Press, 2004, p. 4, available at http://amap.no/acia/; Oran R. Yong and Niels Einarsson, *Arctic Human Development Report*, Stefansson Arctic Institute, Akureyri, Iceland, 2004, pp. 17-18, available at http://www.svs.is/AHDR/AHDR%20chapters/English%20version/Chapters%20PDF.htm; and Hugo Ahlenius, editor in chief et al., *Vital Arctic Graphics, People and Global Heritage on Our Last Wild Shores*, UNEP/GRID-Arendal, Arendal, Norway, p. 6, available at http://www.grida.no/publications/vg/arctic/.

[9] A map showing the line that results from 10o isotherm definition is available at https://www.cia.gov/library/publications/the-world-factbook/reference_maps/pdf/arctic.pdf.

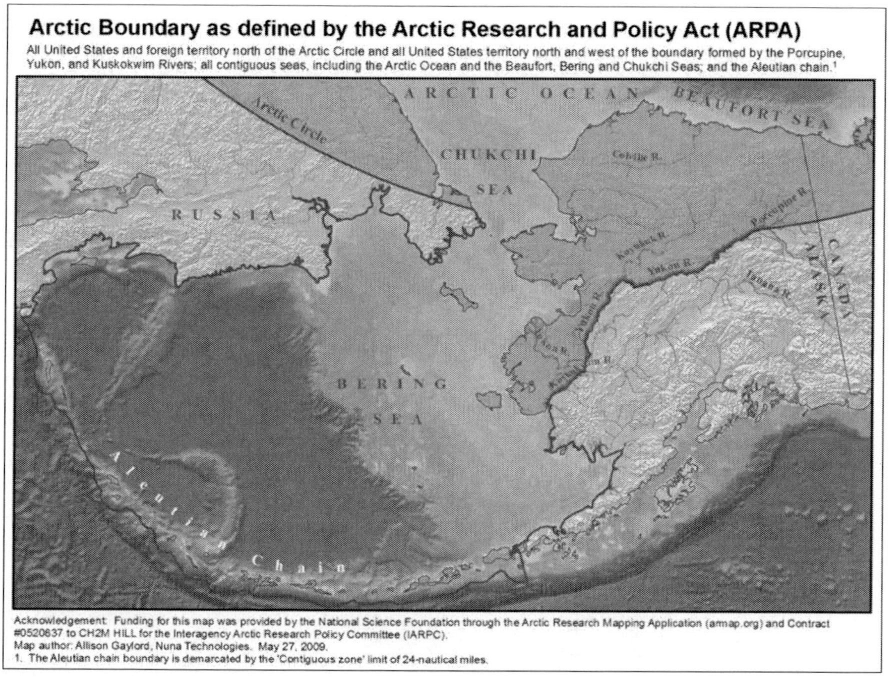

Source: U.S. Arctic Research Commission (http://www.arctic.gov/maps/ARPA_Alaska_only_150dpi.jpg, accessed on December 23, 2011).

Figure 1. Arctic Area of Alaska as Defined by ARPA.

The definition of the Arctic adopted by the Arctic Monitoring and Assessment Programme (AMAP)—a working group of the Arctic Council—"essentially includes the terrestrial and marine areas north of the Arctic Circle (66°32' N), and north of 62° N in Asia and 60° N in North America, modified to include the marine areas north of the Aleutian chain, Hudson Bay, and parts of the North Atlantic, including the Labrador Sea." The AMAP website includes a map showing the Arctic Circle, 10o C isotherm, tree line, and AMAP definitions of the Arctic.[10]

Some observers use the term "high north" as a way of referring to the Arctic. Some observers make a distinction between the "high Arctic"—meaning, in general, the colder portions of the Arctic that are closer to the

10 See also map posted by the Arctic Council at https://arctic-council.org/images/PDF_attachments/Maps/ boundaries.pdf.

North Pole—and other areas of the Arctic that are generally less cold and further away from the North Pole, which are sometimes described as the low Arctic or the subarctic.

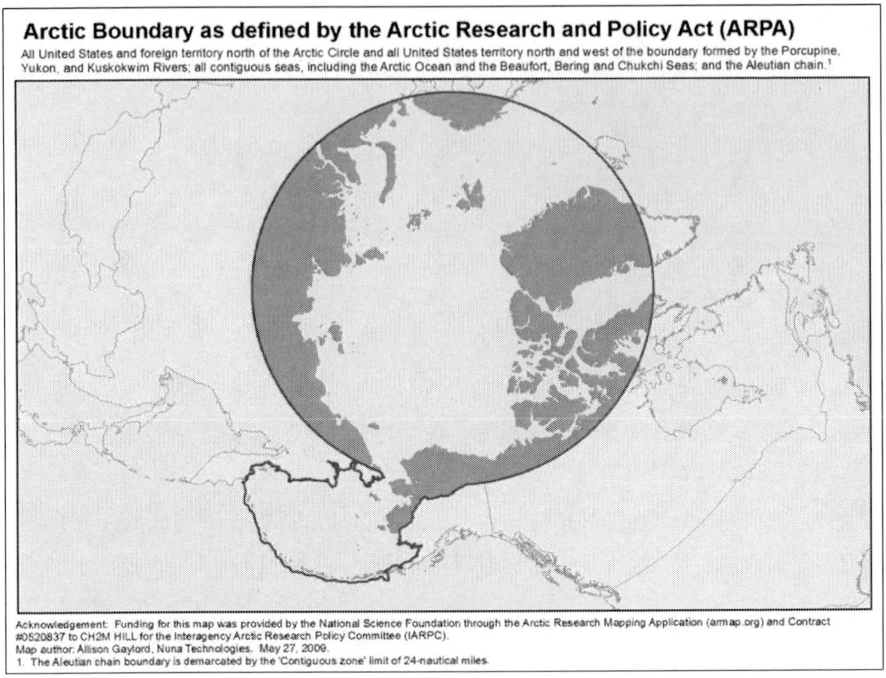

Source: U.S. Arctic Research Commission (http://www.arctic.gov/maps/ARPA_Polar_150dpi.jpg, accessed on December 23, 2011).

Figure 2. Entire Arctic Area as Defined by ARPA.

Eight Arctic States

Eight countries have territory north of the Arctic Circle: the United States (Alaska), Canada, Russia, Norway, Denmark (by virtue of Greenland, a member country of the Kingdom of Denmark), Finland, Sweden, and Iceland. These eight countries are often referred to as the Arctic countries or Arctic States, and they are the member states of the Arctic Council, which is discussed further below.

A subset of the eight Arctic countries are the five countries that are considered Arctic coastal states because they have mainland coasts that front onto waters north of the Arctic Circle: the United States, Canada, Russia, Norway, and Denmark (by virtue of Greenland).[11]

U.S. Identity as an Arctic Nation

As mentioned earlier, the United States, by virtue of Alaska, is an Arctic country and has substantial political, economic, energy, environmental, and other interests in the region. Even so, Alaska is geographically separated and somewhat distant from the other 49 states, and relatively few Americans—fewer than 68,000 as of July 1, 2017—live in the Arctic part of Alaska as shown in Figure 2.[12] A November 8, 2018, research paper on the Arctic in U.S. national identity, based on data collected in online surveys conducted in October and December 2017, stated the following:

> We found that Americans on average continue mildly to disagree with the canonical assertion of U.S. Arctic identity and interests as articulated in government policy. On a scale from 1 to 7, with higher numbers indicating stronger agreement, Americans' average rating was 3.51, up slightly from 3.16 in 2015, but still below the scale midpoint [of 4.0]. A plurality of respondents (27%) answered with a score of one, indicating the strongest disagreement. Men and older individuals showed greater inclination to agree with the assertion of Arctic identity and interests than women or younger respondents, a pattern also observed in 2015. No region of the country showed particularly greater inclination to

---

[11] The northern coast of mainland Iceland is just south of the Arctic Circle. The Arctic Circle passes through Grimsey Island, a small offshore island of Iceland that is about 25 miles north of the northern coast of mainland Iceland. See "Is Iceland in the Arctic Circle?" Iceland Unlimited, January 2017, accessed July 30, 2019, at https://icelandunlimited.is/blog/is-iceland-in-the-arctic-circle/.

[12] Source for figure of fewer than 68,000: CRS analysis of data presented in Table 3.1, entitled Alaska Population by Region, Borough, and Census Area, 2017 to 2045, in Alaska Department of Labor and Workforce Development, Research and Analysis Section, *Alaska Population Projections: 2017 to 2045*, June 2018, p. 26. The table shows that of Alaska's estimated population as of July 1, 2017 of 737,080, a total of 589,680, of about 80%, resided in the Anchorage/Matanuska-Susitna region (401,649), the Fairbanks North Star Borough (97,738), the Kenai Peninsula Borough (58,024), and Juneau (32,269).

agree or disagree, except Alaskans[, who] showed substantially greater agreement.

We also conducted a series of comparative surveys and found that Canadians, with an average rating of 4.87, had a much greater sense of being an Arctic nation than did Americans. American respondents, however, did register somewhat higher agreement than British and Australians in judging their country an Arctic nation with strong Arctic interests. In a separate comparative survey, Americans indicated a stronger sense of being a Pacific nation than an Arctic one.[13]

U.S. Arctic Research

Arctic Research and Policy Act (ARPA) of 1984, as Amended

The Arctic Research and Policy Act (ARPA) of 1984 (Title I of P.L. 98-373 of July 31, 1984)[14] "provide[s] for a comprehensive national policy dealing with national research needs and objectives in the Arctic."[15] The act, among other things

- made a series of findings concerning the importance of the Arctic and Arctic research;
- established the U.S. Arctic Research Commission (USARC) to promote Arctic research and recommend Arctic research policy;
- designated the National Science Foundation (NSF) as the lead federal agency for implementing Arctic research policy;
- established the Interagency Arctic Research Policy Committee (IARPC) to develop a national Arctic research policy and a five-

[13] Zachary D. Hamilla, *The Arctic in U.S. National Identity (2017)*, Arctic Studio, November 8, 2018, p. 1. See also Yereth Rosen, "Americans Still Aren't Fully Embracing the Nation's Arctic Identity, a New Survey Finds," *Arctic Today*, November 13, 2018.

[14] Title II of P.L. 98-373 is the National Critical Materials Act of 1984.

[15] These words are taken from the official title of P.L. 98-373. (Arctic Research and Policy Act of 1984 is the short title of Title I of P.L. 98-373.) The remainder of P.L. 98-373's official title relates to Title II of the act, the short title of which is the National Critical Materials Act of 1984.

year plan to implement that policy, and designated the NSF representative on the IARPC as its chairperson;[16] and

- defined the term "Arctic" for purposes of the act.

The Arctic Research and Policy Act of 1984 was amended by P.L. 101-609 of November 16, 1990. For the texts of the Arctic Research and Policy Act of 1984 and P.L. 101-609, see Appendix A and Appendix B, respectively.

FY2020 NSF Budget Request for Arctic Research

Office of Polar Programs (OPP)

NSF—the lead federal agency for implementing Arctic research policy—carries out Arctic research activities through its Office of Polar Programs (OPP), which operates as part of NSF's Directorate for Geosciences (GEO). NSF is requesting a total of $403.4 million for OPP for FY2020, a decrease of 24.5% from the $534.5 million requested for FY2019, and a decrease of % from the $501.7 million actual for FY2018. Within the $403.4 million requested for OPP for FY2020 is $80.2 million requested for research in both the Arctic and Antarctic, a decrease of 29.4% from the $113.6 million requested for FY2019 and a decrease of 32.7% from the $119.2 million actual for FY2018. Also within the $403.4 million requested for OPP for FY2020 is $43.0 million requested for Arctic research and support logistics, an increase of 10.3% over the $39.3 million requested for FY2019, and a decrease 5.9% from the $45.7 million actual for FY2018.[17] Regarding its FY2020 budget request for OPP, NSF states that

[16] The IARPC currently includes more than a dozen federal agencies, departments, and offices. Additional information on the IARPC is available at http://www.nsf.gov/od/opp/arctic/iarpc/start.jsp.

[17] National Science Foundation, *FY 2020 Budget Request to Congress*, March 18, 2019, p. OPP-1; and National Science Foundation, *FY 2019 Budget Request to Congress*, February 28, 2018, p. OPP-1. The dollar figures in this paragraph have been rounded to the nearest tenth of a million.

The Office of Polar Programs is the primary U.S. supporter of fundamental research in the polar regions. In the Arctic, NSF helps coordinate research planning as directed by the Arctic Research Policy Act of 1984, and the NSF Director chairs the Interagency Arctic Research Policy Committee (IARPC) created for this purpose….

OPP supports investments in research and education and provides support for research infrastructure, such as permanent stations and temporary field camps in the Antarctic and the Arctic, including support for [NSF Big Ideas—NNA, WoU, and URoL.[18] OPP's FY 2020 Budget Request is influenced by three key priorities: (1) supporting critical facilities that enable frontier research in the Earth's polar regions; (2) maintaining strong disciplinary programs that provide a base for our investments in cross-disciplinary system science programs; and (3) maintaining U.S. research community activities in polar system science. These priorities reflect opportunities for fundamental scientific discovery uniquely possible in polar regions, as well as studies to investigate the causes and future trajectory of environmental and ecosystem changes now being observed at the poles that could impact global systems. This work will implement the Foundation's lead-agency role in facilitating the Nation's investment in polar science.

In addition to shared cross-directorate basic research objectives, OPP investments will be guided by recent sponsored studies to identify priority areas and ensure effective polar research programs:

- For the Arctic, IARPC's Arctic Research Plan: FY 2017-20211, and the World Meteorological Organization's Year of Polar Prediction Implementation Plan inform science investment priorities. Efforts to build an integrated research capacity to address the potential opportunities and challenges of Arctic change for the Nation's security and economics and well-being of Arctic residents will continue…

Major Investments

- OPP will support science in both polar regions…

[18] NNA, WoU and URoL are three of NSF's 10 Big Ideas (i.e., its 10 "research agendas that identify areas at the frontiers of science and engineering which promise to be among the most transformative in the coming decade.") NNA, WoU, and URoL stand for Navigating the New Arctic, Windows on the Universe: The Era of Multi-messenger Astrophysics, and Understanding the Rules of Life: Predicting Phenotype, respectively.)

- In FY 2020, OPP will support research at $80.16 million. OPP will continue to support existing priority commitments, including… Arctic observing that received strong endorsement at the Arctic Science Ministerial, as well as new awards….

OPP provides 46 percent of the federal funding for basic research at academic institutions in the polar sciences.[19]

Navigating the New Arctic (NNA)

NSF states in the overview of its FY2020 budget request that "NSF continues to emphasize its 10 Big Ideas, research agendas that identify areas at the frontiers of science and engineering which promise to be among the most transformative in the coming decade. FY 2020 investments will build on progress made in prior years to seed or establish these Big Ideas."[20] Among the 10 big ideas, NSF states in its overview that number 6 is

> Navigating the New Arctic (NNA)—Establishing an observing network of mobile and fixed platforms and tools, including cyber tools, **across the Arctic to document and understand the Arctic's rapid** biological, physical, chemical, and social changes, in partnership with other agencies, countries, and native populations.[21]

For FY2020, NSF is requesting $43.0 million for NNA.[22] NSF states that NNA

> seeks innovations in Arctic observational networks and fundamental convergence research across the social, natural, environmental, computing and information sciences and engineering that address the intersection of natural, social, and built systems. By drawing upon

[19] National Science Foundation, *FY 2020 Budget Request to Congress*, March 18, 2019, p. OPP-1 to OPP-2.

[20] National Science Foundation, *FY 2020 Budget Request to Congress*, March 18, 2019, p. Overview-2.

[21] National Science Foundation, *FY 2020 Budget Request to Congress*, March 18, 2019, p. Overview-4. Emphasis as in original.

[22] National Science Foundation, *FY 2020 Budget Request to Congress*, March 18, 2019, p. NSF-Wide Investments–10.

expertise from across the agency, NNA investments will accelerate research needed to inform decisions regarding the economy, security, and resilience of the U.S. as an Arctic Nation. On the observing front, NNA will advance and address key gaps in the existing array of observational networks, leveraging resources with the Mid-scale RI Big Idea as appropriate. NNA will also support knowledge co-production with indigenous and local people to enhance observations in key areas. Interagency, state government, and international partnerships will be further developed in order to achieve pan-Arctic and Arctic-global perspectives.

NNA will provide support for research activities that advance understanding of the social changes underway, including shifting demographics, changing ways of life, and traditional ecological and other knowledge in danger of being lost. Researchers will also examine economic, cultural, and social impacts on groups and communities living in regions impacted by Arctic changes in order to better understand how the social, economic, and governance systems interact with infrastructure such as the delivery of services and access to resources. Findings from this research can help to anticipate social changes that are likely to emerge from increased economic activity.

NNA-supported data efforts will enable major advances in modeling and prediction of Arctic systems and their global interactions. Strong coupling of observations, computation, and theory will be supported to ensure progress. NNA will leverage investments through the HDR Big Idea by federating data systems supported by NSF and interagency and international partners so that data can be readily discovered, accessed, and interoperated to facilitate more rapid theory and idea development and testing.

NNA will build on NSF's STEM investments and the NSF INCLUDES Big Idea to encourage innovative and appropriately evaluated education and public engagement efforts that leverage exciting NNA science and inspire diverse participation in STEM. NSF plans to invest in NNA through FY 2023.

Goals

1) Improved understanding of Arctic change and its local and global effects that capitalize on innovative and optimized observation infrastructure, advances in understanding of fundamental processes, and new approaches to modeling interactions among the natural environment, built environment, and social systems.

2) New enhanced research communities that are diverse, integrative, and well-positioned to carry out productive research at the intersections of Arctic natural and built environments and social systems.
3) Research outcomes that inform U.S. national security and economic development needs in the Arctic and enable resilient, sustainable Arctic communities.

In FY 2017, NSF issued a Dear Colleague Letter (DCL) on the GCR Big Idea (NSF17-065) to explore Convergence approaches within four of the research-focused NSF Big Ideas including NNA. This DCL requested proposals for Research Coordination Networks (RCNs), workshops, and activities to enhance Arctic observational systems. In FY 2018, NSF issued a DCL on Stimulating Research Related to NNA (NSF18-048), requesting proposals for workshops, RCNs, and proposals be submitted to the Arctic Sciences program in OPP. NSF awarded 25 new projects under these two DCLs and related opportunities. In FY 2019, NSF issued a solicitation for NNA (NSF19-511) and plans to make approximately 25 awards to support research grants, and planning grants that develop convergence research teams.

FY 2020 Investments

NSF's NNA activities in FY 2020 will focus on enabling advances in priority areas, which will be developed by building on outcomes from FY 2017 – FY 2019 activities. In 2019, NNA focused on advances in observing technologies and convergent social/built/natural environment systems science. FY 2020 support for this investment will continue, and NSF expects to issue another solicitation in FY 2020.

NSF will also continue coordination and leveraging of NNA-related activities with external stakeholders, including other federal agencies through the Interagency Arctic Research Policy Committee chaired by the NSF Director, local residents and indigenous peoples through state and local governance structures of Alaska, as well as with international partners through fora such as the biannual International Arctic Science Ministerial. All of the identified FY 2020 NNA activities will support goals 1, 2, and 3.[23]

[23] National Science Foundation, *FY 2020 Budget Request to Congress*, March 18, 2019, pp. NSF-Wide Investments–10 to NSF-Wide Investments–11.

Major U.S. Policy Documents Relating to the Arctic

January 2009 Arctic Policy Directive (NSPD 66/HSPD 25)

On January 12, 2009 (i.e., eight days before its final day in office), the George W. Bush Administration released a presidential directive establishing a new U.S. policy for the Arctic region. The directive, dated January 9, 2009, was issued as National Security Presidential Directive 66/Homeland Security Presidential Directive 25 (NSPD 66/HSPD 25). The directive was the result of an interagency review, and it superseded for the Arctic (but not the Antarctic) a 1994 presidential directive on Arctic and Antarctic policy. The directive, among other things

- states that the United States is an Arctic nation, with varied and compelling interests in the region;
- sets forth a six-element overall U.S. policy for the region;
- describes U.S. national security and homeland security interests in the Arctic; and
- discusses a number of issues as they relate to the Arctic, including international governance; the extended continental shelf and boundary issues; promotion of international scientific cooperation; maritime transportation; economic issues, including energy; and environmental protection and conservation of natural resources.

For the text of NSPD 66/HSPD 25, see Appendix C.

May 2013 National Strategy for Arctic Region

On May 10, 2013, the Obama Administration released a document entitled *National Strategy for the Arctic Region.*[24] The document appears to supplement rather than supersede the January 2009 Arctic policy directive (NSPD 66/HSPD 25) discussed above.[25] The executive summary of

---

[24] *National Strategy for the Arctic Region*, May 2013, 11 pp.; accessed May 14, 2013, at https://obamawhitehouse.archives.gov/sites/default/files/docs/nat_arctic_strategy.pdf. The document includes a cover letter from President Obama dated May 10, 2013.

[25] *National Strategy for the Arctic Region* states on page 6 that the "lines of effort" it describes are to be undertaken "[t]o meet the challenges and opportunities in the Arctic region, and in

National Strategy for the Arctic Region begins by quoting the above statement from the previous May 2010 national security strategy document, and then states the following:

> The *National Strategy for the Arctic Region* sets forth the United States Government's strategic priorities for the Arctic region. This strategy is intended to position the United States to respond effectively to challenges and emerging opportunities arising from significant increases in Arctic activity due to the diminishment of sea ice and the emergence of a new Arctic environment. It defines U.S. national security interests in the Arctic region and identifies prioritized lines of effort, building upon existing initiatives by Federal, state, local, and tribal authorities, the private sector, and international partners, and aims to focus efforts where opportunities exist and action is needed. It is designed to meet the reality of a changing Arctic environment, while we simultaneously pursue our global objective of combating the climatic changes that are driving these environmental conditions. Our strategy is built on three lines of effort:
>
> 1) Advance United States Security Interests – We will enable our vessels and aircraft to operate, consistent with international law, through, under, and over the airspace and waters of the Arctic, support lawful commerce, achieve a greater awareness of activity in the region, and intelligently evolve our Arctic infrastructure and capabilities, including ice- capable platforms as needed. U.S. security in the Arctic encompasses a broad spectrum of activities, ranging from those supporting safe commercial and scientific operations to national defense.
> 2) Pursue Responsible Arctic Region Stewardship – We will continue to protect the Arctic environment and conserve its resources; establish and institutionalize an integrated Arctic management framework; chart the Arctic region; and employ scientific research and traditional knowledge to increase understanding of the Arctic.
> 3) Strengthen International Cooperation – Working through bilateral relationships and multilateral bodies, including the Arctic Council, we will pursue arrangements that advance

---

furtherance of established Arctic Region Policy," at which point there is a footnote referencing the January 2009 Arctic policy directive.

collective interests, promote shared Arctic state prosperity, protect the Arctic environment, and enhance regional security, and we will work toward U.S. accession to the United Nations Convention on the Law of the Sea (Law of the Sea Convention).

Our approach will be informed by the following guiding principles:

- Safeguard Peace and Stability – Seek to maintain and preserve the Arctic region as an area free of conflict, acting in concert with allies, partners, and other interested parties. Support and preserve: international legal principles of freedom of navigation and overflight and other uses of the sea and airspace related to these freedoms, unimpeded lawful commerce, and the peaceful resolution of disputes for all nations.
- Make Decisions Using the Best Available Information – Across all lines of effort, decisions need to be based on the most current science and traditional knowledge.[26]
- Pursue Innovative Arrangements – Foster partnerships with the state of Alaska, Arctic states, other international partners, and the private sector to more efficiently develop, resource, and manage capabilities, where appropriate and feasible, to better advance our strategic priorities in this austere fiscal environment.
- Consult and Coordinate with Alaska Natives – Engage in a consultation process with Alaska Natives, recognizing tribal governments' unique legal relationship with the United States and providing for meaningful and timely opportunity to inform Federal policy affecting Alaskan Native communities.[27]

[26] A footnote in the document at this point states the following: "Traditional knowledge refers to a body of evolving practical knowledge based on observations and personal experience of indigenous communities over an extensive, multigenerational time period." (BOEM Ocean Science, Vol. 9, Issue 2, May/April/June 2012, page 4).

[27] *National Strategy for the Arctic Region*, May 2013, pp. 2-3.

January 2014 Implementation Plan for National Strategy for Arctic Region

On January 30, 2014, the Obama Administration released an implementation plan for the May 2013 national strategy for the Arctic region.[28] The plan states that it

> complements and builds upon existing initiatives by Federal, State, local, and tribal authorities, the private sector, and international partners, and focuses efforts where opportunities exist and action is most needed. The Implementation Plan reflects the reality of a changing Arctic environment and upholds national interests in safety, security, and environmental protection, and works with international partners to pursue global objectives of addressing climatic changes.
>
> This Implementation Plan follows the structure and objectives of the Strategy's three lines of effort and is consistent with the guiding principles. The lines of effort of the Strategy and the Implementation Plan are as follows:
>
> - Advance United States Security Interests
> - Pursue Responsible Arctic Region Stewardship
> - Strengthen International Cooperation
>
> These lines of effort and guiding principles are meant to be implemented as a coherent whole.[29]

The plan also states the following:

> Climate change is already affecting the entire global population, and Alaska residents are experiencing the impacts in the Arctic. To ensure a cohesive Federal approach, implementation activities must be aligned with the Executive Order on Preparing the United States for the Impacts

[28] *Implementation Plan for The National Strategy for the Arctic Region*, January 2014, 32 pp. The news release announcing the implementation plan is posted at http://www.whitehouse.gov/blog/2014/01/30/white-house-releases- implementation-plan-national-strategy-arctic-region. The document itself is posted at https://obamawhitehouse.archives.gov/sites/default/files/docs/implementation_plan_for_the_national_strategy_for_the_arctic_region_-_fi...pdf.

[29] *Implementation Plan for The National Strategy for the Arctic Region*, January 2014, p. 1.

of Climate Change[30] while executing the Strategy. In addition to the guiding principles, the following approaches are important in implementing the activities across all of the lines of effort:

- Foster Partnerships with Arctic Stakeholders. As outlined in the Strategy, all lines of effort must involve Arctic partners, particularly the State of Alaska and Alaska Natives in the Arctic region. Federal agencies, the State of Alaska, tribal communities, local governments, and academia will work with other nations, industry stakeholders, non-governmental organizations, and research partners to address emerging challenges and opportunities in the Arctic environment. The Federal Government should strive to maintain the free flow of communication and cooperation with the State of Alaska to support national priorities.
- Coordinate and Integrate Activities across the Federal Government. Multiple Federal bodies currently have authority for Arctic policy (e.g., the National Ocean Council (NOC), Arctic Policy Group, and Interagency Arctic Research Policy Committee (IARPC)). The National Security Council Staff will develop an Executive Order through the interagency process to maximize efficiency, align interagency initiatives, and create unity of effort among all Federal entities conducting activities in the Arctic.[31]

The plan outlines about 36 specific initiatives. For each, it presents a brief statement of the objective, a list of next steps to be taken, a brief statement about measuring progress in achieving the objective, and the names of the lead and supporting federal agencies to be involved.

---

[30] The passage contains a footnote at this point stating that this executive order was signed by the President on November 1, 2013. It is Executive Order 13653. The text of the order is posted at http://www.gpo.gov/fdsys/pkg/FR- 2013-11-06/pdf/2013-26785.pdf and http://www.whitehouse.gov/the-press-office/2013/11/01/executive-order- preparing-united-states-impacts-climate-change. A fact sheet about it is posted at http://www.whitehouse.gov/the-press-office/2013/11/01/fact-sheet-executive-order-climate-preparedness.

[31] *Implementation Plan for The National Strategy for the Arctic Region*, January 2014, p. 4.

On March 9, 2016, the Obama Administration released three documents discussing the implementation of the national strategy for the Arctic:[32] (1) a report entitled *2015 Year in Review—Progress Report on the Implementation of the National Strategy for the Arctic Region*;[33] (2) an appendix to that report entitled *Appendix A, Implementation Framework for the National Strategy for the Arctic Region*:[34] and (3) another appendix to that report entitled *Appendix B, Interagency Arctic Research Policy Committee 5-Year Plan Collaboration Teams: 2015 Summary of Accomplishments and 2016 Priorities*.[35]

January 2015 Executive Order for Enhancing Coordination of Arctic Efforts

On January 21, 2015, then-President Obama issued Executive Order 13689, entitled "Enhancing Coordination of National Efforts in the Arctic." The order states the following in part:

> As the United States assumes the Chairmanship of the Arctic Council, it is more important than ever that we have a coordinated national effort that takes advantage of our combined expertise and efforts in the Arctic region to promote our shared values and priorities.
>
> As the Arctic has changed, the number of Federal working groups created to address the growing strategic importance and accessibility of this critical region has increased. Although these groups have made

[32] For the text of the Obama Administration's announcement releasing these documents, see "Advancing Implementation of the National Strategy for the Arctic Region," March 9, 2016, accessed March 25, 2016, at https://www.whitehouse.gov/blog/2016/03/09/advancing-implementation-national-strategy-arctic-region.

[33] *2015 Year in Review—Progress Report on the Implementation of the National Strategy for the Arctic Region*, Prepared by the Arctic Executive Steering Committee, March 2016, 35 pp., accessed March 25, 2016, at https://obamawhitehouse.archives.gov/sites/whitehouse.gov/files/documents/Progress%20Report%20on%20the%20Implementation%20of%20the%20National%20Strategy%20for%20the%20Arct ic%20Region.pdf.

[34] *Appendix A, Implementation Framework for the National Strategy for the Arctic Region*, March 2016, 33 pp., accessed March 25, 2016, at https://obamawhitehouse.archives.gov/sites/whitehouse.gov/files/documents/National%20Strategy%20for%20the%20Arctic%20Region%20Implementation%20Framework%20%28Appendix%20A%29%20Final.pdf.

[35] *Appendix B, Interagency Arctic Research Policy Committee 5-Year Plan Collaboration Teams: 2015 Summary of Accomplishments and 2016 Priorities*, undated, 13 pp., accessed March 25, 2016, at https://obamawhitehouse.archives.gov/sites/whitehouse.gov/files/documents/Appendix%20B%20IARPC%202015%20Annual%20Report.pdf.

significant progress and achieved important milestones, managing the broad range of interagency activity in the Arctic requires coordinated planning by the Federal Government, with input by partners and stakeholders, to facilitate Federal, State, local, and Alaska Native tribal government and similar Alaska Native organization, as well as private and nonprofit sector, efforts in the Arctic....

There is established an Arctic Executive Steering Committee (Steering Committee), which shall provide guidance to executive departments and agencies (agencies) and enhance coordination of Federal Arctic policies across agencies and offices, and, where applicable, with State, local, and Alaska Native tribal governments and similar Alaska Native organizations, academic and research institutions, and the private and nonprofit sectors....

... the Steering Committee will meet quarterly, or as appropriate, to shape priorities, establish strategic direction, oversee implementation, and ensure coordination of Federal activities in the Arctic....

The Steering Committee, in coordination with the heads of relevant agencies and under the direction of the Chair, shall:

(a) provide guidance and coordinate efforts to implement the priorities, objectives, activities, and responsibilities identified in National Security Presidential Directive 66/Homeland Security Presidential Directive 25, Arctic Region Policy, the National Strategy for the Arctic Region and its Implementation Plan, and related agency plans;

(b) provide guidance on prioritizing Federal activities, consistent with agency authorities, while the United States is Chair of the Arctic Council, including, where appropriate, recommendations for resources to use in carrying out those activities; and

(c) establish a working group to provide a report to the Steering Committee by May 1, 2015, that:

(i) identifies potential areas of overlap between and within agencies with respect to implementation of Arctic policy and strategic priorities and provides recommendations to increase coordination and reduce any duplication of effort, which may include ways to increase the effectiveness of existing groups; and

(ii) provides recommendations to address any potential gaps in implementation....

It is in the best interest of the Nation for the Federal Government to maximize transparency and promote collaboration where possible with the State of Alaska, Alaska Native tribal governments and similar Alaska Native organizations, and local, private-sector, and nonprofit-sector stakeholders. To facilitate consultation and partnerships with the State of Alaska and Alaska Native tribal governments and similar Alaska Native organizations, the Steering Committee shall:

(a) develop a process to improve coordination and the sharing of information and knowledge among Federal, State, local, and Alaska

> Native tribal governments and similar Alaska Native organizations, and private-sector and nonprofit-sector groups on Arctic issues;
>
> (b) establish a process to ensure tribal consultation and collaboration, consistent with my memorandum of November 5, 2009 (Tribal Consultation). This process shall ensure meaningful consultation and collaboration with Alaska Native tribal governments and similar Alaska Native organizations in the development of Federal policies that have Alaska Native implications, as applicable, and provide feedback and recommendations to the Steering Committee;
>
> (c) identify an appropriate Federal entity to be the point of contact for Arctic matters with the State of Alaska and with Alaska Native tribal governments and similar Alaska Native organizations to support collaboration and communication; and
>
> (d) invite members of State, local, and Alaska Native tribal governments and similar Alaska Native organizations, and academic and research institutions to consult on issues or participate in discussions, as appropriate and consistent with applicable law.[36]

As stated in the above-quoted passage, Executive Order 13689, among other things, established an Arctic Executive Steering Committee (AESC) to "provide guidance to executive departments and agencies (agencies) and enhance coordination of Federal Arctic policies across agencies and offices, and, where applicable, with State, local, and Alaska Native tribal governments and similar Alaska Native organizations, academic and research institutions, and the private and nonprofit sectors." Regarding the AESC, a February 28, 2019, press report states the following: "Although the [executive] order has not been rescinded, the Trump administration has left the committee dormant for the past two years."[37]

December 2017 National Security Strategy Document

The National Security Strategy document released by the Trump Administration in December 2017 mentions the term Arctic once, stating that that "A range of international institutions establishes the rules for how states, businesses, and individuals interact with each other, across land and

---

[36] Executive order, "Enhancing Coordination of National Efforts in the Arctic," January 21, 2015, accessed January 21, 2015, at http://www.whitehouse.gov/the-press-office/2015/01/21/executive-order-enhancing-coordination-national- efforts-arctic.

[37] Melody Schreiber, "Two U.S. Bills Could Advance American Presence in the Arctic," *Arctic Today*, February 28, 2019.

sea, the Arctic, outer space, and the digital realm. It is vital to U.S. prosperity and security that these institutions uphold the rules that help keep these common domains open and free."[38]

U.S. Special Representative for Arctic (Vacant Since January 2017)

On July 16, 2014, during the Obama Administration, then-Secretary of State John Kerry announced the appointment of retired Coast Guard Admiral Robert J. Papp Jr., who served as Commandant of the Coast Guard from May 2010 to May 2014, as the first U.S. Special Representative for the Arctic.[39] Under the Obama Administration, the duties of this position involved, among other things, interacting with ambassadors to the Arctic region from other countries. Papp served as the U.S. Special Representative until January 20, 2017, the final day of the Obama Administration and the first day of the Trump Administration; the position has been unfilled since then.

Arctic Council

Overview

The Arctic Council is the leading international forum for addressing issues relating to the Arctic. It was created in September 1996, following a series of meetings initiated by Finland in 1989.[40] Its founding document is the Ottawa Declaration of September 19, 1996, a joint declaration (not a

---

[38] *National Security Strategy of the United States of America*, December 2017, p. 40.

[39] See "Retired Admiral Robert Papp to Serve as U.S. Special Representative for the Arctic," Press Statement, John Kerry, Secretary of State, Washington, DC, July 16, 2014, accessed July 17, 2014, at http://www.state.gov/secretary/ remarks/2014/07/229317.htm.

[40] See Arctic Council, "Finland," updated September 10, 2015, accessed July 30, 2019, at https://arctic-council.org/ index.php/en/about-us/member-states/finland. The council states that "The establishment of the Arctic Council was preceded by the Arctic Environmental Protection Strategy (June 1991), a declaration on the protection of the Arctic environment." (Arctic Council, "Frequently Asked Questions," updated June 6, 2018, accessed July 30, 2019, at https://arctic-council.org/index.php/en/about-us/arctic-council/faq).

treaty) signed by representatives of the eight Arctic states. The declaration states that the council "is established as a high level forum to," among other things, "provide a means for promoting cooperation, coordination and interaction among the Arctic States, with the involvement of the Arctic indigenous communities and other Arctic inhabitants on common Arctic issues, in particular issues of sustainable development and environmental protection in the Arctic…."[41]

The council describes itself on its website as "the leading intergovernmental forum promoting cooperation, coordination and interaction among the Arctic States, Arctic indigenous communities and other Arctic inhabitants on common Arctic issues, in particular on issues of sustainable development and environmental protection in the Arctic."[42] The State Department describes the council as "the preeminent intergovernmental forum for addressing issues related to the Arctic Region…. The Arctic Council is not a treaty-based international organization but rather an international forum that operates on the basis of consensus, echoing the peaceful and cooperative nature of the Arctic Region.[43]

Thematic areas of work addressed by the council include environment and climate, biodiversity, oceans, Arctic peoples, and agreements on Arctic scientific cooperation, cooperation on marine oil pollution preparedness and response in the Arctic, and cooperation on aeronautical and maritime search and rescue in the Arctic.[44] The Ottawa Declaration states explicitly that "The Arctic Council should not deal with matters related to military security."

---

[41] Arctic Council, "Ottawa Declaration (1996)," accessed July 30, 2019, at https://oaarchive.arctic-council.org/handle/11374/85.

[42] Arctic Council, "The Arctic Council: A Backgrounder," updated September 13, 2018, accessed July 30, 2019, at https://arctic-council.org/index.php/en/about-us.

[43] State Department, "Arctic Region," undated, accessed July 31, 2019, at https://www.state.gov/key-topics-office-of- ocean-and-polar-affairs/arctic/.

[44] Arctic Council, "Thematic Work," undated, accessed July 30, 2019, at https://arctic-council.org/index.php/en/our-work.

Organization and Operations

Eight Member States and Their Senior Arctic Officials (SAOs)

The Arctic Council's membership consists of the eight Arctic states. Each member state is represented by a Senior Arctic Official (SAO), who is usually drawn from that country's foreign ministry. The SAOs meet at least twice per year.[45]

Indigenous Permanent Participants

In addition to the council's eight member states, "six organizations representing Arctic indigenous peoples have status as Permanent Participants. The category of Permanent Participant was created to provide for active participation and full consultation with the Arctic indigenous peoples within the Council. They include: the Aleut International Association, the Arctic Athabaskan Council, Gwich'in Council International, the Inuit Circumpolar Council, Russian Association of Indigenous Peoples of the North and the Saami Council."[46]

Observers

Thirteen non-Arctic states have been approved as observers to the council: Germany, the Netherlands, Poland, and the United Kingdom (approved in 1998); France (2000); Spain (2006); China, India, Italy, Japan, Singapore, and South Korea (2013); and Switzerland (2017). A November 22, 2019, press report states that "in 2015, uncertainty about

---

[45] Arctic Council, "Frequently Asked Questions," updated June 6, 2018, accessed July 30, 2019, at https://arctic- council.org/index.php/en/about-us/arctic-council/faq.

[46] Arctic Council, "The Arctic Council: A Backgrounder," updated September 13, 2018, accessed July 30, 2019, at https://arctic-council.org/index.php/en/about-us. The council states further that "Out of a total of 4 million inhabitants of the Arctic, approximately 500,000 belong to indigenous peoples. Indigenous peoples' organizations have been granted Permanent Participants status in the Arctic Council. The Permanent Participants have full consultation rights in connection with the Council's negotiations and decisions. The Permanent Participants represent a unique feature of the Arctic Council, and they make valuable contributions to its activities in all areas." (Arctic Council, "Permanent Participants," updated March 22, 2017, accessed July 30, 2019, at https://arctic-council.org/index.php/en/about-us/ permanent-participants.)

their role led to a hiatus in observers being admitted."[47] Estonia has announced it will apply to be admitted as an observer in 2021.[48] In addition to state observers, 14 intergovernmental and interparliamentary organizations and 12 nongovernmental organizations have been approved as observers, making for a total of 39 observer states and organizations.[49] The most recently added observer was the International Maritime Organization (IMO), which was added to the list of intergovernmental and interparliamentary observer organizations in 2019.

Working Groups

The council's work is carried out primarily in six working groups that focus on Arctic contaminants; Arctic monitoring and assessment; conservation of Arctic flora and fauna; emergency prevention, preparedness and response; protection of the Arctic marine environment; and sustainable development.[50] The council may also establish task forces or expert groups for specific projects.

Secretariat

The council's standing Secretariat formally became operational in June 2013 in Tromsø, Norway.[51]

Chairmanship

The council has a two-year chairmanship that rotates among the eight member states. The United States held the chairmanship from April 24, 2015, to May 11, 2017, a period which began during the Obama Administration and continued into the first 16 weeks of the Trump

---

[47] Kevin McGwin, "Iceland Begins Its Arctic Council Chairmanship with a Focus on Observers," *Arctic Today*, November 22, 2019.

[48] Kevin McGwin, "Estonia Is Applying to Become an Arctic Council Observer," *Arctic Today*, November 20, 2019.

[49] Arctic Council, "Observers," updated May 23, 2019, accessed July 30, 2019, at https://arctic-council.org/index.php/en/about-us/arctic-council/observers.

[50] For more on the working groups, see Arctic Council, "Working Groups," updated September 10, 2015, accessed July 30, 2019, at https://arctic-council.org/index.php/en/about-us/working-groups.

[51] For more on the Secretariat, see Arctic Council, "The Arctic Counciul Secretariat," updated November 4, 2016, accessed July 30, 2019, at https://arctic-council.org/index.php/en/about-us/arctic-council/the-arctic-council-secretariat.

Administration.[52] The United States had previously held the chairmanship from 1998 to 2000, and will next hold it in 2031-2033.[53] On May 11, 2017, the chairmanship was transferred from the United States to Finland. On May 7, 2019, it was transferred from Finland to Iceland, which will hold the position until May 2021.[54]

Decisionmaking

The council states that "All decisions of the Arctic Council and its subsidiary bodies are by consensus of the eight Arctic Member States."[55] More specifically, the council states that "Decisions at all levels in the Arctic Council are the exclusive right and responsibility of the eight Arctic States with the involvement of the Permanent Participants,"[56] and that

---

[52] See, for example, Yereth Rosen, "US Ends Arctic Council Chairmanship With Reluctance on Climate Action," *Alaska Dispatch News*, May 11, 2017.

[53] During the Obama Administration's portion of the period of U.S. chairmanship, the U.S. chairmanship team was led by then-Secretary of State John Kerry. The State Department during the Obama Administration listed nine additional members of the team, including retired Coast Guard Admiral Robert J. Papp Jr., U.S. Special Representative for the Arctic. ("Meet the U.S. Chairmanship Team," accessed September 29, 2015, at http://www.arctic-council.org/index.php/en/about-us/arctic-council/u-s-chairmanship.) The other eight members of the team were the Honorable Fran Ulmer, Special Advisor to the U.S. Secretary of State on Arctic Science and Policy; Ambassador David Balton, Chair of the Senior Arctic Officials; Julia Gourley, U.S. Senior Arctic Official (SAO); Dr. Nikoosh Carlo, Senior Advisor to the SAO Chair; Dr. Adrianna Muir, Deputy Senior Arctic Official; Nomi Seltzer, Arctic Affairs Advisor; Erin Robertson, Arctic Press and Public Affairs Officer; and Matthew Kastrinsky, Administrative Officer.

[54] See Arctic Council, "Arctic Council Ministers meet, pass Chairmanship from Finland to Iceland, Arctic States conclude Arctic Council Ministerial meeting by signing a joint statement," May 7, 2019, updated June 28, 2019, accessed July 30, 2019, at https://arctic-council.org/index.php/en/our-work2/8-news-and-events/521-arctic-council-ministers-meet-pass-chairmanship-from-finland-to-iceland.

The council states that "The theme of the Arctic Council Chairmanship program for [the Iceland chairmanship period of] 2019-2021 reflects Iceland's commitment to the principle of sustainable development and refers to the necessity of close cooperation between the states and peoples of the region and beyond. With sustainable development as an overarching theme, Iceland will highlight four priorities: The Arctic Marine Environment, Climate and Green Energy Solutions, People and Communities of the Arctic, and a Stronger Arctic Council." (Arctic Council, "Iceland's Chairmanship 2019-2021," undated, accessed July 30, 2019, at https://arctic-council.org/index.php/en/about-us/arctic- council/iceland-chairmanship.)

[55] Arctic Council, "Working Groups," updated September 10, 2015, accessed July 30, 2019, at https://arctic-council.org/index.php/en/about-us/working-groups.

[56] Arctic Council, "Observers," updated May 23, 2019, accessed July 30, 2019, at https://arctic-council.org/index.php/ en/about-us/arctic-council/observers.

"Arctic Council assessments and recommendations are the result of analysis and efforts undertaken by the Working Groups. Decisions of the Arctic Council are taken by consensus among the eight Arctic Council States, with full consultation and involvement of the Permanent Participants."[57]

Limits of Arctic Council as a Governing Body

Regarding the limits of the Arctic Council as a governing body, the council states that

> The Arctic Council is a forum; it has no programming budget. All projects or initiatives are sponsored by one or more Arctic States. Some projects also receive support from other entities. The Arctic Council does not and cannot implement or enforce its guidelines, assessments or recommendations. That responsibility belongs to each individual Arctic State. The Arctic Council's mandate, as articulated in the Ottawa Declaration, explicitly excludes military security.[58]

The Arctic and the U.N. Convention on Law of the Sea (UNCLOS)[59]

Overview of UNCLOS

The United Nations Convention on the Law of the Sea (UNCLOS) "lays down a comprehensive regime of law and order in the world's oceans and seas[,] establishing rules governing all uses of the oceans and their resources."[60] It builds on four 1958 law of the sea conventions to which the

---

[57] Arctic Council, "The Arctic Council: A Backgrounder," updated September 13, 2018, accessed July 30, 2019, at https://arctic-council.org/index.php/en/about-us.

[58] Arctic Council, "The Arctic Council: A Backgrounder," updated September 13, 2018, accessed July 30, 2019, at https://arctic-council.org/index.php/en/about-us.

[59] Parts of this section were prepared by Marjorie Ann Browne, who was a Specialist in International Relations, Foreign Affairs, Defense, and Trade Division until her retirement from CRS on October 10, 2015.

[60] United Nations, "United Nations Convention on the Law of the Sea of 10 December 1982, Overview and full text," updated June 28, 2019, accessed August 2, 2019, at

United States, following Senate consent to ratification, became a party in 1961, and which entered force between 1962 and 1966.[61] All four treaties remain in force for the United States.[62] UNCLOS was adopted in 1982 as the "culmination of more than 14 years of work involving participation by more than 150 countries representing all regions of the world, all legal and political systems and the spectrum of socio/economic development."[63] The treaty was modified in 1994 by an agreement relating to the implementation of Part XI of the treaty, which relates to the seabed and ocean floor and subsoil thereof that are beyond the limits of national jurisdiction.

> UNCLOS entered into force in November 1994. As of April 8, 2019, 168 nations were party to the treaty.[64] As discussed later in more detail, the United States is not a party to the treaty.

https://www.un.org/depts/los/convention_agreements/convention_overview_convention.htm.

[61] These are the Convention on the Territorial Sea and Contiguous Zone, which entered into force on September 10, 1964, the Convention on the Continental Shelf, which entered into force on 10 June 10, 1964, the Convention on the High Seas, which entered into force on September 30, 1962, and the Convention on Fishing and Conservation of Living Resources of the High Seas, which entered into force on March 20, 1966. The four 1958 treaties resulted from the first Conference on the Law of the Sea (UNCLOS I), which took place in 1958. (For additional discussion, see United Nations, "United Nations Conference on the Law of the Sea," undated, accessed August 2, 2019, at http://legal.un.org/diplomaticconferences/1958_los/, and United Nations, "1958 Geneva Conventions on the Law of the Sea," undated, accessed August 2, 2019, at http://legal.un.org/avl/ha/gclos/gclos.html).

[62] See Department of State, *Treaties in Force, Section 2, Multilateral Treaties in Force as of January 1, 2019*, pp. 526, 501, 525, and 516, respectively.

[63] United Nations, "United Nations Convention on the Law of the Sea of 10 December 1982, Overview and full text," updated June 28, 2019, accessed August 2, 2019, at https://www.un.org/depts/los/convention_agreements/convention_overview_convention.htm. More specifically, the treaty resulted from the Third United Nations Conference on the Law of the Sea (UNCLOS III), which took place between 1973 and 1982. For additional discussion, see United Nations, "Third United Nations Conference on the Law of the Sea," undated, accessed August 2, 2019, at http://legal.un.org/diplomaticconferences/1973_los/.

[64] Chronological lists of ratifications of, accessions and successions to the Convention and the related Agreements as of February 3, 2017, accessed December 13, 2018, at http://www.un.org/Depts/los/reference_files/chronological_lists_of_ratifications.htm#.

Part VI of UNCLOS and Commission on Limits of Continental Shelf

Part VI of UNCLOS (consisting of Articles 76 through 85), which covers the continental shelf, and Annex II to the treaty, which established a Commission on the Limits of the Continental Shelf, are particularly pertinent to the Arctic, because Article 77 states that "The coastal State exercises over the continental shelf sovereign rights for the purpose of exploring it and exploiting its natural resources," and that these natural resources include, among other things, "mineral and other non-living resources of the seabed and subsoil," including oil and gas deposits.[65]

Article 76 states that "The continental shelf of a coastal State comprises the seabed and subsoil of the submarine areas that extend beyond its territorial sea throughout the natural prolongation of its land territory to the outer edge of the continental margin, or to a distance of 200 nautical miles" if the outer edge of the continental margin does not extend up to that distance. Article 76 states that "the coastal State shall establish the outer edge of the continental margin wherever the margin extends beyond 200 nautical miles," and that "Information on the limits of the continental shelf beyond 200 nautical miles... shall be submitted by the coastal State to the Commission on the Limits of the Continental Shelf set up under Annex II … The Commission shall make recommendations to coastal States on matters related to the establishment of the outer limits of their continental shelf. The limits of the shelf established by a coastal State on the basis of these recommendations shall be final and binding."

Under Annex II, the commission reviews the information submitted by a coastal state and, by a two-thirds majority, approves its recommendations to the submitting state. If the submitting state disagrees with the commission's recommendations, it "shall, within a reasonable time, make a revised or new submission to the Commission."

---

[65] Other parts of UNCLOS relevant to the Arctic include those relating to navigation and high-seas freedoms, fisheries, and exclusive economic zones.

U.S. Not a Party to UNCLOS

As noted earlier, the United States is not a party to UNCLOS.[66] Although the United States is not a party to UNCLOS, the United States accepts and acts in accordance with the non-seabed mining provisions of the treaty, such as those relating to navigation and overflight, which the United States views as reflecting customary international law of the sea.[67]

The United States did not sign UNCLOS when it was adopted in 1982 because the United States objected to the seabed mining provisions of Part XI of the treaty. Certain other countries also expressed concerns about these provisions.[68] The United Nations states that "To address certain difficulties with the seabed mining provisions contained in Part XI of the Convention, which had been raised, primarily by the industrialized countries, the Secretary-General convened in July 1990 a series of informal consultations which culminated in the adoption, on 28 July 1994, of the Agreement relating to the implementation of Part XI of the United Nations Convention on the Law of the Sea of 10 December 1982. The Agreement entered into force on 28 July 1996."[69]

---

[66] The United States is not a signatory to the treaty. On July 29, 1994, the United States became a signatory to the 1994 agreement relating to the implementation of Part XI of the treaty. The United States has not ratified either the treaty or the 1994 agreement.

[67] In a March 10, 1983, statement on U.S. oceans policy, President Reagan stated, President Reagan stated that "the United States is prepared to accept and act in accordance with the [treaty's] balance of interests relating to traditional uses of the oceans—such as navigation and overflight. In this respect, the United States will recognize the rights of other states in the waters off their coasts, as reflected in the Convention, so long as the rights and freedoms of the United States and others under international law are recognized by such coastal states." (Ronald Reagan Presidential Library & Museum, "Statement on United States Oceans Policy," undated, accessed August 2, 2019, at https://www.reaganlibrary.gov/research/speeches/31083c).

[68] In a March 10, 1983, statement on U.S. oceans policy, President Reagan stated, "Last July, I announced that the United States will not sign the United Nations Law of the Sea Convention that was opened for signature on December 10. We have taken this step because several major problems in the Convention's deep seabed mining provisions are contrary to the interests and principles of industrialized nations and would not help attain the aspirations of developing countries. The United States does not stand alone in those concerns. Some important allies and friends have not signed the convention. Even some signatory states have raised concerns about these problems." (Ronald Reagan Presidential Library & Museum, "Statement on United States Oceans Policy," undated, accessed August 2, 2019, at https://www.reaganlibrary.gov/research/speeches/31083c).

[69] United Nations, "Agreement relating to the implementation of Part XI of the United Nations Convention on the Law of the Sea of 10 December 1982," updated September 2, 2016,

The United States signed the 1994 agreement on July 29, 1994, and U.S. administrations since then have supported the United States becoming a party to UNCLOS. The United Nations includes the United States on a list of countries for which the 1994 agreement is in a status of "provisional application," as of November 16, 1994, by virtue of its signature.[70]

The 1982 treaty and the 1994 agreement were transmitted to the Senate on October 6, 1994, during the 103rd Congress, becoming Treaty Document 103-39. Subsequent Senate action on Treaty Document 103-39, as presented at Congress.gov,[71] can be summarized as follows:

- In 2004, during the 108th Congress, the Senate Foreign Relations Committee held hearings on Treaty Document 103-39 and reported it favorably with a resolution of advice and consent to ratification with declarations and understandings. No further action was taken during the 108th Congress, and the matter was re-referred to the committee at the sine die adjournment of the 108th Congress.
- In 2007, during the 110th Congress, the committee held hearings on Treaty Document 103-39 and reported it favorably with a resolution of advice and consent to ratification with declarations, understandings, and conditions. No further action was taken during the 110th Congress, and the matter was re-referred to the committee at the sine die adjournment of the 110th Congress.
- In 2012, during the 112th Congress, the committee held hearings on Treaty Document 103-39. No further action was taken during the 112th Congress.

---

accessed August 1, 2019, at https://www.un.org/depts/los/convention_agreements/convention_overview_part_xi.htm.

70 United Nations, "Agreement relating to the implementation of Part XI of the United Nations Convention on the Law of the Sea of 10 December 1982," status as of August 1, 2019, accessed August 1, 2019, at https://treaties.un.org/Pages/ ViewDetails.aspx?src=TREATY&mtdsg_no=XXI-6-a&chapter=21&clang=_en.

71 United Nations Convention on the Law of the Sea, Senate Consideration of Treaty Document 103-39, accessed July 31, 2019, at https://www.congress.gov/treaty-document/103rd-congress/39. For a timeline of selected key events relating to the treaty, see Department of State, "Law of the Sea Convention," March 7, 2019, accessed July 30, 2019, at https://www.state.gov/law-of-the-sea-convention/.

The full Senate to date has not voted on the question of whether to give its advice and consent to ratification of Treaty Document 103-39. The latest Senate action regarding Treaty Document 103-39 recorded at Congress.gov is a hearing held by the Senate Foreign Relations Committee on June 28, 2012.

Supporters of the United States becoming a party to UNCLOS argue in connection with the Arctic that changing circumstances in the Arctic strengthen the case for the United States becoming a party, on the grounds that it would improve the ability of the United States to protect its interests in the Arctic, particularly in relation to navigation rights and the continental shelf.[72] Opponents of the United States becoming a party to UNCLOS argue in connection with the Arctic that "The U.S. does not need to join the convention in order to access oil and gas resources on its extended continental shelf, in the Arctic, or in the Gulf of Mexico. To the extent necessary, the U.S. can and should negotiate bilateral treaties with neighboring nations to demarcate the limits of its maritime and continental shelf boundaries.... The U.S. has successfully protected its interests in the Arctic since it acquired Alaska in 1867 and has done so during the more than 30 years that the convention has existed. The harm that would be caused by the convention's controversial provisions far outweighs any intangible benefit that allegedly would result from U.S. accession."[73]

The Obama Administration's January 2014 implementation plan for its national strategy for the Arctic region (see earlier section) included, as one of its 36 or so initiatives, one entitled "Accede to the Law of the Sea Convention." Under this initiative, the State Department and other federal agencies are to "continue to seek the Senate's advice and consent to accede to the Law of the Sea Convention." The document states that "the [Obama]

[72] See, for example, Dermot Cole, "Scare Tactics on Sea Treaty Put US Arctic Interests at Risk," *Arctic Today*, June 17, 2019; Steve Haycox, "Why Signing the Law of the Sea Treaty Would Mean a Tougher U.S. Position in the Arctic," *Anchorage Daily News*, March 22, 2018; Ben Werner, "Zukunft: U.S. Presence in Arctic Won't Stop Chinese, Russian Encroachment Without Law of the Sea Ratification," *USNI News*, August 1, 2017; John Grady, "Senators Renew Call to Ratify Law of the Sea Treaty to Help Chart Future of the Arctic," *USNI News*, July 19, 2017.

[73] Theodore R. Bromund, James Jay Carafano, and Brett D. Schaefer, "7 Reasons U.S. Should Not Ratify UN Convention on the Law of the Sea," Heritage Foundation, June 4, 2018.

Administration is committed, like the last three Administrations, to pursuing accession to the Convention on the Law of the Sea and will continue to place a priority on attaining Senate advice and consent to accession."[74]

The State Department's web page for UNCLOS provides a timeline of events relating to the law of the sea from 1958 through 2012, and a list 60 organizations and companies that have expressed support for the United States becoming a party to UNCLOS.[75]

Potential Alternatives or Supplements to UNCLOS

Some observers over the years have occasionally suggested that a separate international legal regime be negotiated to address the changing circumstances in the Arctic. They argue that these changing circumstances were not envisioned at the time UNCLOS was negotiated. Other observers have occasionally suggested that the Arctic region above a certain parallel be designated a wilderness area. As precedent, they cite Article 4 of the Antarctic Treaty, under which any current claims to sovereign territory are frozen and

> No acts or activities taking place while the present Treaty is in force shall constitute a basis for asserting, supporting or denying a claim to territorial sovereignty in Antarctica or create any rights of sovereignty in Antarctica. No new claim, or enlargement of an existing claim, to territorial sovereignty in Antarctica shall be asserted while the present Treaty is in force.

House and Senate Arctic Caucuses

In the House, an Arctic Working Group Caucus is co-chaired by Representative Rick Larsen and Representative Don Young.[76] In the

---

[74] *Implementation Plan for The National Strategy for the Arctic Region*, January 2014, p. 29.

[75] Department of State, "Law of the Sea Convention," accessed January 23, 2020, at https://www.state.gov/law-of-the- sea-convention/.

[76] Source: United States House of Representatives, Committee on House Administration, *Congressional Member Organizations (CMOs)*, 116th Cong., revised June 2019, p. 7,

Senate, Senator Lisa Murkowski and Senator Angus King announced on March 4 and 5, 2015, the formation of a Senate Arctic Caucus.[77]

ISSUES FOR CONGRESS

Climate Change and Loss of Arctic Sea Ice[78]

Record low extents of Arctic sea ice in 2012 and 2007 have focused scientific and policy attention on climate changes in the high north, and on the implications of projected ice-free[79] seasons in the Arctic within decades. The Arctic has been projected by several scientists to be ice-free in most late summers as soon as the 2030s.[80] This opens opportunities for transport through the Northwest Passage and the Northern Sea Route,

accessed July 31, 2019, at https://cha.house.gov/sites/democrats.cha.house.gov/files/documents/cmo_cso_docs/116th%20CMOs_06-03-2019.pdf.

[77] Press release from the office of Senator Angus King, "King, Murkowski Announce U.S. Senate Arctic Caucus," March 4, 2015, accessed March 17, 2015, at http://www.king.senate.gov/newsroom/press-releases/king-murkowski- announce-us-senate-arctic-caucus. See also press release from the office of Senator Lisa Murkowski, "Senators Murkowski, King Announce U.S. Senate Arctic Caucus," March 5, 2015, accessed March 17, 2015, at http://www.murkowski.senate.gov/public/index.cfm/pressreleases?ID=1ce5edcb-540d-4c43-b264-56bdbb570755, which includes a similar phrase.

[78] This section prepared by Jane Leggett, Specialist in Energy and Environmental Policy, Resources, Science, and Industry Division.

[79] In scientific analyses, "ice-free" does not necessarily mean "no ice." The definition of "ice-free" or sea ice "extent" or "area" varies across studies. Sea ice "extent" is one common measure, equal to the sum of the area of grid cells that have ice concentration of less than a set percentage—frequently 15%. For more information, see the National Snow and Ice Data Center, http://nsidc.org/seaice/data/terminology.html.

[80] Muyin Wang and James E. Overland, "A Sea Ice Free Summer Arctic within 30 Years?," *Geophysical Research Letters* 36, no. L07502 (April 3, 2009): 10.1029/2009GL037820; Marika Holland, Cecilia M. Bitz, and Bruno Tremblay, "Future abrupt reductions in the summer Arctic sea ice," *Geophysical Research Letters* 33, no. L23503 (2006); But see also Julien Boé, Alex Hall, and Xin Qu, "Sources of spread in simulations of Arctic sea ice loss over the twenty-first century," *Climatic Change* 99, no. 3 (April 1, 2010): 637-645; I. Eisenman and J. S. Wettlaufer, "Nonlinear threshold behavior during the loss of Arctic sea ice," *Proceedings of the National Academy of Sciences* 106, no. 1 (January 6, 2009): 28-32; Dirk Notz, "The Future of Ice Sheets and Sea Ice: Between Reversible Retreat and Unstoppable Loss," *Proceedings of the National Academy of Sciences* 106, no. 49 (December 8, 2009): 20590-20595.

extraction of potential oil and gas resources, and expanded fishing and tourism (Figure 3).

More broadly, physical changes in the Arctic include warming ocean, soil, and air temperatures; melting permafrost; shifting vegetation and animal abundances; and altered characteristics of Arctic cyclones. All these changes are expected to affect traditional livelihoods and cultures in the region and survival of polar bear and other animal populations, and raise risks of pollution, food supply, safety, cultural losses, and national security. Moreover, linkages ("teleconnections") between warming Arctic conditions and extreme events in the mid-latitude continents are increasingly evident, identified in such extreme events as the heat waves and fires in Russia in 2010; severe winters in the eastern United States and Europe in 2009/2010 and in Europe in 2011/2012;[81] and Indian summer monsoons and droughts. Hence, changing climate in the Arctic suggests important implications both locally and across the Hemisphere.

Like the rest of the globe, temperatures in the Arctic have varied[82] but show a significant warming trend since the 1970s, and particularly since 1995.[83] The annual average temperature for the Arctic region (from 60o to 90° N) is now about 1.8°F warmer than the "climate normal" (the average

---

[81] Overland et al., state that "a warm Arctic-cold continent pattern represents a paradox of recent global warming: there is not a uniform pattern of temperature increases" due to a set of newly recognized processes described in Overland, J. E, K. R Wood, and M. Wang. "Warm Arctic-cold Continents: Climate Impacts of the Newly Open Arctic Sea." *Polar Research* 30 (2011). The authors raise a critical, unanswered question, "Is the observed severe mid-latitude weather in two adjacent years simply due to an extreme in chaotic processes alone, or do they included a partial but important Arctic forcing and connection due to recent changing conditions?" In other words, are recent patterns random anomalies, or might we expect more of the same?; among other examples, see also Lim, Young-Kwon, and Siegfried D. Schubert. "The Impact of ENSO and the Arctic Oscillation on Winter Temperature Extremes in the Southeast United States." *Geophysical Research Letters* 38, no. 15 (August 11, 2011): L15706.

[82] There was a regionally warm period in the Arctic from the mid-1920s to around 1940, which scientists have assessed to have been driven by natural climate variability. They have found that period to be distinctly different from the recent multi-decadal warming, in part because the early 20th century warming was concentrated in the northern high latitudes. See, for example, Figure 2, upper left graphic, in Geophysical Fluid Dynamics Laboratory, "Simulatoin of Early 20th Century Warming," at http://www.gfdl.noaa.gov/early-20th-century-global-warming.

[83] Steele, Michael, Wendy Ermold, and Jinlun Zhang. "Arctic Ocean Surface Warming Trends over the Past 100 Years." *Geophysical Research Letters* 35, no. 2 (January 29, 2008): L02614.

from 1961 to 1990). Temperatures in October-November are now about 9° F above the seasonal normal.

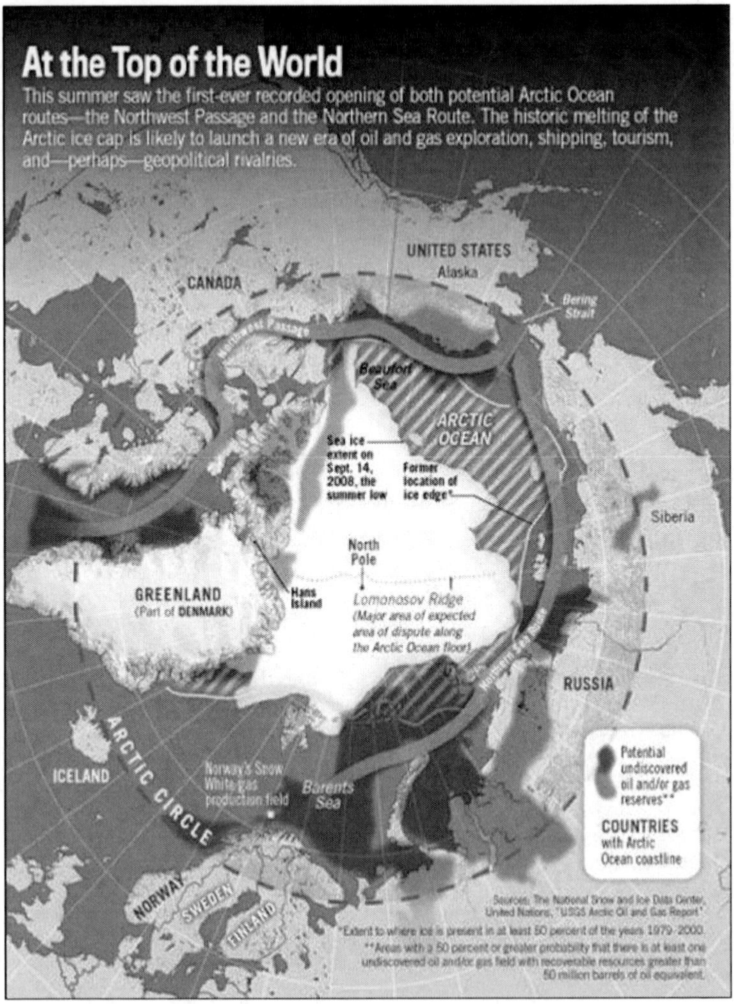

Source: Graphic by Stephen Rountree at *U.S. News and World Report*, http://www.usnews.com/articles/news/world/2008/10/09/global-warming-triggers-an-international-race-for-the-artic/photos/#1.

Figure 3. Arctic Sea Ice Extent in September 2008, Compared with Prospective Shipping Routes and Oil and Gas Resources.

Scientists have concluded that most of the *global* warming of the last three decades is very likely caused by human-related emissions of greenhouse gases (GHG, mostly carbon dioxide); they expect the GHG-induced warming to continue for decades, even if, and after, GHG concentrations in the atmosphere have been stabilized. The extra heat in the Arctic is amplified by processes there (the "polar amplification") and may result in irreversible changes on human timescales.

The observed warmer temperatures along with rising cyclone size and strength in the Arctic have reduced sea ice extent, thickness, and ice that persists year-round ("perennial ice"); natural climate variability has likely contributed to the record low ice extents of 2007 and 2012. The 2007 minimum sea ice extent was influenced by warm Arctic temperatures and warm, moist winds blowing from the North Pacific into the central Arctic, contributing to melting and pushing ice toward and into the Atlantic past Greenland. Warm winds did not account for the near-record sea ice minimum in 2008.[84] In early August 2012, an unusually large storm with low pressure developed over the Arctic, helping to disperse the already weak ice into warmer waters and accelerating its melt rate. By August 24, 2012, sea ice extent had shrunk below the previous observed minimum of late September 2007.[85]

Modeling of GHG-induced climate change is particularly challenging for the Arctic, but it consistently projects warming through the 21st century, with annual average Arctic temperature increases ranging from +1° to +9.0°C (+2° to +19.0°F), depending on the GHG scenario and model used. While such warming is projected by most models throughout the Arctic, some models project slight cooling localized in the North Atlantic Ocean just south of Greenland and Iceland. Most warming would

---

[84] J. Overland, J. Walsh, and M. Wang, *Arctic Report Card - Atmosphere* (NOAA Arctic Research Program, October 6, 2008), ftp://ftp.oar.noaa.gov/arctic/documents/ArcticReportCard_full_report2008.pdf.

[85] National Snow and Ice Data Center, "Arctic sea ice extent breaks 2007 record low" (August 27, 2012); Japanese Aerospace Exploration Agency, "A new record minimum of the Arctic sea ice extent was set on 24 August 2012"; Arctic ROOS (Norway), "Daily Updated Time series of Arctic sea ice area and extent derived from SSMI data provided by NERSC," at http://arctic-roos.org/observations/satellite-data/sea-ice/ice-area-and-extent-in-arctic.

occur in autumn and winter, "with very little temperature change projected over the Arctic Ocean" in summer months.[86]

Due to observed and projected climate change, scientists have concluded that the Arctic will have changed from an ice-covered environment to a recurrent ice-free[87] ocean (in summers) as soon as the late 2030s. The character of ice cover is expected to change as well, with the ice being thinner, more fragile, and more regionally variable. The variability in recent years of both ice quantity and location could be expected to continue.

Geopolitical Environment[88]

Shift to Era of Renewed Great Power Competition

A principal factor affecting the geopolitical environment for the Arctic is the shift that has occurred in recent years from the post-Cold War era that began in the late 1980s and early 1990s, also sometimes known as the unipolar moment (with the United States as the unipolar power), to a new and different international security environment that features, among other things, renewed great power competition with China and Russia and challenges by these two countries and others to elements of the U.S.-led international order that has operated since World War II.[89] This shift in the international security environment, combined with the diminishment of Arctic ice and the resulting increase in human activities in the Arctic, has several potential implications for the geopolitical environment for the Arctic, which are discussed in the following sections.

---

[86] William L. Chapman and John E. Walsh, "Simulations of Arctic Temperature and Pressure by Global Coupled Models," *Journal of Climate* 20, no. 4 (February 1, 2007): 609-632.

[87] See footnote 79. Also, although one Canadian scientist has predicted that recurrent ice-free summers may begin sometime between 2013 and 2020, this is not consistent with other climate models' projections.

[88] This section was prepared by Ronald O'Rourke, Specialist in Naval Affairs, Foreign Affairs, Defense, and Trade Division. It incorporates material prepared by Kristin Archick, Specialist in European Affairs, Foreign Affairs, Defense, and Trade Division, and Derek E. Mix, Analyst in European Affairs, Foreign Affairs, Defense, and Trade Division.

[89] For more on this shift, see CRS Report R43838, *Renewed Great Power Competition: Implications for Defense— Issues for Congress*, by Ronald O'Rourke.

Arctic Tradition of Cooperation and Low Tensions

The renewal of great power competition has raised a basic question as to whether the Arctic in coming years will continue to be a region generally characterized by cooperation and low tensions, as it was during the post-Cold War era, or instead become a region characterized at least in part by competition and increased tensions, as it was during the Cold War. In this regard, the renewal of great power competition poses a potential challenge to the tradition of cooperation, low tensions, peaceful resolution of disputes, and respect for international law—sometimes referred to as the "Arctic spirit"—that has characterized the approach used by the Arctic states, particularly since the founding of the Arctic Council in 1996, for managing Arctic issues.[90]

Some observers argue that the Arctic states and other Arctic stakeholders should attempt to maintain the region's tradition of cooperation and low tensions, and work to prevent the competition and tensions that have emerged in Europe, Asia, and elsewhere in recent years from crossing over into the Arctic. These observers argue that the Arctic tradition of cooperation and low tensions has proven successful in promoting the interests of the Arctic states and other Arctic stakeholders on a range of issues, that it has served as a useful model for other parts of the world to follow, and that in light of tensions and competition elsewhere in the world, this model is needed more now than ever.[91]

---

[90] See, for example, Joshua Tallis, "As 'Arctic Exceptionalism' Melts Away, the US Isn't Sure What It Wants Next," *Defense One*, January 22, 2020; Timo Koivurova, "How US Policy Threatens Existing Arctic Governance," *Arctic Today*, January 17, 2020; Melody Schreiber, "As the Arctic Changes, International Cooperation May Be Put to the Test," *Arctic Today*, July 25, 2018; Stephanie Pezard, Abbie Tingstad, and Alexandria Hall, *The Future of Arctic Cooperation in a Changing Strategic Environment*, RAND Europe (PE-268RC), 2018, 18 pp.; Geoff Ziezulewicz, "As Arctic Waters Open, Nations Plant Their Flags," *Navy Times*, April 8, 2018; James Stavridis, "Avoiding a Cold War in the High North," *Bloomberg*, May 4, 2018; Kristina Spohr, "The Race to Conquer the Arctic—the World's Final Frontier," *New Statesman*, March 12, 2018.

[91] See, for example, Alexandra Brzozowski, "Avoid Politicising Arctic Issues, Expert Warns," *Euractiv*, February 8, 2019; Kevin McGwin, "More Military Activity May Spoil the Arctic's Atmosphere of Collaboration, Warns a Danish Report," *Arctic Today*, December 6, 2018; Hsin Hsuan Sun, "Arctic Council Sells Itself As a Model for International Cooperation at the UN," *Arctic Today*, July 23, 2018; John Grady, "Panel: Cooperation, Not Conflict Key to Future of the Arctic," *USNI News*, April 9, 2018; Levon Sevunts, "Arctic Nations Develop Coast Guard Co-operation," *Barents Observer*, March 13, 2018; Doug

Other observers could argue that, notwithstanding the efforts of Arctic states and other Arctic stakeholders to maintain the Arctic as a region of cooperation and low tensions, it is unreasonable to expect that the Arctic can be kept fully isolated from the competition and tensions that have arisen in other parts of the world. As a consequence, these observers could argue, the Arctic states and other Arctic stakeholders should begin taking steps to prepare for increased competition and higher tensions in the Arctic, precisely so that Arctic issues can continue to be resolved as successfully as conditions may permit, even in a situation of competition and increased tensions.

Still other observers might argue that a policy of attempting to maintain the Arctic as a region of cooperation and low tensions, though well-intentioned, could actually help encourage aggressive behavior by Russia or China in other parts of the world by giving those two countries confidence that their aggressive behavior in other parts of the world would not result in punitive costs being imposed on them in the Arctic. These observers might argue that maintaining the Arctic as a region of cooperation and low tensions in spite of aggressive Russian or Chinese actions elsewhere could help legitimize those aggressive actions and provide little support to peaceful countries elsewhere that might be attempting to resist them. This, they could argue, could facilitate a divide-and-conquer strategy by Russia or China in their relations with other countries, which in the long run could leave Arctic states with fewer allies and partners in other parts of the world for resisting unwanted Russian or Chinese actions in the Arctic.

Still others might argue that there is merit in some or all of the above perspectives, and that the challenge is to devise an approach that best mixes the potential strengths of each perspective.

In a May 6, 2019, speech in Finland that was given prior to the start of formal discussions at an Arctic Council ministerial meeting, Secretary of State Michael Pompeo stated

---

Tsuruoka, "Despite US-Russia Tensions, Arctic Powers Eye Cooperation," *Asia Times*, March 29, 2017.

For reasons I'll explain in a moment, the region [i.e., the Arctic] has become an arena for power and for competition. And the eight Arctic states must adapt to this new future.

In its first two decades, the Arctic Council has had the luxury of focusing almost exclusively on scientific collaboration, on cultural matters, on environmental research— all important themes, very important, and we should continue to do those.

But no longer do we have that luxury of the next hundred years….

To leverage the Arctic's—the Arctic continental, all nations, including non-Arctic nations, should have a right to engage peacefully in this region. The United States is a believer in free markets. We know from experience that free and fair competition, open, by the rule of law, produces the best outcomes.

But all the parties in the marketplace have to play by those same rules. Those who violate those rules should lose their rights to participate in that marketplace. Respect and transparency are the price of admission.

And let's talk about China for a moment. China has observer status in the Arctic Council, but that status is contingent upon its respect for the sovereign rights of Arctic states. The U.S. wants China to meet that condition and contribute responsibly in the region. But China's words and actions raise doubts about its intentions.

Beijing claims to be a "Near-Arctic State," yet the shortest distance between China and the Arctic is 900 miles. There are only Arctic States and Non-Arctic States. No third category exists, and claiming otherwise entitles China to exactly nothing.

That's not to say Chinese investment is unwelcome—indeed, quite the opposite. The United States and Arctic nations welcome transparent Chinese investments that reflect economic interest and national security ambitions. Between 2012 and 2017, China invested in the Arctic nearly $90 billion. It's planning to build infrastructure from Canada, to the Northwest Territories, to Siberia.

Just last month, Russia announced plans to connect the Northern Sea Route with China's Maritime Silk Road, which would develop a new shipping channel from Asia to northern Europe. Meanwhile, China is already developing shipping lanes in the Arctic Ocean.

This is part of a very familiar pattern. Beijing attempts to develop critical infrastructure using Chinese money, Chinese companies, and Chinese workers—in some cases, to establish a permanent Chinese security presence.

Our Pentagon warned just last week that China could use its civilian research presence in the Arctic to strengthen its military presence, including our deployment of submarines— including deployment of submarines to the region as a deterrent against nuclear attack.

We need to examine these activities closely, and we need—and we keep the experience we have learned of other nations in mind. China's pattern of aggressive behavior elsewhere in the—excuse me—aggressive behavior elsewhere should inform what we do and how it might treat the Arctic.

Let's just ask ourselves: Do we want Arctic nations broadly, or indigenous communities specifically, to go the way of former government in Sri Lanka or Malaysia, ensnared by debt and corruption? Do we want crucial Arctic infrastructure to end up like Chinese- constructed roads in Ethiopia, crumbling and dangerous after only a few years? Do we want the Arctic Ocean to transform into a new South China Sea, fraught with militarization and competing territorial claims? Do we want the fragile Arctic environment exposed to the same ecological devastation caused by China's fishing fleet in the seas off its coast, or unregulated industrial activity in its own country? I think the answers are pretty clear.

Then there's Russia. As a fellow Arctic Council member, Russia – the other Arctic states have fruitfully cooperated in a number of areas—expansive conservation efforts. Those are to be applauded. We want cooperation to continue. But we can't have one side cooperate, and the other side derogate its duties.

We're concerned about Russia's claim over the international waters of the Northern Sea Route, including its newly announced plans to connect it with China's Maritime Silk Road. In the Northern Sea Route, Moscow already illegally demands other nations request permission to pass, requires Russian maritime pilots to be aboard foreign ships, and threatens to use military force to sink any that fail to comply with their demands.

These provocative actions are part of a pattern of aggressive Russian behavior here in the Arctic. Russia is already leaving snow prints in the form of army boots. Russia formally announced its intent to increase its military presence in the region in 2014, when it re- opened a Cold War Arctic military base.

Since then, thanks in part to its large icebreaker fleet, Russia has been able to renovate old bases and infrastructure. It claims to have built 475 new military sites, including bases north of the Arctic Circle, as well as 16 new deep-water ports. It secures this presence through sophisticated new air defense systems and anti-ship missiles.

No one denies Russia has significant Arctic interests. We recognize that Russia is not the only nation making illegitimate claims. The U.S. has a long-contested feud with Canada over sovereign claims through the Northwest Passage.

But Russia is unique. Its actions deserve special attention, special attention of this Council, in part because of their sheer scale. But also because we know Russian territorial ambitions can turn violent. 13,000 people have been killed due to Russia's ongoing aggressive action in Ukraine.

And just because the Artic is a place of wilderness does not mean it should become a place of lawlessness. It need not be the case. And we stand ready to ensure that it does not become so.[92]

---

[92] State Department, "Looking North: Sharpening America's Arctic Focus, Remarks, Michael R. Pompeo, Secretary of State, Rovaniemi, Finland, May 6, 2019," accessed August 23, 2019 at: https://www.state.gov/looking-north- sharpening-americas-arctic-focus/.

For press reports about the speech, including reports of reactions by observers in other countries, see Blake Hounshell, "Pompeo Aims to Counter China's Ambitions in the Arctic," *Politico*, May 6, 2019; Courtney McBride, "Pompeo Issues Warning to China, Russia on Arctic," *Wall Street Journal*, May 6, 2019; Simon Johnson, "Pompeo: Russia Is 'Aggressive' in Arctic, China's Work There Also Needs Watching," *Reuters*, May 6, 2019; Eilis Quinn (Radio Canada International), "U.S. Stuns Audience by Tongue-Lashing China, Russia on Eve of Arctic Council Ministerial," *Barents Observer*, May 6, 2019; Amund Trellevik, "USA Launches Head-On Verbal Attack on Russia and China," *High North News*, May 7, 2019; Amund Trellevik, "MP Argues That Pompeo Provoked Unnecessarily—Fears for High North Cooperation," *High North News*, May 7, 2019; Victoria Herrmann, "Pompeo: Melting Sea Ice Presents Trade Opportunities," *CNN*, May 10, 2019; Mia Bennett, "China Reveals Arctic Geopolitics to Be Above Pompeo's Pay Grade," *Arctic Today*, May 11, 2019; Dermot Cole, "Pompeo's Bombastic Arctic Council Performance Could Prove an Aberration—Or A New Normal," *Arctic Today*, May 13, 2019; Siri Gulliksen Tømmerbakke, "Strong Chinese Reactions to Pompeo's Speech," *High North News*, May 16, 2019; Arne O. Holm, "Peave and Love, My Ass," *High North News*, May 16, 2019; Siri Gulliksen Tømmerbakke, "China Will Seek to Treat The Pompeo Speech As Simply 'Road Damage' to Be Routed Around," *High North News*, May 16, 2019; Maiya Clark, "Pompeo Seeks New Strategic Agenda for Arctic, Cautions Allies on China's Intentions, *Daily Signal*, May 17, 2019.

The United States believes that the part of the Northwest Passage that runs through the Canadian archipelago is an international strait; Canada believes it is internal Canadian waters. Since 1985, when the use of the waterway by a U.S. polar icebreaker led to a diplomatic dispute between the United States and Canada, the two countries have agreed to disagree on the issue, and the issue had remained largely dormant. Secretary Pompeo's mention of the issue in his May 6, 2019, speech put a spotlight back on the matter. For press reports discussing this part of the speech, see Lloyd Alter, "American Secretary of State Calls Canadian Claim to Northwest Passage 'Illegitimate,'" *Tree Hugger*, May 7, 2019; Jane George, "Canadian Inuit Challenge U.S. Stance on Northwest Passage," *Nunatsiaq News*, May 9, 2019; Siri Gulliksen Tømmerbakke, "Canada Is Pushing Back at Pompeo's Assertion That Canada's Claim to the Northwest Passage is 'Illegitimate,'" *High North News*, May 16, 2019; Robert Huebert, "Winning at All Costs: The Bizarre Position of the Trump Administration on the Northwest Passage," *Globe and Mail*, May 20, 2019; Dermot Cole, "Insulting Canada Won't Help the US with Its Arctic Ambitions," *Arctic Today*, June 3, 2019; Alastair Allan, "Canada Needs to Cut a Deal with the U.S. to Control the Northwest Passage," *National Post (Canada)*, August 8, 2019. See also Melody Schreiber, "Russia and China May Lose Their Legal Claim to Arctic Seaways as Ice Melts, Experts Say," *Arctic Today*, April 4, 2019.

Arctic Governance

Spotlight on Arctic Governance and Limits of Arctic Council

The renewal of great power competition has put more of a spotlight on the issue of Arctic governance and the limits of the Arctic Council as a governing body.[93] As noted earlier in this report, regarding the limits of the Arctic Council as a governing body, the council states that "The Arctic Council does not and cannot implement or enforce its guidelines, assessments or recommendations. That responsibility belongs to each individual Arctic State. The Arctic Council's mandate, as articulated in the Ottawa Declaration, explicitly excludes military security.[94]

During the post-Cold War era—the period when the Arctic Council was established and began operating—the limits of the Arctic Council as a governing body may have been less evident or problematic, due to the post-Cold War era's general situation of lower tensions and reduced overt competition between the great powers. In the new situation of renewed great power competition, however, it is possible that these limits could become more evident or problematic, particularly with regard to addressing Arctic-related security issues.

If the limits of the Arctic Council as a governing body are judged as having become more evident or problematic, one option might be to amend the rules of the council to provide for some mechanism for enforcing its guidelines, assessments, or recommendations. Another option might be to

---

[93] See, for example, Roman Chuffart, "Is the Arctic Council a Paper Polar bear?" *High North News*, November 29, 2019; David Balton and Fran Ulmer, *A Strategic Plan for the Arctic Council: Recommendations for Moving Forward*, The Arctic Initiative (Belfer Center for Science and International Affairs, Harvard Kennedy School) and The Polar Institute (Woodrow Wilson International Center for Scholars), June 2019, 15 pp.; Mckenna Coffey, "Concerns Rise Over Governance Gap in Arctic," *New Security Beat (Wilson Center)*, August 5, 2019; Melody Schreiber, "Will the Arctic Council Begin Addressing National Security?" *Arctic Today*, May 17, 2019; Arne O. Holm, "Does the Arctic Council Have Enough Power in It to Keep the Arctic a Peaceful Region?" High North News, April 27, 2019; Gosia Smieszek, "Costs and Reality of Reforming the Arctic Council," *Arctic Institute*, April 9, 2019; John Grady, "Old Treaties Called Into Question as Arctic Competition Increases," *USNI News*, December 20, 2016.

[94] Arctic Council, "The Arctic Council: A Backgrounder," updated September 13, 2018, accessed August 2, 2018, at https://arctic-council.org/index.php/en/about-us.

expand the council's mandate to include an ability to address military security issues.

Supporters of such options might argue that they could help the council adapt to the major change in the Arctic's geopolitical environment brought about the shift in the international security environment, and thereby help maintain the council's continued relevance in coming years. They might also argue that continuing to exclude military security from the council's mandate risks either leaving Arctic military security issues unaddressed, or shifting them to a different forum that might have traditions weaker than those of the Arctic Council for resolving disputes peacefully and with respect for international law.

Opponents of such options might argue that they could put at risk council's ability to continue addressing successfully nonmilitary security issues pertaining to the Arctic. They might argue that there is little evidence to date that the council's limits as a governing body have become problematic, and that in light of the council's successes since its founding, the council should be viewed as an example of the admonition, "if it isn't broke, don't fix it."[95]

China and Arctic Governance

China—which is not one of the eight Arctic states and consequently does not have a decisionmaking role in the Arctic Council—has begun to raise questions as to whether the Arctic Council as currently constituted and the current broader legal framework for the Arctic should continue to be the principal means for addressing issues relating to the Arctic, and has begun to use other approaches for influencing Arctic governance.[96] In May

---

[95] For further discussion of the limits of the Arctic Council and what, if anything, to do about it, see, for example, Heather Exner-Pirot et al., "Form and Function: The Future of the Arctic Council," *Arctic Institute*, February 5, 2019. See also Marc Lanteigne, "The Growing Role of 'Track II' Organisations in the Arctic," Over the Circle, May 23, 2018.

[96] See, for example, Pan Yixuan, "Global Governance Needed for Arctic Affairs," *China Daily*, May 10, 2019; Zhang Yao, "Ice Silk Road Framework Welcomed by Countries, Sets New Direction for Arctic Cooperation," *Global Times*, April 7, 2019; Liu Caiyu, "China's Role in Arctic Governance 'Cannot Be Ignored,'" *Global Times*, November 22, 2018; Harriet Moynihan, "China Expands Its Global Governance Ambitions in the Arctic," Chatham House, October 15, 2018; Sydney J. Freedberg Jr., "China & Russia In The Arctic: Axis Of Ambivalence," *Breaking Defense*, July 6, 2018; Nengye Liu and Michelle Lim, "How

2019, a U.S. official stated that the United States "reject[s] attempts by non-Arctic states to claim a role" in Arctic governance.[97]

Arctic and World Order

Another potential implication for the Arctic of renewal of great power competition concerns associated challenges to elements of the U.S.-led international order that has operated since World War II. One element of the U.S.-led international order that has come under challenge is the principle that force or threat of force should not be used as a routine or first-resort measure for settling disputes between countries. Another is the principle of freedom of the seas (i.e., that the world's oceans are to be treated as an international commons).[98] If either of these elements of the U.S.-led international order is weakened or overturned, it could have potentially major implications for the future of the Arctic, given the Arctic's tradition of peaceful resolution of disputes and respect for international law and the nature of the Arctic as a region with an ocean at its center that washes up against most of the Arctic states.

More broadly, some observers assess that the U.S.-led international order in general may be eroding or collapsing, and that the nature of the successor international order that could emerge in its wake is uncertain.[99] An erosion or collapse of the U.S.-led international order, and its replacement by a new international order of some kind, could have significant implications for the Arctic, since the Arctic's tradition of cooperation and low tensions, and the Arctic Council itself, can be viewed as outgrowths of the U.S.-led order.

Arctic Governance Could Become a Testing Ground for Sino-US Relations," *The Conversation*, March 29, 2017.

[97] Reuters, "US Rejects Interference by Non-Arctic Countries in Polar Region: Official," *Arctic Today*, May 3, 2019.

[98] For additional discussion, see CRS Report R43838, *Renewed Great Power Competition: Implications for Defense— Issues for Congress*, and, regarding the principle of freedom of the seas, CRS Report R42784, *U.S.-China Strategic Competition in South and East China Seas: Background and Issues for Congress*.

[99] For additional discussion, see CRS Report R44891, *U.S. Role in the World: Background and Issues for Congress*.

Relative Priority of Arctic in U.S. Policymaking

The renewal of great power competition has raised a question concerning the priority that should be given to the Arctic in overall U.S. policymaking. During the post-Cold War era, when the Arctic was generally a region of cooperation and low tensions, there may have been less need to devote U.S. policymaker attention and resources to the Arctic. Given how renewed great power competition and challenges to elements of the U.S.-led international order might be expressed in the Arctic in terms of issues like resource exploration, disputes over sovereignty and navigation rights, and military forces and operations, it might be argued that there is now, other things held equal, more need for devoting U.S. policymaker attention and resources to the Arctic.[100]

On the other hand, renewed great power competition and challenges to elements of the U.S.-led international order are also being expressed in Europe, the Middle East, the Indo-Pacific, Africa, and Latin America. As a consequence, it might be argued, some or all these other regions might similarly be in need of increased U.S. policymaker attention and resources. In a situation of constraints on total U.S. policymaker attention and resources, the Arctic competes against these other regions for U.S. policymaker attention and resources. As one expression of this issue, it was reported in January 2020 that 3,000 of a planned force of about 10,500 U.S. military personnel scheduled to participate in a cold-weather exercise in Norway in March 2020 were to be diverted to perform missions elsewhere.[101]

---

[100] For an article bearing on this issue, see Heather A. Conley and Matthew Melino, "The Implications of U.S. Policy Stagnation toward the Arctic Region," Center for Strategic and International Studies, May 2019, 5 pp.

[101] John Vandiver, "US Troops Slated for Big Arctic Exercise Get Shifted to Other Missions," *Stars and Stripes*, January 17, 2020; Alte Staalesen, "American Forces Withdraw from Allied Exercise in Arctic Norway," *Arctic Today*, January 17, 2020.

U.S., Canadian, and Nordic Relations with Russia in Arctic

Overview

The renewal of great power competition raises a question for U.S., Canadian, and Nordic policymakers regarding the mix of cooperation and competition to pursue (or expect to experience) with Russia in the Arctic. In considering this question, points that can be noted include the following:

- Geographically, Russia is the most prominent of the eight Arctic states. According to one assessment, Russia "has at least half of the Arctic in terms of area, coastline, population and probably mineral wealth."[102] About 20% of Russia's land mass is north of the Arctic Circle.[103] Russia has numerous cities and towns in its Arctic, uses its coastal Arctic waters as a maritime highway for supporting its Arctic communities, is promoting the Northern Sea Route that runs along Russia's Arctic coast for use by others, and is keen to capitalize on natural resource development in the region, both onshore and offshore. A substantial fraction of Russia's oil and gas production and reserves are in the Arctic. In this sense, of all the Arctic states, Russia might have the most at stake in the Arctic in absolute terms.[104]
- The Arctic is a top strategic priority for Russia. In 2008, 2013, 2014, and 2017, the Russian government adopted strategy documents outlining plans to bolster the country's Arctic military capabilities, strengthen territorial sovereignty, and develop the

---

[102] "The Arctic: Special Report," *The Economist,* June 16, 2012, p. 11. As noted in the "Background" section of this CRS report, of the 4 million or so people who live above the Arctic Circle, roughly half are in Russia.

[103] Testimony of Admiral Charles W. Ray, Coast Guard Vice Commandant, on "Expanding Opportunities, Challenges, and Threats in the Arctic: a Focus on the U.S. Coast Guard Arctic Strategic Outlook" before the Senate Commerce, Science, & Transportation Security Subcommittee, December 12, 2019, p. 3.

[104] See also Stephanie Pezard, *The New Geopolitics of the Arctic, Russia's and China's Evolving Role in the Region*, RAND (Testimony presented before the Standing Committee on Foreign Affairs and International Development of the Canadian House of Commons on November 26, 2018), pp. 1-2.

region's resources and infrastructure.[105] Over the past several years, Russia has invested in the construction of ports and search-and-rescue facilities, some of which are referred to as dual use (civilian-military) facilities. Russia also has reactivated and modernized military bases that fell into disuse with the end of the Cold War.

- Arctic ice is diminishing more rapidly or fully on the Russian side of the Arctic than it is on the Canadian side. Consequently, the Northern Sea Route along Russia's coast is opening up more quickly for trans-Arctic shipping than is the Northwest Passage through the Canadian archipelago.

On the one hand, the United States, Canada, and the Nordic countries continue cooperate with Russia on a range of issues in the Arctic, including, to cite just one example, search and rescue (SAR) under the May 2011 Arctic Council agreement on Arctic SAR (see "Search and Rescue (SAR)"). More recently, the United States and Russia cooperated in creating a scheme for managing two-way shipping traffic through the Bering Strait and Bering Sea.[106] Observers see possibilities for further U.S.-Russian cooperation in the Arctic.[107] On the other hand, as discussed later in this report, a significant increase in Russian military capabilities

---

[105] Nazrin Mehdiyeva, *Russia's Arctic Papers: The Evolution of Strategic Thinking on the High North*, NATO Defense College, November 19, 2018.

[106] See, for example, U.S. Coast Guard, "U.S., Russia Propose Bering Strait Ship Traffic Routing Measures," January 25, 2018; Amy Midgett, "U.S., Russia Jointly Propose Bering Strait Routing Measures," *Coast Guard Maritime Commons*, January 25, 2018; Amy Midgett, "IMO Approves U.S.-Russian Federation Proposal for Bering Strait Routing Measures," *Coast Guard Maritime Commons*, May 25, 2018; Yereth Rosen, "With Marine Traffic Growing, International Shipping Agency Approves US-Russia Plan for Bering Strait Shipping Lanes," *Arctic Today*, May 26, 2018; Associated Press, "Maritime Organization Approves Two-Way Shipping Routes in Bering Strait," *CBC*, May 27, 2018; "U.S., Russia Propose Bering Strait Traffic Routing," *Maritime Executive*, May 27, 2018; Margaret Kriz Hobson, "Amid Ice Melt, New Shipping Lanes Are Drawn Up off Alaska," *E&E News (Scientific American)*, May 29, 2018.

[107] See, for example, Levon Sevunts (Radio Canada International), "In the Arctic As in Space, Russia and West Can Look Past Differences to Pursue Common Goals: Study," *Eye on the Arctic (Radio Canada International)*, June 14, 2019; Mike Sfraga and Lawson Brigham, "US, Russia Can Look North to the Arctic to Find Common Ground," *The Hill*, July 17, 2018. See also Martin Breum "Why Russia Is Likely to Remain Cooperative in the Arctic," *Arctic Today*, November 22, 2018.

and operations in the Arctic in recent years has prompted growing concerns among U.S., Canadian, and Nordic observers that the Arctic might once again become a region of military tension and competition, as well as concerns about whether the United States, Canada, and the Nordic countries are adequately prepared militarily to defend their interests in the region.

Russian actions outside the Arctic could affect relations between Russia and the other Arctic states. For example, in protest of Russia's forcible occupation and annexation of Crimea and its actions elsewhere in Ukraine, Canada announced that it would not participate in an April 2014 working-level-group Arctic Council meeting in Moscow. In addition, former Secretary of State Hillary Clinton reportedly stated that Arctic cooperation may be jeopardized if Russia pursues expansionist policies in the high north.[108] Economic sanctions that the United States imposed on Russia in response to Russian actions in Ukraine could affect Russian Arctic offshore oil exploration.[109]

Northern Sea Route

Another concern for U.S. policymakers in connection with Russia in the Arctic relates to the Northern Sea Route (NSR)—the Arctic shipping route linking Europe and Asia via waters running along Russia's Arctic coast. Russia considers certain parts of the NSR to be internal Russian waters and has asserted a right to regulate commercial shipping passing through these waters[110]—a position that creates a source of tension with the

---

108 Naval Exercise on Hold Due to Ukraine Tensions. *Navy Times*. March 13, 2014. Hillary Warns Against Russia in Arctic. Barents Observer. April 3, 2014. http://Barents observer.com/en/arctic/2014/04/hillary-warns-against-russia- arctic-03-04 Canada Takes Principled Stand on Arctic Council Meetings. Ministry of Foreign Affairs, Trade, and Development Canada. April 15, 2014, http://www.international. gc.ca/media/arctic-arctique/news-communiques/2014/ 04/15a.aspx?lang=eng, Crimea Crises Changes Nordic Defense Perspective.

109 See, for example, Reuters staff, "Expanded U.S. Sanctions May Affect Russia's Foreign Expansion in Oil and Gas" *Reuters*, November 19, 2017.

110 See, for example, James Foggo III, "Russia, China Offer Challenges in the Arctic," *Defense One*, July 10, 2019; Dmitry Sudakov, "Russia Closes Northern Sea Route for Foreign Warships," *Pravda Report*, May 30, 2019; Nastassia Astrasheuskaya and Henry Foy, "Polar Powers: Russia's Bid for Supremacy in the Arctic Ocean," *Financial Times*, April 27, 2019; James R. Holmes, "Don't Let Russia Create a 'Caribbean' in the Arctic," *The Hill*, March 28, 2019; Marex (Maritime Executive), "Russia Tightens Control Over Northern Sea

United States, which considers those waters to be international waters. The U.S.-Russian dispute over this issue could have implications not only for U.S.-Russian relations and the Arctic, but for other countries and other parts of the world as well, since international law is universal in its application, and a successful challenge to international waters in one part of the world can serve as a precedent for challenging it in other parts of the world.

The issue of the U.S.-Russian dispute over the international legal status of the NSR was largely dormant for many years. In March 2019, however, Russia announced that

> The Russian government has elaborated a set of rules for foreign naval vessels' sailing on the Northern Sea Route, [the Russian newspaper] Izvestia informs. The newspaper has obtained a copy of the document that states that all vessels are obliged to comply.
>
> The foreign state must send a notification about the voyage at least 45 days ahead of its start. Included will have to be the name of the ship, its objective, route and period of sailing, as well as ship characteristics such as length, width, deadweight, draft and type of engine power. Also the name of the ship captain must be listed.
>
> The ships must also have on board a Russian maritime pilot.
>
> In case the voyage is not conducted in line with the regulations, Russia will have the right to take extraordinary measures including its forced halt, arrest and in extreme cases elimination, Izvestia writes.[111]

In September 2019, it was reported that Russia had used military commandos to board a Russian- flag commercial ship operating in the NSR that Russian authorities suspected of violating certain regulations.[112]

Route," *Maritime Executive*, March 8, 2019; Atle Staalesen, "Russia Sets Out Stringent New Rules for Foreign Ships on the Northern Sea Route," *Barents Observer*, March 8, 2019; Will Stewart, "Moscow Threatens to Sink Foreign Ships Using Arctic Sea Route That Links Atlantic to the Pacific Unless It Is Given 45 Days Notice of Voyages and Vessels Take a Russian Pilot on Board," *Daily Mail*, March 6, 2019; "Russia Will Restrict Foreign Warships in Arctic Ocean, Defense Official Says," *Moscow Times*, November 30, 2018.

[111] Atle Staalesen, "Izvestia: This Is What Awaits Foreign Military Vessels on Northern Sea Route," *Barents Observer*, March 7, 2019. (A similar version published as Atle Staalesen, "Russia Sets Out Stringent New Rules for Foreign Ships on the Northern Sea Route," *Arctic Today*, March 8, 2019.) See also Maritime Executive, "Russia Tightens Control Over Northern Sea Route," *Maritime Executive*, March 8, 2019; Will Stewart, "Russia Warning: Moscow Could Sink or Detain Foreign Ships in Arctic Waters under New Rules," Express (UK), March 6, 2019.

NATO and the Arctic

Five of the eight Arctic states—the United States, Canada, Denmark, Iceland, and Norway—are members of NATO. The renewal of great power competition has led to a renewal of NATO interest in NATO's more northerly areas. During the Cold War, NATO member Norway and its adjacent sea areas were considered to be the northern flank of NATO's defensive line against potential aggression by the Soviet-led Warsaw Pact alliance. With the end of the Cold War and the shift to the post-Cold War era, NATO planning efforts shifted away from defending against potential aggression by Russia, which was considered highly unlikely, and toward other concerns, such as the question of how NATO countries might be able to contribute to their own security and that of other countries by participating in out-of-area operations, meaning operations in areas outside Europe.

With the renewal of great power competition, NATO is now once again focusing more on the question of how to deter potential Russian aggression against NATO countries, including in the Arctic.[113] As one consequence of that, Norway and its adjacent sea areas are once again receiving more attention in NATO planning.[114] For example, a NATO exercise called Trident Juncture 18 that was held from October 25 to November 7, 2018, in Norway and adjacent waters of the Baltic and the Norwegian Sea, with participation by all 29 NATO members plus Sweden and Finland, was described as NATO's largest exercise since the Cold War, and featured a strong Arctic element, including the first deployment of a U.S. Navy aircraft carrier above the Arctic Circle since 1991.[115] The

[112] See Atle Staalesen, "Navy Commandos Board Cargo vessel on Northern Sea Route," *Barents Observer*, September 13, 2019.

[113] See, for example, Rebecca Pincus, "NATO North? Building a Role for NATO in the Arctic," *War on the Rocks*, November 6, 2019; Anna Wieslander, "It's Time for NATO to Engage in the Arctic," *Defense One*, September 16, 2019; Tyler Cross, "The NATO Alliance's Role in Arctic Security," *Maritime Executive*, July 19, 2019.

[114] See, for example, Teri Schultz, "NATO and Washington Worry About Russian Subs in the High North," *Deutsche Welle*, April 26, 2018.

[115] See, for example, Kevin McGwin, "The EU Moves Toward a New Arctiv Straegy—and a More Independent Role I the Region," *Arctic Today*, December 11, 2019; Christopher Woody, "'We Can Do Better': The Navy's Newest Fleet Commander Says US Ships and Sailors Got 'Beat Up' During NATO's Biggest Exercise Since the Cold War," *Business Insider*, December 4, 2018; Levon Sevunts, "NATO's Arctic Dilemma; Two Visions of the

question of NATO's overall involvement in the Arctic, however, has been a matter of debate within NATO.[116]

European Union (EU) and the Arctic

Three of the eight Arctic states—Denmark, Finland, and Sweden—are members of the European Union (EU), and two other Arctic states—Iceland and Norway—have close ties to the EU as members of the European Economic Area. The EU is showing increased interest in the Arctic,[117] and the European Parliament (EP) supports an active EU role in the Arctic.[118] The EU is considered an "observer in principle" to the Arctic

---

Arctic Collide as NATO and Russia Flex Muscles," *Radio Canada International*, December 3, 2018; Megan Eckstein, "Truman CSG: Arctic Strike Group Operations Required Focus on Logistics, Safety," *USNI News*, November 6, 2018; Mary Thompson-Jones, "NATO's Arctic Exercise is a Good Start to Standing Up to Russian Militarization of the High North," *National Interest*, November 6, 2018; Pierre-Henry Deshayes, "Antifreeze and Balaclavas: NATO Troops in cold War Games," *Military.com*, November 2, 2018; Vasco Cotovio and Frederik Pleitgen, "Submarines a Centerpiece of Russia's Navy," *CNN*, November 19, 2018; Shawn Snow, "The Corps' Armor Makes a Big Showing in Norway as Marines Test Future Force," *Marine Corps Times*, October 25, 2018; Terje Solsvik, "As Winter Comes, NATO Kicks Off Largest Maneuvers Since Cold War," *Reuters*, October 25, 2018; Kyle Rempfer, "US Breaches the Arctic with Marines, Fighter Jets and Aircraft Carriers," *Military Times*, October 23, 2018; Megan Eckstein, "Truman Carrier Strike Group Operating North of Arctic Circle; First Time for US Navy Since 1991," *USNI News*, October 19, 2018; Thomas Gooley, "HST Strike Group Enters Arctic Circle, Prepares for NATO Exercise," *DVIDS (Defense Visual Information Distribution Service)*, October 19, 2018; Thomas Nilsen, "US Marines Launch Exercise in Northern Norway Ahead of Trident Juncture," *Barents Observer*, October 15, 2018.

[116] See, for example, Luke Coffey and Daniel Kochis, "Brussels NATO Summit 2018: Time to Get Serious About the Arctic," Heritage Foundation, June 27, 2018 (Issue Brief 4875), p. 1.

[117] See, for example, Romain Chuffart and Andreas Raspotnik, "The EU and Its Arctic Spirit: Solving Arctic Climate Change from Home?" European View, Vol. 18, No. 2, 2019: 156-162; Kevin McGwin, "The EU Is Poised to Take a Broader—and More Proactive—Role in the Arctic," *Arctic Today*, November 27, 2019; Adam Stepien and Andreas Raspotnik, "Can the EU's Arctic Policy Find True North?" Centre for European Policy Studies (CEPS), September 11, 2019; Wesley Morgan, "Politico Pro Q&A: EU Ambassador at Large for the Arctic Marie-Anne Coninsx," *Politico Pro*, August 6, 2019; Adam Stepien and Andreas Raspotnik, "The EU's Arctic Policy: Between Vision and Reality," *College of Europe Policy Brief*, August 2019, 5 pp.; Martin Breum, "Spurred by Chinese and Russian Activity, EU President Juncker Is Making the Arctic More Central to EU Policy," *Arctic Today*, February 20, 2019.

[118] In March 2017, the EP adopted a resolution largely endorsing the 2016 joint communication on an integrated EU policy for the Arctic (by 483 votes to 100, with 37 abstentions). (European Parliament, *Resolution on An Integrated EU Policy for the Arctic*, P8_TA(2017)0093, March 16, 2017.) The resolution advocated keeping the Arctic a low-tension area, recognized the important role of the Arctic Council in maintaining

Council, but to date has been denied full observer status at the council, alternately by Canada (because of Canadian Inuit objections to the EU's ban on the import of seal products) and Russia (following heightened EU-Russian tensions since Russia's 2014 invasion of Ukraine).[119]

In 2016, the European Commission (the EU's executive) and the EU's High Representative for Foreign Affairs and Security Policy issued a joint communication (or policy paper), *An Integrated European Union Policy for the Arctic*, that states that a "safe, stable, sustainable, and prosperous Arctic" is important for the region, the EU, and the world, and that "the EU has a strategic interest in playing a key role in the Arctic region."[120] The policy outlined in the document seeks to boost the EU's profile in the region and focuses on three broad themes—climate change and safeguarding the environment, sustainable development in the Arctic, and international cooperation on Arctic issues.

In 2017, the EU appointed its first Ambassador-at-Large for the Arctic, and in October 2019, the EU held its first-ever Arctic Forum, a high-level conference in northern Sweden focused on promoting EU efforts in the Arctic.[121] The EU is also a major financial contributor to Arctic research, providing around €200 million in the past decade under the Horizon 2020 Research and Innovation Program.[122] Some analysts contend, however, that the EU's policy statements on the Arctic have yet to coalesce into a clearly defined narrative with concrete goals; the European Commission's in-house think tank argues that the EU must develop a more comprehensive

---

constructive cooperation and stability in the region, and called upon the EU to develop a more concrete EU Arctic strategy and action plan.

[119] Kamrul Hossain, "EU Engagement in the Arctic: Do the Policy Responses from the Arctic States Represent the EU as a Legitimate Stakeholder?," *Arctic Review on Law and Politics*, Vol. 6, No. 2, November 2015.

[120] European Commission and High Representative of the Union for Foreign Affairs and Security Policy, Joint Communication to the European Parliament and the Council, *An Integrated European Union Policy for the Arctic*, April 27, 2016, p. 2.

[121] Martin Breum, "EU Plays Catch-up with US, China, Russia in Arctic," EUObserver.com, October 3, 2019.

[122] European Commission and High Representative of the Union for Foreign Affairs and Security Policy, Joint Communication to the European Parliament and the Council, *An Integrated European Union Policy for the Arctic*, April 27, 2016, p. 2.

strategy that balances protecting the Arctic environment with facilitating the sustainable economic and social development of the region.[123]

China in the Arctic

China's Growing Activities in Arctic

China's activities in the Arctic have grown steadily in recent years, and have emerged as a major topic of focus for the Arctic in a context of renewed great power competition. As noted earlier in this report, China was one of six non-Arctic states that were approved for observer status by the Arctic Council in 2013.[124] China in recent years has engaged in growing diplomatic activities with the Nordic countries, and has increased the size of its diplomatic presences in some of them. China has also engaged in growing economic discussions with Iceland and with Greenland, a territory of Denmark that might be moving toward eventual independence.[125]

China has a polar-capable icebreaker, *Xue Long* (Snow Dragon), that in recent years has made several transits of Arctic waters—operations that China describes as research expeditions.[126] A second polar-capable icebreaker (the first that China has built domestically), named *Xue Long 2*, entered service in 2019.[127] China in 2018 announced an intention to build a

---

[123] European Political Strategy Centre/European Commission, *Walking on Thin Ice: A Balanced Arctic Strategy for the EU*, July 2019; Adam Stepien and Andreas Raspotnik, "The EU's Arctic Policy: Between Vision and Reality," College of Europe Policy Brief, August 2019.

[124] The other five were India, Italy, Japan, Singapore, and South Korea. For the full list of observers and when they were approved for observer status, see Arctic Council, "Observers," accessed July 20, 2018, at https://arctic- council.org/index.php/en/about-us/arctic-council/observers.

[125] See, for example, Marc Lanteigne, "Stumbling Block: China-Iceland Oil Exploration Reaches an Impasse," *Over the Circle*, January 24, 2018. "Greenland Plans Office in Beijing to Boost Trade Ties with China," *Reuters*, July 18, 2018.

[126] See, for example, "Icebreaker Sets Sail on China's 9th Arctic Research Expedition," *Xinhua*, July 20, 2018; "China Begins 9th Arctic Expedition to Help Build 'Polar Silk Road,'" *Global Times*, July 20, 2018.

[127] "China Delivers First Self-Built Icebreaking Research Vessel," *People's Daily Online*, July 12, 2019; Kamaran Malik, "China Launches First Locally-made Icebreaker," *Asia Times*, July 11, 2019.

30,000-ton (or possibly 40,000-ton) nuclear-powered icebreaker,[128] which would make China only the second country (following Russia) to operate a nuclear-powered icebreaker. In December 2019, however, it was reported that China's third polar-capable icebreaker might instead be built as a 26,000-ton, conventionally powered ship.[129] (By way of comparison, the new polar icebreakers being built for the U.S. Coast Guard are to displace 22,900 tons each.) Like several other nations, China has established a research station in the Svalbard archipelago.

China in January 2018 released a white paper on China's Arctic policy that refers to China as a "near-Arctic state."[130] (China's northernmost

---

[128] See Trym Aleksander Eiterjord, "Checking in on China's Nuclear Icebreaker," *Diplomat*, September 5, 2019; Thomas Nilsen, "Details of China's Nuclear-Powered Icebreaker Revealed," *Barents Observer*, March 21, 2019; Zhao Yusha, "China One Step Closer to Nuke-Powered Aircraft Carrier with Cutting-Edge Icebreaker Comes on Stream," ChinaMil.com, June 23, 2018; China Daily, "China's 1st Nuclear-Powered Icebreaker in the Pipeline," *People's Daily Online*, June 25, 2018; Kyle Mizokami, "China Is Planning a Nuclear-Powered Icebreaker," *Popular Mechanics*, June 25, 2018; *China Military Online*, "Why Is China Building a 30,000-Ton Nuclear-Powered Icebreaker?" ChinaMil.com, June 30, 2018.

[129] Malte Humpert, "China Reveals Details of a Newly Designed Heavy Icebreaker," *Arctic Today*, December 17, 2019.

[130] "Full Text: China's Arctic Policy," *Xinhua*, January 26, 2018, accessed July 20, 2018, at http://www.xinhuanet.com/ english/2018-01/26/c_136926498.htm. The white paper states that "China is an important stakeholder in Arctic affairs. Geographically, China is a 'Near-Arctic State', one of the continental States that are closest to the Arctic Circle. The natural conditions of the Arctic and their changes have a direct impact on China's climate system and ecological environment, and, in turn, on its economic interests in agriculture, forestry, fishery, marine industry and other sectors. China is also closely involved in the trans-regional and global issues in the Arctic, especially in such areas as climate change, environment, scientific research, utilization of shipping routes, resource exploration and exploitation, security, and global governance. These issues are vital to the existence and development of all countries and humanity, and directly affect the interests of non-Arctic States including China."

Somewhat similarly, France's June 2016 national roadmap for the Arctic refers to France as a "polar nation." (Republique Francaise, Ministere des Affaires Etrangeres et du Developpement International, *The Great Challenge of the Arctic, National Roadmap for the Arctic*, June 2016, 60 pp.) The document states on page 9 that "France has established itself over the last three centuries as a polar nation, with a strong tradition of expeditions and exploration, and permanent research bases at the poles," and on page 17 that "[b]uilding on its long-standing tradition of exploration and expeditions in high latitudes, France has carved out its place as a polar nation over the last three centuries. France has permanent scientific bases in the Arctic and in Antarctica." It can also be noted that the northernmost part of mainland France, next to Belgium and across the Strait of Dover from England, is almost as far north as the more southerly parts of the Aleutian Islands.

Also somewhat similarly, a November 2018 UK parliamentary report refers to the UK as a "near-Arctic neighbour." The report states the following: "While the UK is not an Arctic territory, northeast of Mongolia, is at about the same latitude as the Aleutian Islands in Alaska, which, as noted earlier in this report, the United States includes in its definition of the Arctic for purposes of U.S. law.) The white paper refers to trans- Arctic shipping routes as the Polar Silk Road, and identifies these routes as a third major transportation corridor for the Belt and Road Initiative (BRI), China's major geopolitical initiative, first announced by China in 2013, to knit Eurasia and other regions together in a Chinese-anchored or Chinese-led infrastructure and economic network.[131]

China appears to be interested in using the NSR to shorten commercial shipping times between Europe and China and perhaps also to reduce China's dependence on southern sea routes (including those going to the Persian Gulf) that pass through the Strait of Malacca—a maritime choke point that China appears to regard as vulnerable to being closed off by other parties (such as the United States) in time of crisis or conflict.[132] China reportedly reached an agreement with Russia on July 4, 2017, to create an "Ice Silk Road,"[133] and in June 2018, China and Russia agreed to a credit agreement between Russia's Vnesheconombank (VEB) and the China Development Bank that could provide up to $9.5 billion in Chinese funds for the construction of select infrastructure projects, including in

---

state, it is a near-Arctic neighbour. The UK's weather system is profoundly affected by changes in the Arctic's climate and sea currents. The UK has been an Observer to the Arctic Council since 1998." (United Kingdom, House of Commons, Environmental Audit Committee, *The Changing Arctic, Twelfth Report of Session 2017-19*, November 29, 2018, p. 3. [Report, together with formal minutes relating to the report, Ordered by the House of Commons to be printed November 6, 2018]. See also pp. 6, 29, and 32.)

See also Eva Dou, "A New Cold War? China Declares Itself a 'Near-Arctic State,' *Wall Street Journal*, January 26, 2018; Grant Newsham, "China As A 'Near Arctic State'—Chutzpah Overcoming Geography," *Asia Times*, January 30, 2018.

[131] See, for example, Zhang Chun, "China's 'Arctic Silk Road,'" *Maritime Executive*, January 10, 2020; Sabena Siddiqui, "Arctic Ambition: Beijing Eyes the Polar Silk Road," *Asia Times*, October 25, 2018. The BRI's other two main corridors, which were announced at the outset of the BRI, are a land corridor that runs east to west across the middle of Eurasia—the "belt" in BRI—and a sea corridor called the Maritime Silk Road that passes through the South China Sea and the Indian Ocean to the Persian Gulf and the Mediterranean Sea—the "road" in BRI. For more on the BRI, see CRS In Focus IF10273, *China's "One Belt, One Road,"* by Susan V. Lawrence and Gabriel M. Nelson. See also Atle Staalesen, "Chinese Money for Northern Sea Route," *Barents Observer*, June 12, 2018. See also Lin Boqiang, "China Can Support Arctic Development as Part of B&R," *Global Times*, August 9, 2018.

[132] See, for example, Jonathan Hall, "Arctic Enterprise: The China Dream Goes North," *Journal of Political Risk*, September 2019.

[133] Xinhua, "China, Russia agree to jointly build 'Ice Silk Road,'" *Xinhuanet*, July 4, 2017.

particular projects along the NSR.[134] In September 2013, the *Yong Shen, a* Chinese cargo ship, became the first commercial vessel to complete the voyage from Asia to Rotterdam via the NSR.[135]

China is interested in oil and gas exploration in the Arctic, and has made significant investments in Russia's Arctic oil and gas industry, including the Yamal natural gas megaproject located on Russia's Yamal Peninsula in the Arctic.[136] China is also interested in mining opportunities in the Arctic seabed and in Greenland.[137] Given Greenland's very small population, China may view Greenland as an entity that China can seek to engage using an approach similar to ones that China has used for engaging with small Pacific and Indian Ocean island states. China may also be interested in Arctic fishing grounds.

China's growing activities in the Arctic may also reflect a view that as a major world power, China should, like other major world powers, be active in the polar regions for conducting research and other purposes. (Along with its growing activities in the Arctic, China has recently increased the number of research stations in maintains in the Antarctic.)

Particularly since China published its Arctic white paper in January 2018, observers have expressed curiosity or concern about China's exact

---

[134] Atle Staalesen, Chinese Money for Northern Sea Route," *Barents Observer*, June 12, 2018. See also Xie Wenwen (Caixin Globus), "For Chinese Companies, Investment I Arctic Infrastructure Offers Both Opportunities and Challenges," *Arctic Today*, June 17, 2019.

[135] "Chinese Make First Successful North Sea Route Voyage," *The Arctic Journal*, September 12, 2013. See also Malte Humpert, "Chinese Shipping Company COSCO To Send Record Number of Ships Through Arctic," *High North News*, June 13, 2019; Marex (Maritime Executive), "Russia and China Sign Arctic Deal," *Maritime Executive*, June 8, 2019.

[136] See, for example, Malte Humpert (High North News), "China Acquires 20 Percent Stake in Novatek's Arctic LNG 2 Project," *Arctic Today*, April 30, 2019; Ernesto Gallo and Giovanni Biava, "A New Energy Frontier Called 'Polar Silk Road,'" *China Daily*, April 12, 2019.

[137] See, for example, Nadia Schadlow, "Why Greenland Is Really About China," *The Hill*, August 28, 2019; Marc Lanteigne and Mingming Shi, "China Steps up Its Mining Interests in Greenland," *Diplomat*, February 12, 2019; John Simpson, "How Greenland Could Become China's Arctic Base," *BBC*, December 18, 2018; Marc Lanteigne, "Do Oil and Water Mix? Emerging China-Greenland Resource Cooperation," *Over the Circle*, November 2, 2018; Martin Breum, "China and the US Both Have Strategic Designs for Greenland," *Arctic Today*, October 17, 2018; Mia Bennett, "The Controversy over Greenland Airports Shows China Isn't Fully Welcome in the Arctic—Yet," *Arctic Today*, September 13, 2018; Mingming Shi and Marc lanteigne, "The (Many0 Roles of Greenland in china's Developing Arctic Policy," *Diplomat*, March 30, 2018; Miguel Martin, "China in Greenland: Mines, Science, and Nods to Independence," *China Brief*, March 12, 2018.

mix of motivations for its growing activities in the Arctic, and about what China's ultimate goals for the Arctic might be.[138]

Arctic States' Response

The renewal of great power competition underscores a question for the Arctic states regarding whether and how to respond to China's growing activities in the Arctic. China's growing activities in the Arctic could create new opportunities for cooperation between China and the Arctic

---

[138] See, for example, Jacquelyn Chorush, "Is China Really Threatening Conquest in the Arctic?" *High North News*, January 6, 2020; Mike Green, "China's Arctic Ambitions Spark Fear Among U.S. Military Leaders," *Washington Times*, December 26, 2019; Ryan D. Martinson, "The Role of the Arctic in Chinese Naval Strategy," *China Brief*, December 20, 2019; Anne-Marie Brady, "Facing Up to China's Military Interests in the Arctic," *China Brief*, December 10, 2019; Jacquelyn Chorush, "Concerns About China in the Arctic Dominate Canadian Security Forum," *High North News*, December 4, 2019; Nikolaj Skydsgaard and Jacob Gronholt-Pedersen, "China Mixing Military and Science in Arctic Push: Denmark," *Reuters*, November 29, 2019; Lyle J. Goldstein, "What Does China Want with the Arctic?" *National Interest*, September 7, 2019; Malte Humpert, Explaining China's Arctic Interests, the U.S.' Efforts to Re-Engage, and the Growing Split Between the Two Countries," *High North News*, September 6, 2019; Malte Humpert, "China Looking to Expand Satellite Coverage in Arctic, Experts Warn Of Military Purpose," *High North News*, September 4, 2019; "Mark Rosen, "Will China Freeze America Out of the Arctic?" *National Interest*, August 14, 2019; Heljar Havnes and Johan Martin Seland, "The Increasing Security Focus in China's Arctic Policy," Arctic Institute, July 16, 2019; Mercy A. Kuo, "The US and China's Arctic Ambitions, Insights from Malte Humpert," *Diplomat*, June 11, 2019; Global Risk Insight, "Why Is China Pouring Money Into The Arctic?" *Oil Price*, June 8, 2019; Somini Sengupta and Steven Lee Myers, "Latest Arena for China's Growing Global Ambitions: The Arctic," *New York Times*, May 24, 2019; Michael Lelyveld, "China's Arctic Investments Generate Heat," *Radio Free Asia*, May 24m, 2019; David Auerswald, "China's Multifaceted Arctic Strategy," *War on the Rocks*, May 24, 2019; John Vandiver, "China Tops Agenda as Military Leaders from 11 Nations Mull Arctic Security," *Stars and Stripes*, May 6, 2019; Sherman Xiaogang Lai, "China's Arctic Policy and Its Potential Impact on Canada's Arctic Security," Center for International Maritime Security (CIMSEC), April 17, 2019; James Stavridis, "China Is Joining the Rush for Arctic Riches," *Bloomberg*, April 16, 2019; Joel Gehrke, "US Wary of Chinese Military Threat in the Arctic," *Washington Examiner*, February 8, 2019; Omair Ahmad, "What Role for China and India in the Arctic?" *Third Pole*, February 4, 2019; Bethlehem Feleke, "Beijing Is Building Influence in the Arctic," *CNN*, December 30, 2018; Emanuele Scimia, "China's Advances in Arctic May Pose Security Threat to Canada," *Asia Times*, December 26, 2018; Fermin Koop, "Reykjavik Meeting Seals China's Commitment to Arctic Expansion," *Chinadialogue Ocean*, November 21, 2018; Rebecca Pincus and Walter A. Berbrick, "Gray Zones in a Blue Arctic: Grappling with China's Growing Influence," *War on the Rocks*, October 24, 2018; Kevin McGwin, "Copenhagen Is Ignoring Russia, Chinese Activity in the Arctic, Lawmakers Warn," *Arctic Today*, October 17, 2018; Trym Aleksander Eiterjord, "The Growing Institutionalization of China's Polar Silk Road; China Is Building an Increasingly Mature Platform to Support Its Scientific and Commercial Interests in the Arctic," *Diplomat*, October 7, 2018.

states.[139] They also, however, have the potential for posing challenges to the Arctic states in terms of defending their own interests in the Arctic.[140]

For U.S. policymakers, a general question is how to integrate China's activities in the Arctic into the overall equation of U.S.-China relations, and whether and how, in U.S. policymaking, to link China's activities in the Arctic to its activities in other parts of the world. Some observers see potential areas for U.S.-Chinese cooperation in the Arctic.[141] Other observers view the Arctic as emerging arena of U.S.-China strategic competition.[142] A specific question could be whether to impose punitive costs on China in the Arctic for unwanted actions that China takes elsewhere. As one potential example of such a cost-imposing action, U.S. policymakers could consider moving to suspend China's observer status on the Arctic Council[143] as a punitive cost-imposing measure for unwanted

---

[139] See, for example, Atle Staalesen, "As Arctic Talks Move to China, Leaders Downplay Divides," *Barents Observer*, May 11, 2019; Thomas Nilsen, "China Seeks a More Active Role in the Arctic," *Barents Observer*, May 11, 2019; Nathan Vanderklippe, "Agreeing on the Arctic: Amid Dispute, Canada Sides with China over the U.S. on How to Manage the North," *Globe and Mail*, May 10, 2019; Nong Hong, "Arctic Ambitions of China, Russia—and Now the US—Need Not Spark a Cold War," *South China Morning Post*, March 11, 2019.

[140] See, for example, Humphrey Hawksley, "Nordic Nations Can Stand Up to China in the Arctic," *Nikkei Asian Review*, June 13, 2019; Thomas Nilsen, "'We Must Be Prepared for Clearer Chinese Presence in Our Neighborhood'; Chief of Norway's Military Intelligence Service, Lieutenant General Morten Haga Lunde, Highlighted Chinese, Russian Arctic Cooperations in His Annual Focus Report," *Barents Observer*, February 11, 2019.

[141] See, for example, Stephen Delaney, "The United States Must Work with China to Ensure Freedom of Navigation in the Arctic," *Global Security Review*, September 6, 2019; Alison McFarland, "Arctic Options: Why America Should Invest in a Future with China," *National Interest*, September 30, 2018.

[142] See, for example, Simone McCarthy, "Trade, tech … and Now the Arctic? The Next Frontier in the China-US Struggle for Global Control," *South China Morning Post*, January 14, 2020 (ellipse as in the article's title); Chen Zinan, "To Keep Hegemony, US Trying to Obstruct China's Rights in Arctic," Global Times, December 25, 2019.

[143] Paragraph 37 of the Arctic Council's rules of procedure states the following:

Once observer status has been granted, Observers shall be invited to the meetings and other activities of the Arctic Council unless SAOs [Senior Arctic Officials] decide otherwise. Observer status shall continue for such time as consensus exists among Ministers. Any Observer that engages in activities which are at odds with the Council's [Ottawa] Declaration [of September 19, 1996, establishing the Council] or these Rules of Procedure shall have its status as an Observer suspended.

Paragraph 5 of Annex II of the Arctic Council's rules of procedure—an annex regarding the accreditation and review of observers—states the following:

Every four years, from the date of being granted Observer status, Observers should state affirmatively their continued interest in Observer status. Not later than 120 days before a

Chinese actions in the South China Sea.[144] As mentioned earlier, in a May 6, 2019 speech in Finland, Secretary of State Pompeo stated (emphasis added):

> The United States is a believer in free markets. We know from experience that free and fair competition, open, by the rule of law, produces the best outcomes.
>
> But all the parties in the marketplace have to play by those same rules. Those who violate those rules should lose their rights to participate in that marketplace. Respect and transparency are the price of admission.
>
> And let's talk about China for a moment. China has observer status in the Arctic Council, but that status is contingent upon its respect for the sovereign rights of Arctic states. The U.S. wants China to meet that condition and contribute responsibly in the region. But China's words and actions raise doubts about its intentions.[145]

In February 2019, it was reported that the United States in 2018 had urged Denmark to finance airports that China had offered to build in Greenland, so as to counter China's attempts to increase its presence and influence there.[146] In May 2019, the State Department announced plan for

---

Ministerial meeting where Observers will be reviewed, the Chairmanship shall circulate to the Arctic States and Permanent Participants a list of all accredited Observers and up-to-date information on their activities relevant to the work of the Arctic Council.

(Arctic Council, *Arctic Council Rules of Procedure*, p. 9. The document was accessed July 26, 2018, at https://oaarchive.arctic-council.org/handle/11374/940.

Paragraph 4.3 of the Arctic Council's observer manual for subsidiary bodies states in part

Observer status continues for such time as consensus exists among Ministers. Any Observer that engages in activities which are at odds with the Ottawa Declaration or with the Rules of Procedure will have its status as an Observer suspended.

(Arctic Council. *Observer Manual for Subsidiary Bodies*, p. 5. The document was accessed July 26, 2018, at https://oaarchive.arctic-council.org/handle/11374/939.)

See also Kevin McGwin, "After 20 years, the Arctic Council Reconsiders the Role of Observers," *Arctic Today*, October 24, 2018.

[144] For more on China's actions in the South China Sea and their potential implications for U.S. interests, see CRS Report R42784, *U.S.-China Strategic Competition in South and East China Seas: Background and Issues for Congress*.

[145] State Department, "Looking North: Sharpening America's Arctic Focus, Remarks, Michael R. Pompeo, Secretary of State, Rovaniemi, Finland, May 6, 2019," accessed August 23, 2019 at: https://www.state.gov/looking-north-sharpening-americas-arctic-focus/.

[146] Drew Hinshaw and Jeremy Page, "How the Pentagon Countered China's Designs on Greenland; Washington Urged Denmark to Finance Airports that Chinese Aimed to Build on North America's Doorstep," *Wall Street Journal*, February 10, 2019. See also Marc Lanteigne, ""Greenland's Airport Saga: Enter the US?" *Over the Circle*, September 18, 2018; Marc Lanteigne, "Greenland's Airports: A Balance between China and Denmark?"

establishing a permanent diplomatic presence in Greenland.[147] Some observers argue that a desire to preclude China (or Russia) from increasing its presence and influence in Greenland may have been one of the reasons why President Trump in August 2019 expressed an interest in the idea of buying Greenland from Denmark.[148]

For Russia, the question of whether and how to respond to China's activities in the Arctic may pose particular complexities. On the one hand, Russia is promoting the NSR for use by others, in part because Russia sees significant economic opportunities in offering icebreaker escorts, refueling posts, and supplies to the commercial ships that will ply the waterway. In that regard, Russia presumably would welcome increased use of the route by ships moving between Europe and China. More broadly, Russia and China have increased their cooperation on security and other issues in recent years, in no small part as a means of balancing or countering the United States in international affairs, and Russian-Chinese cooperation in the Arctic can both reflect and contribute to that cooperation.[149]

Over the Circle, June 15, 2018; Arne Finne (translated by Elisabeth Bergquist), "Intense Airport Debate in Greenland," *High North News*, May 30, 2018.

[147] State Department, "Secretary Pompeo Postpones Travel to Greenland, Press Statement, Morgan Ortagus, Department Spokesperson, May 9, 2019. See also Krestia DeGeorge, "US State Department Announces Plans for a Diplomatic Presence in Greenland," Arctic Today, May 9, 2019; Morten Soendergaard Larsen and Robbie Gramer, "Trump Puts Down New Roots in Greenland," *Foreign Policy*, November 8, 2019.

[148] See, for example, Stuart Lau, "Did China's Growing Presence in Arctic Prompt Donald Trump's Offer to Buy Greenland?" *South China Morning Post*, September 1, 2019; Nadia Schadlow, "Why Greenland Is Really About China," *The Hill*, August 28, 2019; Daniel Lippman, "Trump's Greenland Gambit Finds Allies Inside Government," *Politico*, August 24, 2019; Seth Borenstein (Associated Press), "Icy Arctic Becomes Hot Property for Rival Powers," *Navy Times*, August 22, 2019; Ragnhild Grønning, "Why Trump Is Looking to Buy Greenland—Even If It's Not for Sale," *High North News*, August 19, 2019.

[149] See, for example, Matteo Giovannini, "China and Russia Strengthen Strategic Partnership Along 'Polar Silk Road,'" *China Daily*, December 6, 2019; Melody Schreiber, "US Has 'Vital National Interests' at Stake in Russia-China Relationship in the Arctic, Expert Says," *Arctic Today*, April 2, 2019; Rebecca Pincus, China and Russia in the Arctic, Testimony before the U.S.-China Economic and Security Review Commission Hearing on "An Emerging China-Russia Axis? Implications for the United States in an Era of Strategic Competition," March 21, 2019; Liz Ruskin, "China, Russia Find Common Cause in Arctic," *Alaska Public Media*, March 21, 2019; Olga Alexeeva and Frederic Lasserre, "An Analysis of Sino-Russian Cooperation in the Arctic in the BRI Era," *Advances in Polar Science*, December 2018: 269-282; Marc Lanteigne, "Northern Crossroads, Sino-Russian Cooperation in the Arctic," *National Bureau of Asian Research*, March 2018, 6 pp. Nicholas Groffman, "Why China-Russia Relations Are Warming Up in the Arctic," *South China Morning Post*, February 17, 2018.

On the other hand, Russian officials are said to be wary of China's continued growth in wealth and power, and of how that might eventually lead to China becoming the dominant power in Eurasia, and to Russia being relegated to a secondary or subordinate status in Eurasian affairs relative to China. Increased use by China of the NSR could accelerate the realization of that scenario: As noted above, the NSR forms part of China's geopolitical Belt and Road Initiative (BRI). Some observers argue that actual levels of Sino-Russian cooperation in the Arctic are not as great as Chinese or Russian announcements about such cooperation might suggest.[150]

Linkages between Arctic and South China Sea

Another potential implication of the renewal of great power competition is a linkage that is sometimes made between the Arctic and the South China Sea relating to international law of the sea or the general issue of international cooperation and competition.[151] One aspect of this linkage relates to whether China's degree of compliance with international law of the sea in the South China Sea has any implications for understanding potential Chinese behavior regarding its compliance with international law of the sea (and international law generally) in the Arctic.

---

[150] See, for example, James Foggo III, "Russia, China Offer Challenges in the Arctic," *Defense One*, July 10, 2019; Anita Parlow, "Does a Russia-China Alignment in the Arctic Have Staying Power?" *Arctic Today*, June 27, 2019; Marc Lanteigne, "Scenes from a Northern Crossroad: China and Russia in the Arctic," *Over the Circle*, February 20, 2019; Marc Lanteigne, "No Borderline: A Norway-Russia Frontier Festival Connects with China," *Over the Circle*, February 16, 2019; Atle Staalesen, "Beijing Finds a Chinatown on NoRway's Arctic Coast,; The Asian Superpower Looks towards the Arctic and Finds a Home in This Year's Barents Spektakel Winter Festival," *Barents Observer*, February 12, 2019; Elizabeth Wishnick, "Russia and the Arctic in China's Quest for Great-Power Status," in Ashley J. Tellis, Alison Szalwinski, and Michael Wills, editors, *Strategic Asia 2019, China's Expanding Strategic Ambitions*, National Bureau of Asian Research, Seattle and Washington, DC, pp. 64-75.

[151] See, for example, Robinson Meyer, "The Next 'South China Sea' Is Covered in Ice," *Atlantic*, May 15, 2019; Justin D. Nankivell, "The Role of History and Law in the South China Sea and Arctic Ocean," *Maritime Awareness Project*, August 7, 2017; Sydney J. Freedberg, "Is The Arctic The next South China Sea? Not Likely," *Breaking Defense*, August 4, 2017; Caroline Houck, "The Arctic Could Be the Next South China Sea, Says Coast guard Commandant," *Defense One*, August 1, 2017; Daniel Thomassen, "Lessons from the Arctic for the South China Sea," Center for International Maritime Security, April 4, 2017.

A second aspect of this linkage, mentioned earlier, is whether the United States should consider the option of moving to suspend China's observer status on the Arctic Council as a punitive cost- imposing measure for unwanted Chinese actions in the South China Sea.

A third aspect of this linkage concerns the question of whether the United States should become a party to UNCLOS: Discussions of that issue sometimes mention both the situation in the South China Sea[152] and the extended continental shelf issue in the Arctic.[153]

Extended Continental Shelf Submissions, Territorial Disputes, and Sovereignty Issues[154]

Extended Continental Shelf Submissions

Motivated in part by a desire to exercise sovereign control over the Arctic region's increasingly accessible oil and gas reserves (see "Oil, Gas, and Mineral Exploration"), the four Arctic coastal states other than the United States—Canada, Russia, Norway, and Denmark (of which Greenland is a territory)—have made or are in the process of preparing submissions to the Commission on the Limits of the Continental Shelf regarding the outer limits of their extended continental shelves.

Russia has been attempting to chart the Arctic Ocean's enormous underwater Lomonosov Ridge in an attempt to show that it is an extension of Russia's continental margin. The ridge spans a considerable distance across the Arctic Ocean. A 2001 submission by Russia was rejected as insufficiently documented. Canada views a portion of the ridge as part of its own continental shelf.[155]

---

[152] For further discussion of this situation, see CRS Report R42784, *U.S.-China Strategic Competition in South and East China Seas: Background and Issues for Congress*.

[153] See, for example, Ben Werner, "Zukunft: U.S. Presence in Arctic Won't Stop Chinese, Russia Encroachment Without Law of the Sea Ratification," *USNI News*, August 1, 2017.

[154] This section was prepared by Carl Ek, who was a Specialist in International Relations, Foreign Affairs, Defense, and Trade Division, until his retirement on April 30, 2014. For questions relating to this section, contact Derek E. Mix, Analyst in European Affairs, Foreign Affairs, Defense, and Trade Division.

[155] "Russia, Canada Make Competing Claims To Arctic Resources," *The Canadian Press*, September 16, 2010.

In August 2007, a Russian submersible on a research expedition deposited an encased Russian Federation flag on the seabed of the presumed site of the North Pole. The action captured worldwide attention, but analysts note that it did not constitute an official claim to the Arctic seabed or the waters above it, that it has no legal effect, and that it therefore was a purely symbolic act.

At a May 2008 meeting in Ilulissat, Greenland, the five Arctic coastal states reaffirmed their commitment to the UNCLOS legal framework for the establishment of extended continental shelf limits in the Arctic.[156] (For further discussion, see "Extent of the Continental Margin" in "Oil, Gas, and Mineral Exploration.")

Territorial Disputes and Sovereignty Issues

In addition to this process, there are four unresolved Arctic territorial disputes:

Scientists have forecast that in coming decades, global warming will reduce the ice pack in Canada's northern archipelago sufficiently to permit ships to use the trans-Arctic shipping route known as the Northwest Passage during the summer months (see "An October 12, 2015, press report states the following:

> More people are wishing to explore icy environments, says Peter Hellberg, manager responsible for the SAR process at the Swedish Maritime Administration. Hellberg is part of an IMO/International Civil Aviation Organization (ICAO) working group that is re- evaluating search and rescue (SAR) operations in Polar waters as a result of this push.
>
> The working group includes both a maritime and aeronautical perspective, and it has identified a need for more detailed guidance for SAR organizations which will be achieved through an update of the International Aeronautical and Maritime Search and Rescue Manual (IAMSAR) planned for 2019.
>
> While the IAMSAR manual is not mandatory, it is followed by most SAR organizations around the world. It provides the framework for setting up a multi-national SAR, giving different parties guidance on the necessary arrangements for Arctic areas.

[156] "5 Countries Agree To Talk, Not Compete, Over the Arctic," *New York Times*, May 29, 2008.

The guidance will be expanded on based on the Polar Code and other recent IMO regulatory updates, and from an aeronautical perspective, from lessons learned after the disappearance of Malaysian Airlines' MH370.

- Commercial Sea Transportation"). The prospect of such traffic raises a major jurisdictional question. Ottawa maintains that such a passage would be an inland waterway, and would therefore be sovereign Canadian territory subject to Ottawa's surveillance, regulation, and control. The United States, the European Union, and others assert that the passage would constitute an international strait between two high seas.
- The United States and Canada are negotiating over a binational boundary in the Beaufort Sea.
- The United States and Russia in 1990 signed an agreement regarding a disputed area of the Bering Sea; the U.S. Senate ratified the pact the following year, but the Russian *Duma* has yet to approve the accord.
- Denmark and Canada disagree over which country has the territorial right to Hans Island, a tiny, barren piece of rock between Greenland and Canada's Ellesmere Island. Some analysts believe the two countries are vying for control over a future sea lane that might be created if the Arctic ice were to melt sufficiently to create a Northwest Passage. Others claim that the governments are staking out territorial claims in the event that future natural resource discoveries make the region economically valuable.[157]

In addition to these disputes, Norway and Russia had been at odds for decades over the boundary between the two in the so-called "Grey Zone" in the Barents Sea, an area believed to hold rich undersea deposits of petroleum. On September 15, 2010, Norwegian Prime Minister Jens Stoltenberg and Russian President Dmitry Medvedev signed an agreement in Murmansk, a Russian city near the Norwegian border. The accord

---

[157] For additional information, see Natalie Mychajlyszyn, *The Arctic: Geopolitical Issues*, Canadian Library of Parliament, October 24, 2008.

awards roughly half of the 175,000-square- kilometer area to each country; it spells out fishing rights, and provides for the joint development of future oil and gas finds that straddle the boundary line. Some observers believe it is noteworthy that Russia would concede sovereignty over such a large, resource-rich area to a small, neighboring country. But others have noted that Moscow may be hoping for Norwegian cooperation in developing offshore resources, and eventually in winning approval when Russia makes its Article 76 UNCLOS submission.[158]

In August 2010, Canadian Foreign Minister Lawrence Cannon announced a new "Statement of Canada's Arctic Policy," which reaffirmed the government's commitment to Canada's sovereignty in the region, to economic and social development, to environmental protection, and to empowerment of the peoples in the north. The statement also emphasized the government's intention to negotiate settlements to its disputes with the United States over the Beaufort Sea boundary, and with Denmark over Hans Island. Minister Cannon declared that "making progress on outstanding boundary issues will be a top priority."[159] Also, despite their dispute over Hans Island, Canada and Denmark have been working together on Arctic issues. In May 2010, the two countries' military chiefs of staffs signed a memorandum of understanding on Arctic Defense, Security, and Operational Cooperation, committing the two countries to "enhanced consultation, information exchange, visits, and exercises."[160]

---

[158] "Russia, Norway Sign Deal On Barents Sea Border, Seek More Development in Mineral-rich Arctic," *Associated Press*, September 15, 2010.

[159] Cannon quoted in "Canada Seeks To Settle Arctic Borders," *Agence France Presse*, August 20, 2010. For additional information concerning Canada's August statement on Arctic policy, see "Statement on Canada's Arctic foreign policy: Exercising sovereignty and promoting Canada's Northern Strategy abroad," Foreign Affairs and International Trade Canada website, http://www.international.gc.ca/polar-polaire/canada_arctic_foreign_policy-la_politique_etrangere_du_canada_pour_arctique.aspx?lang=eng.

[160] "Canada and Denmark Sign Arctic Cooperation Arrangement," *Targeted News Service*, May 17, 2010.

U.S. Military Forces and Operations[161]

Overview

U.S. military forces, particularly the Navy and Coast Guard, have begun to pay more attention to the Arctic in their planning and operations. The Department of Defense (DOD), the Navy, and the Coast Guard have all issued Arctic strategy documents in recent years. Whether DOD and the military services are adequately resourcing these strategies and taking appropriate and adequate actions for defending U.S. interests in the Arctic has emerged as a topic of debate and congressional oversight.[162]

During the Cold War, the Arctic was an arena of military competition between the United States and the Soviet Union, with both countries, for example, operating nuclear-powered submarines, long-range bombers, and tactical combat aircraft in the region. The end of the Cold War and the collapse of most elements of the Russian military establishment following the dissolution of the Soviet Union in December 1991 greatly reduced this competition and led to a reduced emphasis on the Arctic in U.S. military planning.

Renewed tensions with Russia following its seizure and annexation of Crimea in March 2014, combined with a significant increase in Russian military capabilities and operations in the Arctic in recent years,[163] have led

---

[161] This section prepared by Ronald O'Rourke, Specialist in Naval Affairs, Foreign Affairs, Defense, and Trade Division.

[162] See, for example, Dermot Cole, "The US Writes, But Does Not Implement, Arctic Strategies," *Arctic Today*, January 13, 2020.

[163] Regarding the increase in Russian military capabilities and operations in the Arctic, see, for example, Malte Humpert, "New Satellite Images Reveal Extent of Russia's Military and Economic Buildup in the Arctic," *High North News*, May 3, 2019, updated December 31, 2019; Malte Humpert, "Satellite Images Reveal New Russian Long-Range Radar in the Arctic," *High North News*, December 17, 2019; Atle Staalesen, "Deployment of S-400 in Arctic Bases Creates Air Defense Shield Over Northern Russia," *Barents Observer*, December 11, 2019; Tom Balmforth, "Russia Plans to Set Up Arctic Air Defense 'Dome' with S-400 Missiles," *Reuters*, December 9, 2019; Joseph Trevithick, "Russia Plans To Set Up Massive New Radar Array To Help 'Control' The Arctic Region," *The Drive*, December 2, 2019; Editorial Team, "Russia Flaunts Increased Military Strength on Norway Border," *Life in Norway*, December 2, 2019; Jamie Dettmer, "Russia Ups the Ante in the Arctic," *Voice of America*, November 6, 2019; Charlie Bradley, "Putin Lays Down Arctic Gauntlet to Trump as Russian Military Starts Nuclear War Games," *Express (UK)*, October 16, 2019; Atle Staalesen, "Russia's New Arctic Missile System Comes to Pechenga Valley," *Barents Observer*, August 27, 2019; Atle Staalesen, "Weapons Arrive to Russia's New Arctic

to growing concerns among U.S. officials and other observers that the Arctic is once again becoming a region of military tension and competition,[164] and to concerns about whether the United States is adequately prepared militarily to defend its interests in the region.[165]

Garrison," *Barents Observer*, August 5, 2019; Atle Staalesen, "Russia Establishes a New Arctic Air Squadron to Protect the Northern Sea Route," *Arctic Today*, July 19, 2019; Atle Staalesen, "Russia's New Arctic Missile Knocks Out Targets in Massive Test Launch," *Barents Observer*, June 14, 2019; Mathieu Boulègue, *Russia's Military Posture in the Arctic, Managing Hard Power in a 'Low Tension' Environment*, Chatham House, June 2019, 44 pp.; Pavel Felgenhauer, "Russia's New Electronic Warfare Capabilities in the Arctic," *Eurasia Daily Monitor*, May 30, 2019; Atle Staalesen, "Russia Says Its Radio-Electronic Shield Now Covers the Arctic," Barents Observer, May 21, 2019; Atle Staalesen, "Navy Is Building a New Major Depot for Arctic Operations," *Barents Observer*, May 2, 2019; Diana Magnay, "Arctic Battleground: Russia's Race to Militarise Resource-Rich Region," *Sky News*, April 6, 2019; Mary Ilyushina and Frederik Pleitgen, "Arctic Ambitions," *CNN*, April 5, 2019; Thomas Nilsen, "Russia Claims to Have Demonstrated Complex Exercise Outside Norway," *Barents Observer*, April 11, 2019; Vladimir Isachenkov, "Russia Revamps Arctic Military Base to Stake Claim on Region," *Associated Press*, April 4, 2019; "Russia Will Deploy New Missiles to Defend Arctic, Admiral Says," *Moscow Times*, March 13, 2019; Nerijus Adomaitis, "Norway Says It Proved Russian GPS Interference During NATO Exercises," *Reuters*, March 18, 2019; Atle Staalesen, "Russia's Navy Is Already Flying to New Bases in the Arctic to Counter Foreign Threats Above and Below the Ice," *Business Insider*, March 15, 2019; Pavel Felgenhauer, "Russia Claims Total Military Superiority in the Arctic," *Eurasia Daily Monitor*, March 14, 2019; Joel Gehrke, "Moscow Warns US: Russian Special Forces Are Training for Arctic Conflict," *Washington Examiner*, February 22, 2019; Kyle Mizokami, "Nearly a Dozen Russian Jets Practiced an Attack on a Norwegian Radar Base," *Popular Mechanics*, February 13, 2019; Thomas Nilsen, "Eleven Russia Fighter Jets Make a Mock Attack on a Norwegian Arctic Radar Installation," Arctic Today, February 13, 2019; "Russian Tor-Family Air Defense System Trains to Defend the Arctic," *Defence Blog*, February 9, 2019; Thomas Nilsen, "Russia Resumes North Pole Patrols with Fighter Jets," *Arctic Today*, February 4, 2019; Atle Staalesen, "New Arctic Naval Base Built in 6 Months," *Barents Observer*, January 30, 2019; Thomas Nilsen, "In Polar Night, Russia Exercises Strategic Bombers and Newest Frigate," *Barents Observer*, January 27, 2019.

164 See, for example, Terri Moon Cronk, "Climate Change Opens Door to Arctic for Competitors, DOD Official Says," *DOD News*, December 13, 2019; Laurence Peter, "Danes See Greenland Security Risk Amid Arctic Tensions," *BBC News*, November 29, 2019; Gwladys Fouche, "On Norway's Icy Border with Russia, Unease over Military Buildup," *Reuters*, November 6, 2019; Jackie Northam, "In A Remote Arctic Outpost, Norway Keeps Watch On Russia's Military Buildup," NPR, November 3, 2019; John Grady, "NORTHCOM Says U.S., Canada Must Maintain 'Clear- Eyed' View of Arctic Threats," *USNI News*, July 23, 2019; Russ Read, "NORTHCOM Commander Warns the Arctic Is An 'Avenue of Approach' for Russia," *Washington Examiner*, July 22, 2019; David Brennan, "Battle for the Arctic:
U.S. Warms Russia 'Is Way Ahead of US' in Race to Control New Frontier," *Newsweek*, July 19, 2019; Geoff Upton, "Great Power Problems: Russia Want Control over the Arctic Region," *National Interest*, April 25, 2019; Scott Waldman, "As Russian Military Moves into Thawing Arctic, U.S. Strategy Shifts," *E&E News (Scientific American)*, March 6 2019; Joel Gehrke, "'It's Nobody's Lake': US Admiral Warms China and Russia over

As noted earlier, Russia since 2008 has adopted a series of strategy documents outlining plans that call for, among other things, bolstering the country's Arctic military capabilities. Russia has reactivated and modernized military bases that fell into disuse with the end of the Cold War, and in December 2014 established a new Arctic Joint Strategic Command at Severomorsk, the home of the Russian navy's Northern Fleet.

U.S. military officials, military officials from other Arctic states, and other observers have stressed the cooperative aspects of how the Arctic states have addressed Arctic issues, and have sometimes suggested that the competitive aspects of the situation have been exaggerated in some press accounts.[166] Some observers argue that that Russia's recent military investment in the Arctic is being exaggerated, or reflects normal modernization of aging capabilities, or is intended partly for domestic

Arctic," *Washington Examiner*, February 20, 2019; Jon Lockett, "Global Warring, Russian and US Warships Converge on Arctic to Build Naval and Air Bases as Melting Ice Leaves Cold War Frontiers Wide Open," *The Sun (UK)*, January 24, 2019; Jonathan Watts, "Military Buildup in Arctic as Melting Ice Reopens Northern Borders," *Guardian*, January 24, 2019.

165 See, for example, Liz Ruskin, "Russia's Military Dominance over Arctic Grows While US Trteads Water, Security Experts Tell Senate Panel," KTOO, December 17, 2019; Shawn Snow, "NORTHCOM Commander Says US Needs Arctic Early Warning System," *Military Times*, November 26, 2019; Christopher Woody, "Russia Is Finding New Islands in the Arctic, While the US Is Still Trying to Figure Out How to Get Up There," *Business Insider*, October 29, 2019; Hal Brands, "America Is Losing the Battle of the Arctic," *Bloomberg*, July 30, 2019; Kyle Rempfer, "Marines, Sailors and Coasties Play Cath-Up in Arctic Warfare," *Military Times*, May 30, 2019; Kyle Rempfer, "US Military Leaders Say the Arctic Now the 'First Line of Defense'—and the US Is Already Behind," *Business Insider*, May 10, 2019; Nico Hines, "Inside the West's Plans for Arctic War Against Russia," *Wired*, May 7, 2019; Patrick Tucker, "Admiral: The US IS 'Operating Blind' Ion the Arctic," *Defense One*, May 6, 2019; Kyle Rempfer, "NORTHCOM: Arctic Now America's 'First Line of Defense,'" *Defense News*, May 6, 2019; Kyle Rempfer, "Northern Border Along Arctic, Not Southern, Is What Worries NORAD Leaders," *Military Times*, February 27, 2019; "Canada, U.S. Must Do More to Check Russia's Military Moves in the Arctic, Says NORAD Chief," *CBC News*, February 12,2019; Liz Ruskin, "Calls for 'Vigilance' on Russian Military Buildup in Arctic," KTOO, February 2, 2019.

166 See, for example, Mia Bennett, "U.S. Rhetoric About the Strategic Importance of the Arctic Is Out of Step with Its Spending Priorities," *Arctic Today*, July 26, 2019; "Arctic Conflict With Russia 'Not Likely In The Short-Term', Analyst Says," *Forces*, January 30, 2019; Danita Catherine Burke, "Why the New Arctic 'Cold War' Is a Dangerous Myth," *The Conversation*, December 13, 2018; Martin Breum, "Why Russia Is Likely to Remain Cooperative in the Arctic," *Arctic Today*, November 22, 2018; John Grady, "Panel: Cooperation, Not Conflict Key to Future of the Arctic," *USNI News*, April 8, 2018; Levon Sevunts," NATO Wants to Keep the Arctic An Area of Low Tensions, Stoltenberg," *Radio Canada International*, April 4, 2018; Levon Sevunts, "Arctic Nations Develop Coast Guard Co- operation," *Independent Barents Observer*, March 13, 2018.

Russian consumption.[167] Even so, U.S. military forces (and U.S. intelligence agencies)[168] are paying renewed attention to the Arctic. This is particularly true in the case of the Navy and Coast Guard, for whom diminishment of Arctic sea ice is opening up new operating areas for their surface ships. The U.S. Air Force, Army, and Marine Corps, too, are now focusing more on Arctic operations.[169] Canada, the UK, and the Nordic countries are taking or contemplating steps to increase their own military presence and operations in the region.[170]

---

[167] See, for example, Elizabeth Buchanan and Mathieu Boulègue, "Russia's Military Exercises in the Arctic Have More Bark Than Bite," *Foreign Policy*, May 20, 2019; Arne F. Finne, "Russia Is a Responsible Actor in the Arctic," *High North News*, January 22, 2019; Stephanie Pezard, *The New Geopolitics of the Arctic, Russia's and China's Evolving Role in the Region*, RAND (Testimony presented before the Standing Committee on Foreign Affairs and International Development of the Canadian House of Commons on November 26, 2018), pp. 2-4 (which presents comments on both sides of the issue of whether other countries should be concerned by Russia's Arctic military capabilities); Thomas Nilsen, "New Weapons Testing Is Worrying, But Does Not Raise Tensions in the North, [Norway's] Defense Minister Says," *Barents Observer*, August 26, 2018.

[168] See Brian Bennett and W.J. Hennigan, "US Builds Up Arctic Spy Network As Russia, China Increase Presence," *Los Angeles Times*, September 7, 2015.

[169] See, for example, Christopher Woody, "The US Air Force Is Fixing Up a Remote Base That Could Help Keep an Eye on Russia," *Business Insider*, January 15, 2020; Hank Davis, "WATCH: Marines, Navy 'Take' Adak Islandas Part of Arctic Training," KTUU.com, September 20, 2019; Thomas Nilsen, "B-2 Stealth Bomber on Mission above Arctic Circle outside Norway," Barents Observer. September 6, 2019; Lolita C. Baldor, "US Steps Up Winter-Warfare Training As Global Threat Shifts," *Associated Press*, February 20, 2019; April Davis, "Soldiers Learn Arctic Survival from Northern Neighbors," U.S. Army, February 7, 2019; Heather Wilson, "Air Power and the Arctic: The Importance of Projecting Strength in the North," *Defense News*, January 9, 2019.

[170] See, for example, "1,000 Royal Marines On Largest Arctic Deployment In 'Many Years,'" *Forces Network*, January 17, 2020; Nikolaj Skydsgaard, "Danish PM Suggests Increasing Arctic Surveillance in Response to Russian Activity," *Reuters*, December 3, 2019; Kevin McGwin, "Denmark Will Triple Arctic Defense Spending," *Arctic Today*, December 3, 2019; Gerard O'Dwyer, "Nordic Countries See Russia Flex Its Missile Muscles," *Defense News*, August 5, 2019; Kevin McGwin, "Keeping Eye on Russian Build-Up, Danes Ready to Send Fighter Jets to Greenland," *Arctic Today*, May 20, 2019; YLE News, "US, Norwegian Troops Join Military Drill in Northern Finland," *Barents Observer*, May 16, 2019; Robbie Gramer, "Vanguards of the Thawing Arctic," *Foreign Policy*, May 4, 2019; Thomas Nilsen, "Norway Rearms Finnmark for New Security Landscape," *Arctic Today*, April 17, 2019; Helene Cooper, "Military Drills in Arctic Aim to Counter Russia, but the First Mission Is to Battle the Cold," *New York Times*, April 12, 2019; Rebecca Ricks, "Night And Day: Conducting Arctic Beach landings With Royal Marines," *Forces*, March 6, 2019; Thomas Nilsen, "7,000 Foreign Soldiers Ready for Exercise in Northern Sweden," *Barents Observer*, February 26, 2019; Rebecca Ricks, "Exercise Clockwork: Taking To The Skies Over The Arctic Circle," *Forces*, January 30, 2019; Rebecca Ricks, "Arctic Strategy: Hundreds Of Royal Marines Prepare For Norway Deployment," *Forces*, January 22, 2019;

DOD in General

January 2014 Implementation Plan for National Strategy for Arctic Region

The Obama Administration's January 2014 implementation plan for its national strategy for the Arctic region (see "Background") makes DOD the lead federal agency for one of the plan's 36 or so specific initiatives, and a supporting agency for 18 others.[171] The initiative for which DOD is designated the lead federal agency is entitled "Develop a framework of observations and modeling to support forecasting and prediction of sea ice."[172]

January 2018 National Defense Strategy

The January 2018 unclassified summary of the 2018 National Defense Strategy[173] does not specifically mention the Arctic.

June 2019 DOD Arctic Strategy

DOD's June 2019 Arctic Strategy—the successor to DOD's 2013 and 2016 Arctic strategies— states

> This document articulates the Department of Defense's (DoD) strategy for the Arctic region in an era of strategic competition. It is informed by the 2017 National Security Strategy and anchored in the priorities of the 2018 National Defense Strategy (NDS) and its focus on competition with China and Russia as the principal challenge to long-term U.S. security and prosperity. This strategy supersedes the 2016 DoD Arctic strategy.
>
> DoD's desired end-state for the Arctic is a secure and stable region where U.S. national interests are safeguarded, the U.S. homeland is defended, and nations work cooperatively to address shared challenges. This strategy sets forth DoD's assessment of the Arctic security environment, risks posed to U.S. national security interests, DoD Arctic

---

Kevin McGwin, "In Face of Uncertainty about Russia, a Nordic Gang of Four Emerges," *Arctic Today*, January 17, 2019.

171 *Implementation Plan for The National Strategy for the Arctic Region*, January 2014, pp. 6-32.

172 *Implementation Plan for The National Strategy for the Arctic Region*, January 2014, pp. 15-16.

173 Department of Defense, *Summary of the 2018 National Defense Strategy of the United States of America: Sharpening the American Military's Competitive Edge*, January 2018, 11 pp.

objectives, and the strategic approach by which DoD will achieve these objectives.

A secure and stable Arctic region benefits the United States and necessitates a rules-based order, reflecting Arctic nations' respect for national sovereignty and constructive engagement to address shared challenges. The network of U.S. allies and partners with shared national interests in this rules-based order is the United States' greatest strategic advantage in the Arctic region, and thus the cornerstone of DoD's Arctic strategy. DoD cooperation with Arctic allies and partners strengthens our shared approach to regional security and helps deter strategic competitors from seeking to unilaterally change the existing rules-based order.

NDS goals and priorities guide DoD's strategic approach to the Arctic. The Joint Force must be able to deter, and if necessary, defeat great power aggression. DoD must prioritize efforts to address the central problem the NDS identifies – i.e., the Joint Force's eroding competitive edge against China and Russia, and the NDS imperative to ensure favorable regional balances of power in the Indo-Pacific and Europe. Developing a more lethal, resilient, agile, and ready Joint Force will ensure that our military sustains its competitive advantages, not only for these key regions of strategic competition, but globally as well. Maintaining a credible deterrent for the Arctic region requires DoD to understand and shape the Arctic's geo-strategic landscape for future operations and to respond effectively to contingencies in the Arctic region, both independently and in cooperation with others. DoD's strategic approach seeks to do so by implementing three ways in support of the desired Arctic end-state (each described in detail in this document):

- Building Arctic awareness;
- Enhancing Arctic operations; and
- Strengthening the rules-based order in the Arctic.[174]

[174] Department of Defense, Office of the Under Secretary of Defense for Policy, Report to Congress, Department of Defense Arctic Strategy, As required by Section 1071 of the John S. McCain National Defense Authorization Act for Fiscal Year 2019 (P.L. 115-232), June 2019, p. 2. For press reports regarding the document, see Connie Lee, "Great Power Competition Extends to Arctic," National Defense Magazine, August 12, 2019; Hal Brands, "America Is Losing the Battle of the Arctic," Bloomberg, July 30, 2019; Melody Schreiber, "U.S. Arctic Defense Strategy Ramps Up Rhetoric Without Committing Resources, Experts Say," Arctic Today, June 11, 2019; Malte Humpert, "New U.S. Department of Defense Arctic Strategy Sees Growing Uncertainty and Tension in Region," High North News, June 7, 2019.

DOD Cooperation with Canada and Other Countries

DOD has been taking a number of steps in recent years to strengthen U.S.-Canadian cooperation and coordination regarding military operations in the Arctic.[175]

FY2020 National Defense Authorization Act

The conference report (H.Rept. 116-333 of December 9, 2019) on S. 1790, the FY2020 National Defense Authorization Act, includes a number of provisions relating to the Arctic, including the following:

- Section 1238, which requires a report on Russian and Chinese military activities in the Arctic;
- Section 1260E, which requires an independent study on Chinese foreign direct investment in the Arctic countries;
- Section 1706, which requires a report on DOD's plans for mass-casualty disaster response operations in the Arctic; and
- Section 1752, which requires a report evaluating potential sites for one or more strategic ports in the Arctic, and authorizes the Secretary of Defense to designate one or more ports in the Arctic as DOD strategic Arctic ports.

Navy and Coast Guard in General

The Navy and Coast Guard are addressing the potential implications of increased human activities in the Arctic for the Navy and Coast Guard, particularly implications for required numbers of ships and aircraft, ship and aircraft characteristics, ship and aircraft operations, new or enlarged Arctic bases, and supporting systems, such as navigation and communication systems. The Coast Guard annually deploys cutters and aircraft into the region to perform missions and better understand the

---

175 See, for example, Murray Brewster, "Canadian, U.S. Military Leaders Agree on Framework to Retool Norad," *CBC News*, August 8, 2019; Geoff Ziezulewicz, "The Navy Sends a Carrier Back to Russia's Arctic Haunts," *Navy Times*, October 20, 2018; "Harry S. Truman Strike Group Enters Arctic Circle, Prepares for NATO Exercise," *Navy News Service*, October 19, 2018; Amy Bushatz, "The Navy Could Send More Sailors to Alaska," *Military.com*, August 16, 2018; Vivienne Machi, "United States, Canada studying Options to Replace Arctic Early Warning Radars," *National Defense*, July 27, 2018.

implications of operating such units there, and the Navy has recently deployed some surface ships to the region.[176] Key points relating to the Navy and Coast Guard in the Arctic include the following:

- The diminishment of Arctic ice is creating new operating areas in the Arctic for Navy surface ships and Coast Guard cutters.
- U.S. national security interests in the Arctic include "such matters as missile defense and early warning; deployment of sea and air systems for strategic sealift, strategic deterrence, maritime presence, and maritime security operations; and ensuring freedom of navigation and overflight."[177]
- In a context of renewed great power competition, discussions of the Arctic have begun to focus more on security issues, and on countering Russian and Chinese activities in the region. Some Members of Congress, in addition to urging DOD to establish a new strategic port in the Arctic, have urged the Navy to bolster its plans for sending surface ships into Arctic waters as part of a broader U.S. effort to counter substantial Russian and growing Chinese activities in the region. Russia's and China's investments in building and operating polar icebreakers have become more prominent in discussions about procuring new polar icebreakers for the Coast Guard.

---

[176] See, for example, Zachariah Hughes, "As Polar Ice Cap Recedes, The U.S. Navy Looks North," *NPR*, June 12, 2019; Megan Eckstein, "Theodore Roosevelt Strike Group in Alaska for High-End, Joint Exercise Northern Edge 2019," *USNI News*, May 14, 2019; Liz Ruskin, "Navy Plans to Be More Active in the Arctic," *Alaska Public Media*, May 2, 2019; Ben Kesling, "U.S. Navy Plans to Extend Its Reach in the Arctic," *Wall Street Journal*, May 1, 2019; Malte Humpert, "U.S. navy Plans to Send Surface Vessels Through Arctic," *High North News*, March 11, 2019; Ben Kesling, "Cold War Games: U.S. Is Preparing to Test the Waters in Icy Arctic," *Wall Street Journal*, January 11, 2019; Yereth Rosen, US Coast Guard Settles into Kotzebue for the Summer," *Arctic Today*, July 2, 2018; David Axe, "Three U.S. and British Submarines Meet at the North Pole," *War Is Boring*, March 22, 2018; Kris Osborn, "The Navy Is Training for War with Russia in the Arctic," *National Interest*, March 20, 2018; Corey B. Barker, "Arctic-Deployed Navy Submarines Participate in Ice Exercise 2018," Defense.gov, March 13, 2018; Elizabeth McLaughlin, "Regular US Navy Arctic Submarine Exercise Takes On New Importance," *ABC News*, March 12, 2018; Submarine Force Pacific, "Navy's Arctic Ice Exercise Features Multinational Participation," Defense.gov, March 8, 2018.

[177] NSPD 66/HSPD 25, Section III B.

- SAR in the Arctic is a mission of increasing importance, particularly for the Coast Guard, and one that poses potentially significant operational challenges (see "Search and Rescue (SAR)" above).
- More complete and detailed information on the Arctic is needed to more properly support expanded Navy and Coast Guard ship and aircraft operations in the Arctic.
- The Navy and the Coast Guard currently have limited infrastructure in place in the Arctic to support expanded ship and aircraft operations in the Arctic.[178]
- Expanded ship and aircraft operations in the Arctic may require altering ship and aircraft designs and operating methods.[179]
- Cooperation with other Arctic countries will be valuable in achieving defense and homeland security goals.

Navy

January 2019 Navy Strategic Outlook for the Arctic

The Navy in recent years has issued a series of strategy and roadmap documents for the Arctic. The latest of these is a January 2019 strategic outlook for the Arctic document, which states

> The Navy will defend the United States from attack and preserve the Nation's strategic influence in the Arctic. Naval forces will operate to deter aggression and enable peaceful resolution of crises on terms acceptable to the U.S. and its allies and partners….
>
> In support of National and Department of Defense objectives for the Arctic, the Navy will pursue the following strategic objectives:
>
> - Defend U.S. sovereignty and the homeland from attack;
> - Ensure the Arctic remains a stable, conflict-free region;

[178] See, for example, Geoff Ziezulewicz, "Welcome to the Arctic: Degraded Radios, Poor Satellite Geometry and Sea Charts Dating Back to Capt. Cook," *Navy Times*, September 19, 2019.

[179] See, for example, Ben Werner, "Arleigh Burke Destroyers Are More Viable Option for Near-Term Navy Presence in Arctic," *USNI News*, September 18, 2019; Megan Eckstein, "CNO: Arctic Operations Limited Now, But Future Ship designs Should Consider Environment," *USNI News*, September 12, 2016.

- Preserve Freedom of the Seas; and
- Promote partnerships within the U.S. Government and with allies and partners to achieve the above objectives….

While there are recognized threats, opportunities, and risks in our return to an era of Great Power Competition, the Arctic is assessed to be low risk for conflict because nations have demonstrated the ability to resolve differences peacefully. Within the Arctic Ocean, nations exercise rights and freedoms consistent with international law, such as navigation and overflight, marine scientific research, the laying and maintenance of undersea pipelines and cables, and fishing. Nations respect the exclusive rights and jurisdiction of Arctic coastal states for exploitation of resources in the EEZ and continental shelf. It remains unlikely that any of the eight Arctic states would risk a large-scale conflict; however, the Navy will be prepared to deter conflict and protect our national interests….

Though the United States has not acceded to the United Nations Convention on the Law of the Sea (UNCLOS), the U.S. Navy has long treated its provisions related to traditional ocean uses as customary international law. The customary international law principles reflected in UNCLOS serve as the legal framework for rights and obligations in the Arctic Ocean including the delineation of the outer limits of the continental shelf, protection of the marine environment, freedom of navigation, military survey, and marine scientific research….

The Navy and U.S. Coast Guard (USCG) collaborate in close partnership during times of war and peace to protect our nation's ports and waterways and to promote U.S. maritime security interests overseas. The combined efforts of the Navy and the Coast Guard in the Arctic Ocean demonstrate a cooperative relationship that acknowledges the distinctive authorities, missions, competencies, and cultures of each service. The Coast Guard has statutory responsibilities, and fulfills most U.S. maritime missions in the Arctic….

The Navy will work with the Royal Canadian Navy to ensure common Arctic Region interests are addressed in a complementary manner. For decades, Canada and the United States have been partners in the defense of North America, cooperating within the framework of the North Atlantic Treaty Organization (NATO) and the North American Aerospace Defense Command (NORAD)….

As the Arctic Ocean continues to become more accessible, the importance of the Bering Strait as a strategic chokepoint will increase….

The Greenland, Iceland, United Kingdom-Norwegian (GIUK-N) Gap is a strategic corridor for naval operations in the high north….

The Northern Sea Route and the Northwest Passage are two recognized Arctic Ocean routes for connecting the Pacific to the Atlantic, and are ice covered the majority of the year. Seasonal ice cover varies along the length of the transit, though the exact location and duration remains difficult to predict. The Navy will consider implications of increasing maritime traffic [along these two routes] in future security planning….

Navy prioritization of leadership and resources on core functions in regions of likely conflict will continue to override competing interests to develop capabilities in peaceful regions such as the Arctic….

In alignment with higher-level guidance, and considering the context described above, the Navy strategic objectives for the Arctic are:

Defend U.S. sovereignty and the homeland from attack…. New opportunities and security challenges in the Arctic may emerge as conditions in the region change.… The Navy will ensure it remains prepared to operate in the Arctic to counter any threats to the homeland that may arise.

Ensure the Arctic remains a stable, conflict-free region…. While Arctic and non-Arctic states have thus far demonstrated a commitment to a stable and conflict-free region, Navy presence in cooperation with international partners will encourage all nations to act responsibly and in accordance with the rights and freedoms provided in international law.

Preserve Freedom of the Seas. Access to the global commons and freedom of the seas are national priorities. U.S. policy since 1983 provides that the U.S. will exercise and assert its navigation and overflight rights and freedoms on a worldwide basis in a manner consistent with the balance of U.S. interests as reflected in UNCLOS….

DoD's FON [Freedom of Navigation] Program employs every branch of military service, including the U.S. Coast Guard. In the Arctic, Navy submarines can conduct FONOPS, either undersea or by surfacing, and Navy surface combatants could conduct FONOPS in open water conditions during the summer melt season….

Promote partnerships within the U.S. Government and with allies and partners to achieve the above objectives. Today's great power competition requires both a renewed focus on the Navy's role as part of the Joint Force and a rededication to engagement. The Navy will engage [in the Arctic] with allies and partners, in order to protect the American homeland, maintain maritime superiority in key maritime regions, and promote U.S. interests globally.

The Navy will collaborate with interagency and international partners and allies to improve all-domain awareness, information sharing, and communications [in the Arctic] through participation in training, operations, and exercises. The Navy's submarine fleet has decades of experience performing missions and exercises under the sea ice….

The Navy continues to make investments in its global operational modeling capability to improve the ability to protect operational forces, installations, and equipment from hazardous conditions of the physical environment. The Navy collaborates with NOAA, DOE, NASA, NSF, and the U.S. Air Force to improve Arctic modeling and forecasting capabilities….

The Navy will continue to play a significant leadership role in the Arctic. Through its global reach capability and worldwide command and control, Navy leadership will support joint and interagency efforts, enhance information sharing, and promote interoperability….

The Arctic provides additional maneuver room for distributed nuclear and conventional forces to operate in support of deterrence….

The Navy will be ready to exercise Sea Control in order to defend United States' maritime access and interests in the Arctic….

The Arctic enables power projection from vectors [i.e., geographic points of origin] [that are] in addition to [those made possible by] traditional operating areas….

Maritime security in the Arctic includes defending sea lines of communication and the homeland from seaborne attacks, maritime domain awareness, and supporting Coast Guard operations as required….

The Navy will continue to evaluate the requirements and priorities associated with the rapid movement of personnel, materiel, and forces to and from or within the Arctic by sea to provide timely movement, positioning, and sustainment of the Joint Force across the range of military operations….

The Navy will provide support as required to search and rescue missions [in the Arctic] conducted and led by the Coast Guard and as directed in support of international partners….

The Navy will remain ready to support missions [in the Arctic] such as pollution response or natural disaster recovery; integrated planning efforts with local, state, federal, and native communities; and interoperability with the Coast Guard and international partners. Mitigating navigation hazards and chemical spills in the Arctic operational environment requires specialized equipment and tactics, techniques and procedures. …

The Navy will continually assess its preparedness and make informed decisions on Arctic operations and planning in response to changes in the Arctic operational environment or changes in its strategic context. As changes to threats, opportunities, and risks in the Arctic security environment may occur, they will be viewed in balance with the global nature of Great Power Competition.[180]

---

[180] U.S. Navy, *Strategic Outlook for the Arctic*, January 2019, pp. 4-13. For a press report about the strategy, see Melody Schreiber, "As Climate Changes, US Admirals See Increasing Need for Arctic Presence, *Arctic Today*, May 21, 2019.

Reestablishment of 2nd Fleet for North Atlantic and Arctic

In May 2018, the Navy announced that it would reestablish the 2nd Fleet, which was the Navy's fleet during the Cold War for countering Soviet naval forces in the North Atlantic. The fleet's formal reestablishment occurred in August 2018. The 2nd Fleet was created in 1950 and disestablished in September 2011. In its newly reestablished form, it is described as focusing on countering Russian naval forces not only in the North Atlantic but in the Arctic as well.[181]

Upcoming Freedom of Navigation (FON) Operation in Arctic

In January 2019, the Navy announced that "in coming months" it will send a Navy warship through Arctic waters on a freedom of navigation (FON) operation to assert U.S. navigational rights under international law in Arctic waters. The U.S. government's FON program was established in 1979 and annually includes multiple U.S. Navy FON operations conducted in various parts of the world.[182] The upcoming FON operation in the Arctic, however, will reportedly be the Navy's first ever FON operation in the Arctic.[183]

---

181 See, for example, Christopher Woody, "The US Navy's Newest Fleet Is Bulking Up for 'Leaner, Agile' Operations to Counter Russia in the Atlantic and the Arctic," *Business Insider*, January 18, 2019; Patricia Kime, "The Navy Isd gearing Up for 'Leaner, Agile' Operations in Arctic, North Atlantic," *Military.com*, January 16, 2019; Rich Abott, "2nd Fleet To Be Fully Operational in 2019, Sees Real Russian Threat," *Defense Daily*, December 3, 2018; Sam LaGrone, "U.S. 2nd Fleet Racing Toward a 2019 Operational Capability," USNI News, November 29, 2018; Paul McLeary, "New Second Fleet To Stay Lean, Unpredictable, Commander Says; & Watching China," *Breaking Defense*, November 29, 2018; Sam LaGrone, "CNO: New 2nd Fleet Boundary Will Extend North to the Edge of Russian Waters," *USNI News*, August 24, 2018.

182 For background information on the FON program, see the section entitled "Freedom of Navigation (FON) Program" in CRS Report R42784, *U.S.-China Strategic Competition in South and East China Seas: Background and Issues for Congress*.

183 See Ben Kesling, "Cold War Games: U.S. Is Preparing to Test the Waters in Icy Arctic," Wall Street Journal, January 11, 2019; Samuel Osborne, "US Navy to Sail Warship Through Arctic Waters in Show of Strength to Russia and China," *Independent (UK)*, January 14, 2019. See also Megan Eckstein "Navy May Deploy Surface Ships to Arctic This Summer as Shipping Lanes Open Up," *USNI News*, January 8, 2019.

Coast Guard

April 2019 Coast Guard Arctic Strategic Outlook

The Coast Guard, like the Navy, has released a series of Arctic-related studies and strategy documents in recent years. In April 2019, the Coast Guard released a new Arctic strategic outlook document as the successor to its previous 2013 Arctic strategy document. The April 2019 strategic outlook document states

> The United States is an Arctic Nation, and the United States Coast Guard has served as the lead federal agency for homeland security, safety, and environmental stewardship in the Arctic region for over 150 years. Since Revenue Cutters first sailed to Alaska in 1867 to establish U.S. sovereignty, the Service's role has expanded, including representing American interests as a leader in the international bodies governing navigation, search and rescue, vessel safety, fisheries enforcement, and pollution response across the entire Arctic. As the region continues to open and strategic competition drives more actors to look to the Arctic for economic and geopolitical advantages, the demand for Coast Guard leadership and presence will continue to grow.
>
> Since the release of the Coast Guard Arctic Strategy in 2013, the resurgence of nation-state competition has coincided with dramatic changes in the physical environment of the Arctic, which has elevated the region's prominence as a strategically competitive space. America's two nearest-peer powers, Russia and China, have both declared the region a national priority and made corresponding investments in capability and capacity to expand their influence in the region. Russia and China's persistent challenges to the rules-based international order around the globe cause concern of similar infringement to the continued peaceful stability of the Arctic region. As the only U.S. Service that combines both military and civil authorities, the Coast Guard is uniquely suited to address the interjurisdictional challenges of today's strategic environment by modeling acceptable behavior, building regional capacity, and strengthening organizations that foster transparency and good governance across the Arctic.
>
> The Arctic's role in geostrategic competition is growing, in large part because reductions in permanent sea ice have exposed coastal borders and facilitated increased human and economic activity. The warming of the Arctic has led to longer and larger windows of reduced ice conditions. From 2006 to 2018, satellite imagery observed the 12 lowest Arctic ice extents on record. This has led to greater access through Arctic shipping routes. While the near-term future of these routes is uncertain, a polar

route has the potential to reduce transit times of traditional shipping routes by up to two weeks. Russia's establishment of a Northern Sea Route Administration, along with the use of high ice-class Liquefied Natural Gas (LNG) tankers built specifically to export natural gas from its Yamal LNG facility, have contributed significantly to the increase in commercial shipping traffic in the Arctic. In addition, opportunities for potential resource extraction and expanding Arctic tourism offer new prospects for some of the Nation's most isolated communities and broader benefits to America. However, changing terrain and subsistence food patterns, as well as the impacts of increasingly frequent and intense winter storms, continue to challenge the communities and increase risk in the maritime domain.

The Coast Guard will adhere to the following principles as it manages these risks and seizes the opportunities created by these changes:

Partnership. The Arctic is an exceptional place that demands collaboration across national boundaries. The Coast Guard will partner with the Arctic Nations, as well as partners and allies with Arctic interests, to contribute to keeping the Arctic a conflict-free region. The Service will continue to dedicate resources to forums, such as the Arctic Council, and to combined operations and exercises to safeguard and secure the Arctic domain.

The unique and valuable relationship the Coast Guard has established with tribal entities builds mutual trust and improves mission capacity and readiness. We will continue to incorporate lessons-learned from engagements with Alaska Native communities, as well as industry and other Arctic residents, in the development and implementation of policy and strategy.

Unity of Effort. The Coast Guard will advance the Nation's strategic goals and priorities in the Arctic and exercise leadership across the Arctic community of federal, state, and local agencies. As a military Service, the Coast Guard will strengthen interoperability with the Department of Defense and complement the capabilities of the other military services to support the National Security Strategy and the National Military Strategy.

A Culture of Continuous Innovation. The Coast Guard cannot meet the challenges of tomorrow's Arctic with today's paradigms. Rapid technological advancements within the maritime industry, combined with robust investments by strategic competitors, have raised the stakes. The Service must take this opportunity to leverage transformative technology and lead the employment of innovative policies to solve complex problems.

While the strategic context has changed, Coast Guard missions in the Arctic are enduring. The Coast Guard will protect the Nation's vital interests by upholding the rules-based order in the maritime domain while cooperating to reduce conflict and risk. We will help safeguard the

Nation's Arctic communities, environment, and economy. The Service will pursue these ends through three complementary lines of effort:

Line of Effort 1: Enhance Capability to Operate Effectively in a Dynamic Arctic

In order to prosecute its missions in the Arctic, the Coast Guard must fully understand and operate freely in this vast and unforgiving environment. Effective capability requires sufficient heavy icebreaking vessels, reliable high-latitude communications, and comprehensive Maritime Domain Awareness. In order to respond to crises in the Arctic, our Nation must also muster adequate personnel, aviation, and logistics resources in the region. The Coast Guard is the sole provider and operator of the U.S. polar capable fleet but currently does not have the capability or capacity to assure access in the high latitudes. Closing the gap requires persistent investment in capabilities and capacity for polar operations, including the Polar Security Cutter. The Coast Guard will pursue this line of effort through three sub-objectives.

- Fill Gaps in the Coast Guard's Arctic Operational Capability and Capacity
- Establish Persistent Awareness and Understanding of the Arctic Domain
- Close the Critical Communications Gap in the Arctic

Line of Effort 2: Strengthen the Rules-Based Order

Actions by strategic competitors will challenge the long-standing norms that have made the Arctic an area of peace and low tension. The institutions contributing to a conflict-free Arctic will face new challenges requiring active and committed American leadership. The Coast Guard is uniquely positioned to provide this leadership in the maritime domain. The Coast Guard is dedicated to strengthening institutions–such as the Arctic Council, the Arctic Coast Guard Forum (ACGF), and the International Maritime Organization (IMO)– and partnerships which reinforce the rules-based order and foster transparency.

Rules and norms endure when nations demonstrate a commitment to upholding them. Working closely with allies and partners, the Coast Guard will deter threats to international maritime norms and America's national interests by conducting operations and exercises along the full spectrum of competition. Working in partnership with the Department of Defense, the Coast Guard will continue to support to the Nation's defense priorities in the Arctic. The Coast Guard will work closely with joint and international partners to build capability and demonstrate resolve in the Arctic.

The Coast Guard will pursue this line of effort through two sub-objectives.

- Strengthen Partnerships and Lead International Forums
- Counter Challenges to the International Rules-Based Order in the Maritime Domain

Line of Effort 3: Innovate and Adapt to Promote Resilience and Prosperity

The tyranny of distance and the harsh Arctic climate pose significant challenges to agencies charged with providing maritime safety and security to all Americans, including the hundreds of villages and thousands of seasonal workers in the U.S. Arctic. Search and rescue, law enforcement, marine safety, waterways management, and other Coast Guard missions are complicated by the Arctic's dynamic and remote operating environment. The Coast Guard will collaborate with stakeholders to develop new practices and technology to serve the maritime community and manage risk in the region.

As the Nation's maritime first responder, the Coast Guard will lead and participate in planning and exercises that include federal, state, tribal, local, international, non- governmental and industry partners to test preparedness and adaptability. During a crisis in the Arctic's maritime domain, the Service will lead an effective, unified response. The Coast Guard will pursue this line of effort through three sub-objectives.

- Support Regional Resilience and Lead in Crisis Response
- Address Emerging Demands in the Maritime Law Enforcement Mission
- Advance and Modernize the Arctic Marine Transportation System

Conclusion. Increased accessibility and activity will create more demand for Coast Guard services in the Arctic maritime domain. While long-term trends point to a more consistently navigable and competitive region, other environmental and economic factors make it difficult to predict the scope and pace of change. Near-term variability in the physical environment exposes mariners and communities to unpredictable levels of risk. As the region attracts increasing attention from both partner and competitor states, America's economic and security interests will become even more closely tied to the Arctic. Each development is significant on its own, but in combination, these trends create a new risk landscape for the Nation and the Coast Guard. This updated strategic outlook reflects a recognition of these realities and outlines the Service's lines of effort to succeed in the new Arctic.[184]

Polar Icebreaking[185]

Polar Icebreaker Operations

Within the U.S. government, the Coast Guard is the U.S. agency responsible for polar icebreaking. U.S. polar ice operations conducted in large part by the Coast Guard's polar icebreakers support 9 of the Coast Guard's 11 statutory missions.[186] The roles of U.S. polar icebreakers can be summarized as follows:

- conducting and supporting scientific research in the Arctic and Antarctic;
- defending U.S. sovereignty in the Arctic by helping to maintain a U.S. presence in U.S. territorial waters in the region;
- defending other U.S. interests in polar regions, including economic interests in waters that are within the U.S. exclusive economic zone (EEZ) north of Alaska;

---

[184] Coast Guard, *Arctic Strategic Outlook*, April 2019, Executive Summary (pp. 4-7). For press reports about the strategy, see, for example, Ragnhild Grønning, "Strategic Competition in Focus in US Coast Guard Arctic Strategy," *High North News*, May 2, 2019; Dermot Cole, "The Coast Guard's New Arctic Strategic Vision Downplays a Big Problem: Budget Constraints," *Arctic Today*, April 29, 2019; Kyle Rempfer, "Great Powers on Ice: Coast Guard Drops Arctic Manifesto," *Military Times*, April 25, 2019; Dan Lamothe, "New Coast Guard Strategy for the Arctic Calls for 'Projecting Sovereignty' to Contest China and Russia," *Washington Post*, April 22, 2019.

[185] This section prepared by Ronald O'Rourke, Specialist in Naval Affairs, Foreign Affairs, Defense, and Trade Division. For more on the Coast Guard's polar icebreakers, see CRS Report RL34391, *Coast Guard Polar Security Cutter (Polar Icebreaker) Program: Background and Issues for Congress*.

[186] The nine missions supported by polar ice operations are search and rescue; maritime safety; aids to navigation; ice operations; marine environmental protection; living marine resources; other law enforcement (protect the exclusive economic zone [EEZ]); ports, waterways and costal security; and defense readiness. The two missions not supported by polar ice operations are illegal drug interdiction and undocumented migrant interdiction. (Department of Homeland Security, *Polar Icebreaking Recapitalization Project Mission Need Statement, Version 1.0*, approved by DHS June 28, 2013, p. 10).

- monitoring sea traffic in the Arctic, including ships bound for the United States; and
- conducting other typical Coast Guard missions (such as search and rescue, law enforcement, and protection of marine resources) in Arctic waters, including U.S. territorial waters north of Alaska.[187]

The Coast Guard's large icebreakers are called polar icebreakers rather than Arctic icebreakers because they perform missions in both the Arctic and Antarctic. Operations to support National Science Foundation (NSF) research activities in both polar regions account for a significant portion of U.S. polar icebreaker operations.

Providing support for NSF's research in the Antarctic focuses on performing an annual mission, called Operation Deep Freeze (ODF), to break through Antarctic sea ice so as to reach and resupply McMurdo Station, the large U.S. Antarctic research station located on the shore of McMurdo Sound, near the Ross Ice Shelf. The Coast Guard states that *Polar Star*, the Coast Guard's only currently operational heavy polar icebreaker, "spends the [northern hemisphere] winter [i.e., the southern hemisphere summer] breaking ice near Antarctica in order to refuel and resupply McMurdo Station. When the mission is complete, the *Polar Star* returns to dry dock [in Seattle] in order to complete critical maintenance and prepare it for the next ODF mission. Once out of dry dock, it's back to Antarctica, and the cycle repeats itself."[188] In terms of the maximum thickness of the ice to be broken, the annual McMurdo resupply mission generally poses the greatest icebreaking challenge for U.S. polar icebreakers, though Arctic ice can frequently pose its own significant icebreaking challenges for U.S. polar icebreakers. The Coast Guard's

---

[187] This passage, beginning with "The roles of …," originated in an earlier iteration of this CRS report and was later transferred by the Government Accountability Office (GAO) with minor changes to Government Accountability Office, *Coast Guard[:]Efforts to Identify Arctic Requirements Are Ongoing, but More Communication about Agency Planning Efforts Would Be Beneficial*, GAO-10-870, September 2010, p. 53.

[188] NyxoLyno Cangemi, "Coast Guard Icebreaker Crew Completes Second Arctic Mission; U.S. Interests in Arctic Domain Depends [sic] on Fleet Recapitalization," *DVIDS (Defense Visual Information Distribution System)*, October 19, 2018.

medium polar icebreaker, *Healy*, spends most of its operational time in the Arctic supporting NSF research activities and performing other operations.

Although polar ice is diminishing due to climate change, observers generally expect that this development will not eliminate the need for U.S. polar icebreakers, and in some respects might increase mission demands for them. Even with the diminishment of polar ice, there are still significant ice-covered areas in the polar regions, and diminishment of polar ice could lead in coming years to increased commercial cargo ship, cruise ship, research ship, and naval surface ship operations, as well as increased exploration for oil and other resources, in the Arctic— activities that could require increased levels of support from polar icebreakers, particularly since waters described as "ice free" can actually still have some amount of ice.[189] Changing ice conditions in Antarctic waters have made the McMurdo resupply mission more challenging since 2000.[190]

Current Polar Icebreaker Fleet

The operational U.S. polar icebreaking fleet currently consists of one heavy polar icebreaker, *Polar Star*, and one medium polar icebreaker, *Healy*. In addition to *Polar Star*, the Coast Guard has a second heavy polar icebreaker, *Polar Sea*. *Polar Sea*, however, suffered an engine casualty in June 2010 and has been nonoperational since then. *Polar Star* and *Polar Sea* entered service in 1976 and 1978, respectively, and are now well beyond their originally intended 30-year service lives. The Coast Guard has used *Polar Sea* as a source of spare parts for keeping *Polar Star* operational.

Polar Security Cutter (PSC) Program

A Department of Homeland Security (DHS) Mission Need Statement (MNS) approved in June 2013 states that "current requirements and future projections ... indicate the Coast Guard will need to expand its icebreaking

---

[189] For more on changes in the Arctic due to diminishment of Arctic ice, see CRS Report R41153, *Changes in the Arctic: Background and Issues for Congress*.

[190] National Research Council, *Polar Icebreakers in a Changing World, An Assessment of U.S. Needs*, Washington, 2007, pp. 6-7, 14, 63.

capacity, potentially requiring a fleet of up to six icebreakers (3 heavy and 3 medium) to adequately meet mission demands in the high latitudes"[191]

The Coast Guard in its FY2013 budget initiated a program to acquire three new heavy polar icebreakers, to be followed years from now by the acquisition of up to three new medium polar icebreakers. The program was originally referred to as the polar icebreaker program but is now referred to as the Polar Security Cutter (PSC) program.

The Coast Guard estimates the total procurement costs of the new three heavy PSCs as $1,039 million (i.e., about $1.0 billion) for the first ship, $792 million for the second ship, and $788 million for the third ship, for a combined estimated cost of $2,619 million (i.e., about $2.6 billion). The first ship will cost more than the other two because it will incorporate design costs for the class and be at the start of the production learning curve for the class.

The PSC program has received a total of $1,034.6 million (i.e., about $1.0 billion) in procurement funding through FY2019, including $300 million provided through the Navy's shipbuilding account in FY2017 and FY2018. The Coast Guard's proposed FY2020 budget requests $35 million in procurement funding for the PSC program, which is enough to cover the PSC program's FY2020 government program-management costs. The Coast Guard's FY2019 budget submission had projected that a total of $125 million in procurement funding would be requested for the PSC program in FY2020.

On April 23, 2019, the Coast Guard-Navy Integrated Program Office for the PSC program awarded a $745.9 million fixed-price, incentive-firm contract for the detail design and construction (DD&C) of the first PSC to VT Halter Marine of Pascagoula, MS, a shipyard owned by Singapore Technologies (ST) Engineering. VT Halter was the leader of one of three industry teams that competed for the DD&C contract. The first PSC is

191 Department of Homeland Security, *Polar Icebreaking Recapitalization Project Mission Need Statement, Version 1.0*, approved by DHS June 28, 2013, p. 9.

scheduled to begin construction in 2021 and be delivered in 2024, though the DD&C contract includes financial incentives for earlier delivery.[192]

SEARCH AND RESCUE (SAR)[193]

Overview

Increasing sea and air traffic through Arctic waters has increased concerns regarding Arctic-area search and rescue (SAR) capabilities.[194] Table 1 presents figures on ship casualties in Arctic Circle waters from 2005 to 2017. As shown in the table, the number of ship casualties in Arctic waters has grown in recent years to an annual total of about 50 to 70, most of which are caused by damage to or failure of ship machinery or the stranding or wrecking or ships. Given the location of current U.S. Coast Guard operating bases, it could take Coast Guard aircraft several hours, and Coast Guard cutters days or even weeks, to reach a ship in distress or a downed aircraft in Arctic waters. The Coast Guard states that "the closest Coast Guard Air Station to the Arctic is located in Kodiak, AK, approximately 820 nautical miles south of Utqiagvik, AK, which is nearly the same distance as from Boston, MA, to Miami, FL."[195] In addition to such long distances, the harsh climate complicates SAR operations in the region.

---

[192] For more on the polar icebreaker program, see CRS Report RL34391, *Coast Guard Polar Security Cutter (Polar Icebreaker) Program: Background and Issues for Congress*.

[193] This section prepared by Ronald O'Rourke, Specialist in Naval Affairs, Foreign Affairs, Defense, and Trade Division.

[194] See, for example, Romain Chuffart, "French and Danish Navies Hold Joint SAR Exercises in Greenland's Waters," *High North News*, September 6, 2019; "Arctic Search and Rescue May Face Challenges," *Cruise Industry News*, June 27, 2019; Robbie Gramer, "Stretched Thin on Thin Ice; With the Arctic Melting and Norther Coast Guards Struggling to Keep Up, the Next Disaster Is a Matter of When, Not If," *Foreign Policy*, September 12, 2018; Edward Struzik, "As Ice Recedes, the Arctic Isn't Prepared for More Shipping Traffic," phys.org, September 5, 2018; Derek Minemyer, "Arctic Council Members Say Alaska Needs Search and Rescue in the Arctic, Now," *KTUU*, August 16, 2018; Timothy William James Smith, *Search and Rescue in the Arctic: Is the U.S. Prepared?* RAND Corporation, 2017, 148 pp. (Dissertation report RGSD-382.)

[195] Coast Guard, *Arctic Strategic Outlook*, April 2019, p. 11.

Table 1. Ship Casualties in Arctic Circle Waters, 2005-2014 (Ships of 100 gross tons or more)

	2005	2006	2007	2008	2009	2010	2011	2012	2013	2014	2015	2016	2017
Machinery damage/ failure	2	3	5	13	14	16	12	13	20	27	46	32	46
Wrecked/ stranded	1	4	10	11	14	9	9	8	10	14	6	11	9
Miscella-neous	0	0	5	1	4	4	2	6	5	5	6	4	6
Collision	0	0	0	1	4	10	4	4	2	0	3	2	4
Fire/ explosion	0	0	3	1	2	6	6	1	4	2	4	1	3
Contact (e.g., harbor wall)	0	0	1	1	2	4	1	3	6	4	5	1	1
Hull damage	0	1	3	1	6	2	2	1	2	1	1	2	2
Foundered (i.e., sunk or submerged)	0	0	1	1	2	0	3	1	1	2	0	1	0
Labor dispute	0	0	0	0	0	0	0	0	0	0	0	1	0
Total	3	8	28	30	48	51	39	37	50	55	71	55	71

Sources: For 2005-2007: Allianz Global Corporate & Specialty, Safety and Shipping Review 2015, p. 28. (Table entitled "Arctic Circle Waters—All Casualties including Total Losses 2005–2014."). For 2008-2017: Allianz Global Corporate & Specialty, Safety and Shipping Review 2018, p. 29. (Table entitled "Arctic Circle Waters— Causes of Casualties (Shipping Incidents) 2008-2017.") The two tables include similar source notes; the one for the second source states "Source: Lloyd's List Intelligence Casualty Statistics; Data Analysis & Graphic: Allianz Global Corporate & Specialty.")

Particular concern has been expressed about cruise ships carrying large numbers of civilian passengers that may experience problems and need assistance.[196] There have already been incidents of this kind with cruise ships in recent years in waters off Antarctica, and a Russian-flagged

[196] See, for example, Jane George, "Exercise Held to Prepare for Arctic Cruise Ship Mishap," *Nunatsiaq News*, April 15, 2019; Brian Castner, "How to Rescue a Cruise Ship in the Northwest Passage," *Motherboard (Vice)*, October 24, 2017; Henry Fountain, "With More Ships in the Arctic, Fears of Disaster Rise," *New York Times*, July 23, 2017; Gwladys Fouche, "Uncharted Waters: Mega-Cruise Ships Sail the Arctic," *Reuters*, October 10, 2016; Abbie Tingstad and Timothy Smith, "Being Safer in the Arctic," *National Interest*, October 3, 2016.

passenger ship with 162 people on board ran aground on Canada's Northwest Passage on August 24, 2018.[197] Coast Guard officials have noted the long times that would be needed to respond to potential emergency situations in certain parts the Arctic. The Coast Guard is participating in exercises focused on improving Arctic SAR capabilities.[198]

Increasing U.S. Coast Guard SAR capabilities for the Arctic could require one or more of the following: enhancing or creating new Coast Guard operating bases in the region; procuring additional Arctic-capable aircraft, cutters, and rescue boats for the Coast Guard; and adding systems to improve Arctic maritime communications, navigation, and domain awareness.[199] It may also entail enhanced forms of cooperation with navies and coast guards of other Arctic countries.

A 2017 survey of Arctic SAR capabilities conducted as part of the Finnish Border Guard's Arctic Maritime Safety Cooperation project in cooperation with the Arctic Coast Guard Forum stated the following:

> The key challenges for Arctic search and rescue identified in this survey include long distances, severe weather, ice and cold conditions, a poor communications network, lack of infrastructure and lack of resource presence in the region. In addition, the capacity to host patients, achieving situational awareness, and unsuitable evacuation and survival equipment pose major challenges for maritime safety and SAR in the Arctic.
>
> The Arctic SAR authorities have recognized a need to further develop advanced information sharing between coast guards, emergency authorities, and other stakeholders involved in SAR operations. In addition, joint training and systematic sharing of lessons learned, as well as technological innovation in communications networks and

---

[197] Malte Humpert, "A Cruise Ship Runs Aground in Canada's Arctic Waters; The Akademik Ioffe's Sister Ship Was Nearby, and Together with Canadian Coast guard Ships, Was Able to Rescue All Passengers," *Arctic Today*, August 28, 2018.

[198] See, for example, Av Ragnhild Grønning, "Mass Exercise Tests Emergency Response in the Arctic," *High North News*, August 219, 2016; Julia Bergman, "Mass Rescue Exercise Tests Response Capabilities in Arctic," *New London Day*, August 24, 2016; Wendy Laursen, "Mass Arctic Rescue Exercise Begins," *Maritime Executive*, August 24, 2016; Tim Ellis, "Cruise-Ship Evacuation Exercise Begins as Luxury Liner Prepares for Arctic Ocean Transit," *KUAC (alaskapublic.org)*, August 23, 2016; Bill Colclough, "USCG to Participate with AK Command in Arctic Chinook Exercises," *Alaska Native News*, August 23, 2016.

[199] For a report assessing certain emergency scenarios in the Arctic, including search and rescue scenarios, see *Opening the Arctic Seas, Envisioning Disasters and Framing Solutions*, Coastal Response and Research Center, University of New Hampshire, report of January 2009, based on conference held March 18-20, 2008, at Durham, NH.

connections, navigation, survival and rescue equipment, and healthcare services are being called for in order to improve SAR capabilities in the Arctic.

The survey recommends enhancing practical cooperation between various stakeholders involved in Arctic SAR such as coast guards, rescue centers, other authorities, industry groups, private operators, academia and volunteer organizations. It encourages further information sharing on infrastructure projects and resource assets, Automatic Identification System and weather data, emergency plans and standard operating procedures, as well as exercises and lessons learned via a common database. Furthermore, developing joint courses specifically intended for Arctic SAR and establishing a working group that examines new innovations and technological developments, are recommended as potential initiatives for improving practical international cooperation.[200]

May 2011 Arctic Council Agreement on Arctic SAR

On May 12, 2011, representatives from the member states of the Arctic Council, meeting in Nuuk, Greenland, signed an agreement on cooperation on aeronautical and maritime SAR in the Arctic. Key features of the agreement include the following:

- Article 3 and the associated Annex to the agreement essentially divide the Arctic into SAR areas within which each party has primary responsibility for conducting SAR operations, stating that "the delimitation of search and rescue regions is not related to and shall not prejudice the delimitation of any boundary between States or their sovereignty, sovereign rights or jurisdiction," and that "each Party shall promote the establishment, operation and maintenance of an adequate and effective search and rescue capability within its area."
- Article 4 and the associated Appendix I to the agreement identify the competent authority for each party. For the United States, the competent authority is the Coast Guard.

---

200 Emmi Ikonen, *Arctic Search and Rescue Capabilities Survey: Enhancing International Cooperation 2017*, Finnish Border Guard, Arctic Maritime Safety Cooperation (SARC) project, August 2017, p. iv. See also Finnish Red Cross, *Red Cross Arctic Disaster Management Study*, August 2018, 71 pp.

- Article 5 and the associated Appendix II to the agreement identify the agencies responsible for aeronautical and maritime SAR for each party. For the United States, those agencies are the Coast Guard and the Department of Defense.
- Article 6 and the associated Appendix III to the agreement identify the aeronautical and/or maritime rescue coordination centers (RCCs) for each party. For the United States, the RCCs are Joint Rescue Coordination Center Juneau (JRCC Juneau) and Aviation Rescue Coordination Center Elmendorf (ARCC Elmendorf).
- Article 12 states that "unless otherwise agreed, each Party shall bear its own costs deriving from its implementation of this Agreement," and that "implementation of this Agreement shall be subject to the availability of relevant resources."[201]

Figure 4 shows an illustrative map of the national areas of SAR responsibility based on the geographic coordinates listed in the Annex to the agreement.

An October 12, 2015, press report states the following:

> More people are wishing to explore icy environments, says Peter Hellberg, manager responsible for the SAR process at the Swedish Maritime Administration. Hellberg is part of an IMO/International Civil Aviation Organization (ICAO) working group that is re- evaluating search and rescue (SAR) operations in Polar waters as a result of this push.
>
> The working group includes both a maritime and aeronautical perspective, and it has identified a need for more detailed guidance for SAR organizations which will be achieved through an update of the International Aeronautical and Maritime Search and Rescue Manual (IAMSAR) planned for 2019.
>
> While the IAMSAR manual is not mandatory, it is followed by most SAR organizations around the world. It provides the framework for

---

[201] Source: Text of final version of agreement made ready for signing and dated April 21, 2011, https://oaarchive.arctic- council.org/bitstream/handle/11374/531/EDOCS-1910-v1-ACMMDK07_Nuuk_2011_Arctic_SAR_Agreement_unsigned_EN.PDF?sequence=8&isAllowed=y. For a State Department fact sheet on the agreement, see "Secretary Clinton Signs the Arctic Search and Rescue Agreement with Other Arctic Nations," May 12, 2011, accessed October 2, 2015, at https://2009-2017.state.gov/r/pa/prs/ps/2011/05/ 163285.htm.

setting up a multi-national SAR, giving different parties guidance on the necessary arrangements for Arctic areas.

The guidance will be expanded on based on the Polar Code and other recent IMO regulatory updates, and from an aeronautical perspective, from lessons learned after the disappearance of Malaysian Airlines' MH370.[202]

Source: "Arctic Search and Rescue Agreement," accessed July 7, 2011, at http://www.arcticportal.org/features/features-of-2011/arctic-search-and-rescue-agreement.

Figure 4. Illustrative Map of Arctic SAR Areas in Arctic SAR Agreement, (Based on geographic coordinates listed in the agreement).

202 Wendy Laursen, "MH370, Polar Code Spark SAR Rethink," *Maritime Executive*, October 12, 2015.

COMMERCIAL SEA TRANSPORTATION[203]

Background

The search for a shorter route from the Atlantic to Asia has been the quest of maritime powers since the Middle Ages. The melting of Arctic ice raises the possibility of saving several thousands of miles and several days of sailing between major trading blocs.[204] If the Arctic were to become a viable shipping route, the ramifications could extend far beyond the Arctic. For example, lower shipping costs could be advantageous for China (at least its northeast region), Japan, and South Korea because their manufactured products exported to Europe or North America could become less expensive relative to other emerging manufacturing centers in Southeast Asia, such as India.[205] Melting ice could potentially open up two trans-Arctic routes (see Figure 3):[206]

- The Northern Sea Route (NSR, a.k.a. the "Northeast Passage"), along Russia's northern border from Murmansk to Provideniya, is about 2,600 nautical miles in length. It was opened by the Soviet Union to domestic shipping in 1931 and to transit by foreign vessels in 1991. This route would be applicable for trade between northeast Asia (north of Singapore) and northern Europe. In recent summers, less than a handful of large, non-Russian-flagged cargo ships have transited the NSR.[207] Russia reportedly seeks to reserve carriage of oil and gas extracted along the NSR to Russian-flagged ships.[208]

---

[203] This section prepared by John Frittelli, Specialist in Transportation Policy, Resources, Science, and Industry Division.

[204] Extended daylight hours in the Arctic during the summer may also be an advantage.

[205] Presentation by Stephen Carmel, Senior Vice President, Maersk Line Ltd., Halifax International Security Forum, *Arctic Security: The New Great Game?* November 21, 2009, available at http://fora.tv/.

[206] A third but more remote possibility is a route directly over the North Pole.

[207] Northern Sea Route Information Office; http://www.arctic-lio.com/.

[208] The Independent Barents Observer, "Icebreaking Tankers Have Course for Russia's New Arctic LNG Plant, And They All Carry Foreign Flags," December 11, 2017;

- The Northwest Passage (NWP) runs through the Canadian Arctic Islands. The NWP actually consists of several potential routes. The southern route is through Peel Sound in Nunavut, which has been open in recent summers and contains mostly one-year ice. However, this route is circuitous, contains some narrow channels, and is shallow enough to impose draft restrictions on ships. The more northern route, through McClure Strait from Baffin Bay to the Beaufort Sea north of Alaska, is much more direct and therefore more appealing to ocean carriers, but more prone to ice blockage.[209] The NWP is potentially applicable for trade between northeast Asia (north of Shanghai) and the northeast of North America, but it is less commercially viable than the NSR.[210] Cargo ship transits have been extremely rare but cruise vessel excursions and research vessels are more common.

Destination Traffic, Not Trans-Arctic Traffic

Most cargo ship activity currently taking place in the Arctic is to transport natural resources from the Arctic or to deliver general cargo and supplies to communities and natural resource extraction facilities. Thus, cargo ship traffic in the Arctic presently is mostly regional, not trans-Arctic.

While there has been a recent uptick in Arctic shipping activity, this activity has more to do with a spike in commodity prices than it does with the melting of Arctic ice. Even so, fewer ships ply the Arctic seas now than in the past. The NSR continues to account for the bulk of Arctic shipping activity.

---

https://thebarentsobserver.com/en/industry-and-energy/2017/12/icebreaking-tankers-have-course-russias-new-arctic-lng-plant-and-they.

209 This was the route pioneered by the *SS Manhattan*, an oil tanker modified for ice breaking in 1969 to carry Alaskan North Slope oil to the Atlantic. This was the first commercial passage through the NWP, but the building of the Alaskan pipeline was found to be the more economical means of transporting oil from the North Slope to the lower 48 states.

210 Although the NWP is often compared to the alternative route through the Panama Canal in terms of distance and sailing days from Asia to the U.S. east coast, another alternative to consider is the shorter and faster transcontinental rail route across Canada or the United States.

Unpredictable Ice Conditions Hinder Trans-Arctic Shipping

Arctic waters do not necessarily have to be ice free to be open to shipping. Multiyear ice can be over 10 feet thick and problematic even for icebreakers, but one-year ice is typically 3 feet thick or less. This thinner ice can be more readily broken up by icebreakers or ice-class ships (cargo ships with reinforced hulls and other features for navigating in ice-infested waters). However, more open water in the Arctic has resulted in another potential obstacle to shipping: unpredictable ice flows. In the NWP, melting ice and the opening of waters that were once covered with one-year ice has allowed blocks of multiyear ice from farther north and icebergs from Greenland to flow into potential sea lanes. The source of this multiyear ice is not predicted to dissipate in spite of climate change. Moreover, the flow patterns of these ice blocks are very difficult to forecast.[211] Thus, the lack of ice in potential sea lanes during the summer months can add even greater unpredictability to Arctic shipping. This is in addition to the extent of ice versus open water, which is also highly variable from one year to the next and seasonally.

The unpredictability of ice conditions is a major hindrance for trans-Arctic shipping in general, but can be more of a concern for some types of ships than it is for others. For instance, it would be less of a concern for cruise ships, which may have the objective of merely visiting the Arctic rather than passing through and could change their route and itinerary depending on ice conditions. On the other hand, unpredictability is of the utmost concern for container ships that carry thousands of containers from hundreds of different customers, all of whom expect to unload or load their cargo upon the ship's arrival at various ports as indicated on the ship's advertised schedule. The presence of even small blocks of ice or icebergs from a melting Greenland ice sheet requires slow sailing and could play havoc with schedules. Ships carrying a single commodity in bulk from one port to another for just one customer have more flexibility in terms of

---

[211] S.E.L. Howell and J.J. Yackel, "A Vessel Transit Assessment of Sea Ice Variability in the Western Arctic, 1969- 2002: Implications for Ship Navigation," *Canadian Journal of Remote Sensing*, vol. 30, no. 2, 2004.

delivery windows, but would not likely risk an Arctic passage under prevailing conditions.

Ice is not the sole impediment to Arctic shipping. The region frequently experiences adverse weather, including not only severe storms, but also intense cold, which can impair deck machinery. During the summer months when sea lanes are open, heavy fog is common in the Arctic.

Commercial ships would face higher operating costs on Arctic routes than elsewhere. Ship size is an important factor in reducing freight costs. Many ships currently used in other waters would require two icebreakers to break a path wide enough for them to sail through; ship owners could reduce that cost by using smaller vessels in the Arctic, but this would raise the cost per container or per ton of freight.[212] Also, icebreakers or ice-class cargo vessels burn more fuel than ships designed for more temperate waters and would have to sail at slower speeds. The shipping season in the Arctic only lasts for a few weeks, so icebreakers and other special required equipment would sit idle the remainder of the year. None of these impediments by themselves may be enough to discourage Arctic passage but they do raise costs, perhaps enough to negate the savings of a shorter route. Thus, from the perspective of a shipper or a ship owner, shorter via the Arctic does not necessarily mean cheaper and faster.[213]

Basic Navigation Infrastructure Is Lacking

Considerable investment in navigation-related infrastructure would be required if trans-Arctic shipping were to become a reality. Channel marking buoys and other floating visual aids are not possible in Arctic waters because moving ice sheets will continuously shift their positions.

Therefore, vessel captains would need to rely on marine surveys and ice charts. For some areas in the Arctic, however, these surveys and charts

---

[212] "Arctic Unlikely to See Major Shipping Growth," *New Zealand Transport and Logistics Business Week*, April 24, 2008.

[213] Stephen M. Carmel, Senior Vice President, Maersk Line Ltd., "The Cold, Hard Realities of Arctic Shipping," *United States Naval Institute, Proceedings*; July 2013, pp. 38-41.

are out of date or not sufficiently accurate.[214] To remedy this problem, aviation reconnaissance of ice conditions and satellite images would need to become readily available for ship operators.[215] Ship-to-shore communication infrastructure would need to be installed where possible. Refueling stations may be needed, as well as, perhaps, transshipment ports where cargo could be transferred to and from ice-capable vessels at both ends of Arctic routes. Shipping lines would need to develop a larger pool of mariners with ice navigation experience. Marine insurers would need to calculate the proper level of risk premium for polar routes, which would require more detailed information about Arctic accidents and incidents in the past.

The U.S. Army Corps of Engineers, along with the state of Alaska, has studied the feasibility of a "deep-draft" port in the Arctic (accommodating ships with a draft of up to 35 feet). The northern and northwestern coastlines of Alaska are exceptionally shallow, generally limiting harbor and near-shore traffic to shallow-draft barges. Coast Guard cutters and icebreakers have drafts of 35 to 40 feet while NOAA research vessels have drafts of 16 to 28 feet, so at present these vessels are based outside the Arctic and must sail considerable distances to reach Arctic duty stations.

Supply vessels supporting offshore oil rigs typically have drafts over 20 feet. A deep-draft port could serve as a base of operations for larger vessels, facilitating commercial maritime traffic in the Arctic.[216] The study concluded that the existing harbors of Nome or Port Clarence on Alaska's west coast may be the most suitable for deepening because of their proximity to the Bering Strait and deeper water.[217] However, at a July 2016

---

[214] In July and August 2010, NOAA surveyed the Bering Straits area in order to update its charts but stated that it will take more than 25 years to map the prioritized areas of navigational significance in U.S. Arctic waters. See http://www.noaanews.noaa.gov/stories2010/20100720_fairweather.html.

[215] Ice reporting that currently exists is intended for scientists not mariners.

[216] For further information, see http://www.poa.usace.army.mil/en/cw/AKPortsStudy.htm, and FY2013 USACE Budget Justification, p. POD-5.

[217] *Alaska Deep-Draft Arctic Port System Study,* March 2013; http://www.poa.usace.army.mil/Library/ReportsandStudies/AlaskaRegionalPortsStudy.aspx. The navigation channel at Nome presently ranges from 10 to 20 feet in depth. Much of the harbor at Port Clarence has a natural depth of 35 to 40 feet; http://www.charts.noaa.gov/ OnLineViewer/AlaskaViewer Table.shtml.

hearing, the Coast Guard indicated its preferred strategy was to rely on mobile assets (vessels and aircraft) and seasonal bases of operation rather than pursue a permanent port in the Arctic.[218]

The U.S. Committee on the Marine Transportation System, a Cabinet-level committee of federal agencies with responsibilities for marine transportation, identified a list of infrastructure improvements for Arctic navigation in a 2013 report.[219] The report prioritizes improvements to information infrastructure (weather forecasting, nautical charting, ship tracking) and emergency response capabilities for ships in distress.

Regulation of Arctic Shipping

Due to the international nature of the shipping industry, maritime trading nations have adopted international treaties that establish standards for ocean carriers in terms of safety, pollution prevention, and security. These standards are agreed upon by shipping nations through the International Maritime Organization (IMO), a United Nations agency that first met in 1959.[220]

Key conventions that the 168 IMO member nations have adopted include the Safety of Life at Sea Convention (SOLAS), which was originally adopted in response to the *Titanic* disaster in 1912 but has since been revised several times; the Prevention of Pollution from Ships (MARPOL), which was adopted in 1973 and modified in 1978; and the Standards for Training, Certification, and Watchkeeping for Seafarers (SCTW), which was adopted in 1978 and amended in 1995. It is up to ratifying nations to enforce these standards. The United States is a party to these conventions, and the U.S. Coast Guard enforces them when it boards and inspects ships and crews arriving at U.S. ports and the very few ships engaged in international trade that sail under the U.S. flag.

---

[218] Oral testimony of Admiral Charles D. Michel, Coast Guard Vice Commandant, House Committee on Transportation and Infrastructure, Subcommittee on Coast Guard and Maritime Transportation, *Coast Guard Arctic Implementation Capabilities*, July 12, 2016.

[219] U.S. Committee on the Marine Transportation System, *U.S. Arctic Marine Transportation System: Overview and Priorities for Action*, 2013; http://www.cmts.gov/downloads/CMTS_Arctic_MTS_Report_Narrative.pdf.

[220] See http://www.imo.org/ for more information.

Like the United States, most of the other major maritime trading nations lack the ability to enforce these regulations as a "flag state" because much of the world's merchant fleet is registered under so-called "flags of convenience." While most ship owners and operators are headquartered in major economies, they often register their ships in Panama, Liberia, the Bahamas, the Marshall Islands, Malta, and Cyprus, among other "open registries," because these nations offer more attractive tax and employment regulatory regimes. Because of this development, most maritime trading nations enforce shipping regulations under a "port state control" regime—that is, they require compliance with these regulations as a condition of calling at their ports. The fragmented nature of ship ownership and operation can be a further hurdle to regulatory enforcement. It is common for cargo ships to be owned by one company, operated by a second company (which markets the ship's space), and managed by a third (which may supply the crew and other services a ship requires to sail), each of which could be headquartered in different countries.

New Arctic Polar Code

While SOLAS and other IMO conventions include provisions regarding the operation of ships in ice-infested waters, they were not specific to the polar regions. To supplement these requirements, a new IMO polar code went into effect on January 1, 2017.[221] The code applies to passenger and cargo ships of 500 gross tons or more engaged in international voyages. It does not apply to fishing vessels, military vessels, pleasure yachts, or smaller cargo ships. The polar requirements are intended to improve safety and prevent pollution in the Arctic, and they include provisions on ship construction, ship equipment related to navigation, and crew training and ship operation. The code requires ships to carry fully or partially enclosed lifeboats. The code requires that the crew have training in ice navigation. Nations can enforce additional requirements on ships arriving at their ports or sailing through their coastal

[221] http://www.imo.org/en/MediaCentre/HotTopics/polar/Pages/default.aspx.

waters. For instance, U.S. Coast Guard regulations largely follow IMO conventions but mandate additional requirements in some areas. U.S. coastal states can require ships calling at their ports to take additional safety and pollution prevention safeguards.[222] Canada and Russia have additional pollution regulations for Arctic waters exceeding MARPOL. The U.S. Coast Guard has studied and has recommended a specific vessel traffic separation scheme for the Bering Strait between Alaska and Russia, which experiences over 400 transits per year.[223] The U.S. Coast Guard is seeking IMO approval of this routing scheme.

Oil, Gas, and Mineral Exploration[224]

Decreases in summer polar ice may alter options for oil, gas, and mineral exploration in Arctic offshore and onshore areas. Offshore of Alaska, the U.S. outer continental shelf (OCS) covers more than 1 billion acres,[225] including some areas with high oil and gas potential. Even with warmer temperatures, exploration and development in the Arctic are still subject to harsh conditions, especially in winter. This makes it costly and challenging to develop the infrastructure necessary to produce, store, and transport oil, gas, and minerals from newly discovered deposits. Severe weather poses challenges to several ongoing offshore operations as well as to new exploration.

Offshore oil and gas exploration is affected by efforts to map the margins of the U.S. OCS. Shrinking sea ice cover in the Arctic has intensified interest in surveying and mapping the continental margins of multiple countries with lands in the Arctic. Delineating the extent of the

---

[222] For example, see Alaska State Legislature, HJR 19, Arctic Marine Safety Agreements; http://www.akleg.gov/basis/ Bill/Detail/30?Root=HJR%2019.

[223] 82 *Federal Register* 11935, February 27, 2017.

[224] This section prepared by Laura Comay, Analyst in Natural Resources Policy, Resources, Science, and Industry Division; Peter Folger, Section Research Manager, Resources, Science, and Industry Division; and Marc Humphries, Analyst in Energy Policy, Resources, Science, and Industry Division.

[225] This region includes some areas within the Arctic boundary as defined by the ARPA (15 U.S.C. 4111; see Figure 1), such as the Beaufort and Chukchi Seas, and some areas outside that boundary, such as Cook Inlet.

continental margins beyond the 200 nautical mile Exclusive Economic Zone (EEZ) could lead to consideration of development on substantial amounts of submerged lands. Mapping projects are underway, by individual countries and through cooperative government studies, to support submissions to the Commission on the Limits of the Continental Shelf, including for areas that may contain large amounts of oil, natural gas, methane hydrates, or minerals.

With respect to onshore development, shrinking glaciers could expose land containing economic deposits of gold, iron ore, or other minerals previously covered by glacial ice. At the same time, warming that causes permafrost to melt could pose challenges to oil, gas, and mineral activities because ground structures, such as pipelines and other infrastructure that depend on footings sunk into the permafrost for support, could be compromised. In addition, warmer temperatures shorten the ice road transport seasons for oil, gas, and mineral development, creating transportation challenges.

Offshore Oil and Gas Exploration

The shrinking Arctic ice cap, or conversely, the growing amount of ice-free ocean in the summertime, has increased interest in exploring for offshore oil and gas in the Arctic. Reduced sea ice in the summer means that ships towing seismic arrays[226] can explore regions of the Arctic Ocean, Chukchi Sea, Beaufort Sea, and other offshore regions for longer periods of time with less risk of colliding with floating sea ice. Less sea ice over longer periods compared to previous decades also means that the seasonal window for offshore Arctic drilling remains open longer in the summer, increasing the chances for making a discovery.

In addition to the improved access to larger portions of the Arctic afforded by shrinking sea ice, interest in Arctic oil and gas was fueled by a 2008 U.S. Geological Survey (USGS) appraisal of undiscovered oil and

---

[226] A seismic array is typically a long string or streamer of geophones—acoustic devices used for recording seismic signals—towed behind a ship while the ship traverses a prospective oil and gas-bearing portion of the seafloor. The seismic signals are processed and interpreted to give a cross-section or three-dimensional image of the subsurface.

gas north of the Arctic Circle.[227] The USGS stated that the "extensive Arctic continental shelves may constitute the geographically largest unexplored prospective area for petroleum remaining on Earth."[228] In the report, the USGS estimated that 90 billion barrels of oil, nearly 1,700 trillion cubic feet of natural gas, and 44 billion barrels of natural gas liquids may remain to be discovered in the Arctic (including both U.S. and international resources north of the Arctic Circle).[229] This would constitute approximately 13% of the world's undiscovered conventional oil resources and 30% of natural gas, according the U.S. Energy Information Administration.[230] In terms of U.S. resources specifically, DOI's Bureau of Ocean Energy Management (BOEM) estimated in 2017 that the Alaska portions of the U.S. OCS contain undiscovered, technically recoverable resources of approximately 27 billion barrels of oil and 132 trillion cubic feet of natural gas (although not all of these resources may be economically viable to recover).[231]

Despite the warming trend in the Arctic, severe weather and sea ice continue to pose challenges to exploration. In addition, any discovery of new oil and gas deposits far from existing storage, pipelines, and shipping facilities could not be developed until infrastructure is built to extract and transport the petroleum.

Some have expressed interest in expanding America's ocean energy portfolio in the region. Currently, among 15 federal planning areas in the

---

227 See USGS Fact Sheet 2008-3049, *Circum-Arctic Resource Appraisal: Estimates of Undiscovered Oil and Gas North of the Arctic Circle*, at http://pubs.usgs.gov/fs/2008/3049/, hereinafter referred to as "USGS 2008 Fact Sheet."

228 USGS 2008 Fact Sheet.

229 USGS 2008 Fact Sheet, p. 1.

230 U.S. Energy Information Administration, "Today in Energy: Arctic Oil and Natural Gas Resources," January 20, 2012, at http://www.eia.gov/todayinenergy/detail.cfm?id=4650.

231 Bureau of Ocean Energy Management, "Assessment of Undiscovered Technically Recoverable Oil and Gas Resources of the Nation's Outer Continental Shelf, 2016a," at https://www.boem.gov/UTRR-Update_BTU/. Although dated 2016, the assessment reflects 2017 updates of nearshore Beaufort Sea resources based on new data from seismic surveys and drilling onshore and in state waters, pursuant to DOI Secretarial Order 3352. Not all of the undiscovered *technically* recoverable resources would necessarily be *economically* recoverable. BOEM's *2019-2024 Outer Continental Shelf Oil and Gas Leasing Draft Proposed Program* (January 2015, chapter 5, at https://www.boem.gov/NP-Draft-Proposed-Program-2019-2024/, hereinafter cited as "BOEM 2019-2024 Draft Proposed Program.") estimates a range of resources that would be economically recoverable under various oil and gas price points and cost conditions.

region, the Beaufort Sea and Cook Inlet are the only two areas with active federal leases,[232] and only the Beaufort Sea has any producing wells in federal waters (from a joint federal-state unit).[233] The Trump Administration has stated its interest in promoting offshore development in the region. In January 2018, the Administration issued a draft 5-year offshore oil and gas leasing program for 2019-2024 that would schedule lease sales in all 15 Alaska planning areas, including three sales in the Beaufort Sea and three in the Chukchi Sea.[234] Current lease sales on the Alaska OCS are governed by the Obama Administration's leasing program for 2017-2022, which includes one lease sale in the Cook Inlet (scheduled for 2021) and none in other Alaska planning areas.[235]

Offshore oil and gas activities in the region have fluctuated as industry weighs changing oil prices, development costs, and regulations. For example, in 2015, Shell Oil Company announced its decision to cease exploration in offshore Alaska for the foreseeable future. Shell cited several reasons for the decision, including insufficient indications of oil and gas at its Burger J well in the Chukchi Sea, the high costs associated with Arctic exploration, and the "challenging and unpredictable" federal regulatory environment.[236] BOEM also reported that, between February and November 2016, companies relinquished more than 90% of leases they had held in the Beaufort and Chukchi Sea planning areas, in the midst of a slump in oil prices.[237] While there were 450 active leases in the Chukchi Sea planning area at the end of 2015, as of November 2019 there were

---

[232] Although part of BOEM's Alaska region, Cook Inlet lies outside the Arctic boundary as defined by the ARPA (15 U.S.C. 4111; see Figure 1).

[233] Bureau of Ocean Energy Management, "BP Exploration (Alaska) (BPXA)—Northstar," at http://www.boem.gov/About-BOEM/BOEM-Regions/Alaska-Region/Leasing-and-Plans/Plans/BP-North-Star.aspx.

[234] BOEM 2019-2024 Draft Proposed Program, p. 8.

[235] Bureau of Ocean Energy Management, *2017-2022 Outer Continental Shelf Oil and Gas Leasing Proposed Final Program*, November 2016, at https://www.boem.gov/2017-2022-OCS-Oil-and-Gas-Leasing-PFP/, hereinafter cited as "BOEM 2017-2022 Proposed Final Program."

[236] Royal Dutch Shell, PLC, "Shell Updates on Alaska Exploration," press release, September 28, 2015, at http://www.shell.com/global/aboutshell/media/news-and-media-releases/2015/shell-updates-on-alaska- exploration.html.

[237] BOEM 2017-2022 Proposed Final Program, p. S-3.

none.[238] In the Beaufort Sea, active leases dropped from 77 at the end of 2015 to 40 in November 2019.[239] Despite these changes, some activities have indicated ongoing industry interest in the region. For example, in November 2017, the Trump Administration approved an application for permit to drill (APD) on a lease in the Beaufort Sea held by the Eni U.S. Operating Company.[240] In October 2018, BOEM issued conditional approval to Hilcorp Alaska LLC for an oil and gas development and production plan in the Beaufort Sea, which would be the region's first production facility entirely in federal waters.[241] Recent discoveries onshore and in state waters on Alaska's North Slope have contributed to ongoing interest in the region.

The evolving federal regulatory environment for Arctic offshore activities has been shaped by concerns about industry's ability to respond to potential oil spills, given the region's remoteness and harsh conditions. The section of this report on "Oil Pollution Implications of Arctic Change" discusses this issue in greater detail. In July 2016, BOEM and the Bureau of Safety and Environmental Enforcement (BSEE) released final safety regulations for Arctic exploratory drilling that include multiple requirements for companies to reduce the risks of potential oil spills—for example, the requirement that companies have a separate rig available at

---

[238] For 2015 data, see BOEM, "Combined Leasing Report, as of January 1, 2016," at https://www.boem.gov/ Combined-Leasing-Reports-2016/. For November 2019 data, see BOEM, "Combined Leasing Report, as of November 1, 2019," at https://www.boem.gov/sites/default/files/documents/oil-gas-energy/leasing/Combined%20 Leasing%20Statistics%20November%202019.pdf.

[239] Ibid.

[240] Bureau of Safety and Environmental Enforcement (BSEE), "BSEE Approves New Drilling Operations in Arctic," press release, November 28, 2017, at https://www.bsee.gov/newsroom/latest-news/statements-and-releases/press-releases/bsee-approves-new-drilling-operations-in. The BSEE Director stated in the press release that "responsible resource development in the Arctic is a critical component to achieving American energy dominance."

[241] Department of the Interior press release, "Interior Approves Long-Awaited First Oil Production Facility in Federal Waters Offshore Alaska," October 24, 2018, at https://www.doi.gov/pressreleases/interior-approves-long-awaited-first- oil-production-facility-federal-waters-offshore. On May 29, 2019, NOAA's National Marine Fisheries Service issued a proposed rule to govern the taking of marine mammals incidental to the construction and operation of the project's drilling and production island (84 *Federal Register* 24926). Public comments were accepted through July 31, 2019 (84 *Federal Register* 32697).

drill sites to drill a relief well in case of a loss of well control.[242] Some Members of Congress and industry stakeholders opposed the regulations as overly prescriptive and unnecessarily burdensome, while other Members and environmental organizations asserted that the rules did not go far enough in protecting the region from potential environmental damage and addressing the potential contributions of Arctic oil and gas activities to climate change.[243] In a 2017 executive order, President Trump directed the Secretary of the Interior to review the Arctic regulations, and in 2018 the Department of the Interior announced work on rule revisions.[244] Legislation was introduced in the 115th Congress both to repeal the Obama Administration's version of the Arctic rule and, conversely, to codify it in law.[245]

Concerns about the impacts of oil and gas activities have led in the past to bans by both Congress and the President on leasing in certain Arctic Ocean areas deemed especially sensitive.[246] For example, congressional and presidential moratoria since the 1980s effectively banned federally regulated planning and permitting in the Bristol Bay area of the North Aleutian Basin. Congress allowed most statutory bans in the region to expire in 2004.[247] President Obama reinstated the moratorium in the North

---

242 Department of the Interior, "Requirements for Exploratory Drilling on the Arctic Outer Continental Shelf," 81 *Federal Register* 46477, July 15, 2016.

243 For differing congressional viewpoints, see, e.g., U.S. Congress, House Committee on Natural Resources, Subcommittee on Energy and Mineral Resources, hearing on Arctic Resources and American Competitiveness, 114th Cong., 1st sess., June 16, 2015, at http://naturalresources.house.gov/calendar/eventsingle.aspx?EventID=398713.

244 Executive Order 13795, "Implementing an America-First Offshore Energy Strategy," April 28, 2017, at https://www.govinfo.gov/content/pkg/FR-2017-05-03/pdf/2017-09087.pdf. For the DOI announcement, see Department of the Interior, "Semiannual Regulatory Agenda," 83 *Fed. Reg*. 58039, November 16, 2018.

245 For example, in the 115th Congress, H.R. 4239, the SECURE American Energy Act, would have provided that the Arctic rule would have no force or effect. Conversely, S. 2720, the Clean Coasts Act, would have enacted the regulation into law. These measures were not enacted, and no similar legislation has been introduced to date in the 116th Congress.

246 Section 12(a) of the Outer Continental Shelf Lands Act (43 U.S.C. §1341(a)) authorizes the President to, "from time to time, withdraw from disposition any of the unleased lands of the outer Continental Shelf."

247 FY2004 DOI Appropriations (P.L. 108-108). Furthermore, the Continuing Appropriations Resolution 2009 (P.L. 110-329) did not extend the annual congressional moratorium on oil and gas leasing activities in the lower 48 states. On March 11, 2009, the Omnibus Appropriations Act, 2009 (P.L. 111-8) was enacted without moratorium provisions, confirming that the congressional oil and gas development bans in federal waters along the

Aleutian Basin, indefinitely withdrawing acreage located in Bristol Bay from eligibility for oil and gas leasing.[248] Also, in December 2016, President Obama indefinitely withdrew from leasing disposition other large portions of the U.S. Arctic, including the entire Chukchi Sea planning area and almost all of the Beaufort Sea planning area.[249] President Obama separately withdrew from leasing consideration planning areas in the North Bering Sea.[250] In April 2017, President Trump issued Executive Order 13795, which modified President Obama's withdrawals so as to open all of these areas for leasing consideration except for the North Aleutian Basin.[251] However, in a March 2019 court decision, the U.S. District Court for the District of Alaska vacated this provision in President Trump's executive order, ruling that the Outer Continental Shelf Lands Act gives the President the authority to make withdrawals, but not to revoke prior presidential withdrawals.[252]

Atlantic and Pacific coasts, parts of Alaska, and the Gulf of Mexico that had been in place since 1982 had not been restored in 2009 appropriations measures.

[248] Presidential Memorandum, "Withdrawal of Certain Areas of the United States Outer Continental Shelf from Leasing Disposition," December 16, 2014, at http://www.whitehouse.gov/the-press-office/2014/12/16/presidential-memorandum-withdrawal-certain-areas-united-states-outer-con. Earlier, President Obama had withdrawn the area from leasing for a time-limited period. Presidential Memorandum, "Withdrawal of Certain Areas of the United States Outer Continental Shelf from Leasing Disposition," March 31, 2010, at http://www.doi.gov/whatwedo/energy/ocs/upload/ 2010alaska-mem-rel.pdf.

[249] Presidential Memorandum, "Withdrawal of Certain Portions of the United States Arctic Outer Continental Shelf from Mineral Leasing," December 20, 2016, at https://www.whitehouse.gov/the-press-office/2016/12/20/presidential-memorandum-withdrawal-certain-portions-united-states-arctic. Earlier, President Obama had indefinitely withdrawn from leasing disposition certain smaller areas in the Beaufort and Chukchi Seas, including the Hanna Shoal region of the Chukchi Sea and other areas. Presidential Memorandum, "Withdrawal of Certain Areas of the United States Outer Continental Shelf Offshore Alaska from Leasing Disposition," January 27, 2015, at http://www.whitehouse.gov/the-press-office/2015/01/27/presidential-memorandum-withdrawal-certain-areas-united-states-outer-con.

[250] Executive Order 13754, "North Bering Sea Climate Resilience," December 9, 2016, at https://www.gpo.gov/fdsys/ pkg/FR-2016-12-14/pdf/2016-30277.pdf.

[251] Executive Order 13795, "Implementing an America-First Offshore Energy Strategy," April 28, 2017, at https://www.gpo.gov/fdsys/pkg/FR-2017-05-03/pdf/2017-09087.pdf. For additional discussion, see CRS Legal Sidebar WSLG1799, *Trump's Executive Order on Offshore Energy: Can a Withdrawal be Withdrawn?*

[252] League of Conservation Voters v. Trump, 363 F.Supp.3d 1013 (D.Alaska 2019). The President's withdrawal authority is contained in Section 12(a) of the Outer Continental Shelf Lands Act (43 U.S.C. Section 1341(a)).

Extent of the Continental Margin

Increased interest in developing offshore resources in the Arctic has sparked efforts by nations bordering the Arctic Ocean to map the extent of their continental margins beyond the 200-mile EEZ limit. As discussed earlier, under Article 76 of UNCLOS, nations can make a submission to the Commission on the Limits of the Continental Shelf (hereinafter referred to as the Commission) concerning the extent of their continental shelves.[253] Under Article 76, the extent of the continental margin beyond the 200-mile limit depends on the position of the foot of the continental slope, the thickness of sediments, and the depth of water. Also, the continental margin could include geologic features that extend from the continent out to sea, which may include undersea ridges continuing for hundreds of miles offshore.

Arctic border countries have conducted complex investigations needed to support submissions to the Commission for an extended continental shelf in the Arctic. Submissions have been made by four countries, including the Russian Federation, which made its initial UNCLOS submission to a portion of the Arctic continental shelf in 2001.[254] Russia's 2001 submission included the Lomonosov Ridge, an undersea feature spanning the Arctic from Russia to Canada, as an extension of its continental margin. The submission demonstrated Russia's bid to extend activities in Arctic regions. The Russian Federation presented a revised submission in 2015 to the Commission that included not only the Lomonosov Ridge but also the Mendeleev Rise—another subsea feature claimed by Russia to be a natural part of its continental margin—as components of the extended Russian continental shelf.[255] The Commission

---

[253] For more information, see the section of this report on "Extended Continental Shelf Submissions, Territorial Disputes, and Sovereignty Issues."

[254] United Nations, Division for Ocean Affairs and the Law of the Sea, http://www.un.org/depts/los/clcs_new/ commission_submissions.htm. Denmark, Norway, and Canada also submitted claims to portions of the Arctic continental shelf (https://www.state.gov/frequently-asked-questions-u-s-extended-continental-shelf-project/). In 2009, the Commission approved Norway's submission.

[255] United Nations, "Partial Revised Submission of the Russian Federation to the Commission on the Limits of the Continental Shelf in Respect of the Continental Shelf of the Russian Federation in the Arctic Ocean," 2015, http://www.un.org/Depts/los/clcs_new/ submissions_files/rus01_rev15/2015_08_03_Exec_Summary_English.pdf.

has not rendered a decision on the revised Russian Federation submission as of late 2019.

The United States has started to gather and analyze data for a potential submission through an initiative called the Extended Continental Shelf (ECS) Project.[256] The U.S. ECS project has also assisted more than 30 countries with their efforts to delineate their extended continental shelves worldwide.[257] Canada and the United States share overlapping regions of the seabed as part of the extended continental margin of both nations. Much of the data to delineate the ECS for both countries was collected in a two-ship operation involving the U.S. Coast Guard Cutter *Healy* and the Canadian Coast Guard ship *Louis S. Saint Laurent*.[258] The two-ship operation collected more than 13,000 linear kilometers (about 8,078 miles) of seismic data over four field seasons in the Arctic beginning in 2007. The data collected will help each country delineate the extent of its own ECS, which should then enable the countries to determine the amount of overlap in the seabed and ultimately establish a maritime boundary in the Arctic.[259]

The United States also has potentially overlapping ECS areas with Russia. Russia (then the Soviet Union) and the United States agreed to a maritime boundary in 1990, and so far Russia has not asserted its ECS in any areas that might be considered part of the U.S. ECS.[260]

---

[256] The purpose of the U.S. Extended Continental Shelf (ECS) Project is to establish the full extent of the continental shelf of the United States, consistent with international law. The work to delineate the ECS is coordinated by the ECS Task Force, located at the National Oceanic and Atmospheric Administration's (NOAA's) National Centers for Environmental Information in Boulder, CO. The Department of State, U.S. Geological Survey (USGS), and NOAA conduct the majority of work on the project. NOAA has the lead in collecting bathymetric data. USGS has the lead in collecting seismic data. For more information, see the project's website at https://www.state.gov/u-s-extended- continental-shelf-project/.

[257] U.S. ECS Project, https://www.state.gov/international-support-and-cooperation-u-s-extended-continental-shelf- project/.

[258] Ibid.

[259] Ibid.

[260] The Senate gave advice and consent to ratify the maritime boundary agreement in 1991. Although the Russian Duma has not approved the agreement, both countries continue to provisionally apply the boundary agreement. See U.S. Extended Continental Shelf Project, https://www.state.gov/frequently-asked-questions-u-s-extended-continental- shelf-project/.

Onshore Mineral Development

A warming Arctic means new opportunities and challenges for mineral exploration and development onshore. Receding glaciers expose previously ice-covered land that could host economic mineral deposits that were previously undetectable and unmineable below the ice.

Longer summers also would extend exploration seasons for areas that are not currently ice- covered, but are only accessible for ground surveys during the warmer months. In some parts of the Arctic, less sea ice allows ships to transport heavy equipment to remote locations, and to convey ore from mines to the market further south. Some railway and mining operators are considering developing railroads and other infrastructure to transport ore year-round. As with onshore oil and gas development, however, mining infrastructure that depends on footings sunk into permafrost could become unstable if the permafrost melts in response to warmer temperatures.[261] Also, mineral deposits that may be technically recoverable with current technology may not be economically profitable.

Some industry commentators suggest that mining might offer better long-term economic development opportunities compared to oil and gas development because of a larger permanent workforce and project lifetimes of several decades.[262] Similar to oil and gas, however, industry observers note that uncertainties and knowledge gaps exist in the understanding of environmental change in the Arctic, and how to deal with the risks associated with significant Arctic industrial activity.[263]

One important part of the current infrastructure in the Arctic that supports oil, gas, and mineral development is the construction and use of ice roads—built and used during the winter, but not passable during the

---

[261] This section does not treat onshore oil and gas development on federal lands in the Arctic region, such as the National Petroleum Reserve-Alaska or the Arctic National Wildlife Refuge. For more information on the oil and gas program for the Arctic National Wildlife Refuge and related issues, see CRS Report RL33872, *Arctic National Wildlife Refuge (ANWR): An Overview.*

[262] See, e.g., Cecelia Jamasmie, "Melting Arctic is nothing but good news for the mining industry: Lloyd's/Chatham House," Mining.com, April 13, 2012, at http://www.mining.com/melting-arctic-is-nothing-but-good-news-for-the-mining-industry-lloydschatam-house/.

[263] Charles Emmerson and Glada Lahn, *Arctic Opening: Opportunity and Risk in the High North*, Chatham House, 2012, at https://www.chathamhouse.org/publications/papers/view/182839.

warmer months. Warmer temperatures are shortening the ice road transport seasons and creating transportation challenges. For example, the opening date for tundra roads in northern Alaska usually occurred in early November prior to 1991 and has shifted to January in recent years.[264]

Oil Pollution and Pollution Response[265]

Oil Pollution Implications of Arctic Change

Climate change impacts in the Arctic, particularly the decline of sea ice and retreating glaciers, have stimulated human activities in the region, many of which have the potential to create oil pollution. A primary concern is the threat of a large oil spill in the area. Although a major oil spill has not occurred in the Arctic region,[266] recent economic activity, such as oil and gas exploration and tourism (cruise ships), increases the risk of oil pollution (and other kinds of pollution) in the Arctic. Significant spills in high northern latitudes (e.g., the 1989 *Exxon Valdez* spill in Alaska and spills in the North Sea) suggest that the "potential impacts of an Arctic spill are likely to be severe for Arctic species and ecosystems."[267]

Risk of Oil Pollution in the Arctic

A primary factor determining the risk of oil pollution in the Arctic is the level and type of human activity being conducted in the region. Although climate changes in the Arctic are expected to increase access to natural resources and shipping lanes, the region will continue to present logistical challenges that may hinder human activity in the region. For example (as discussed in another section of this report),[268] the unpredictable ice conditions may discourage trans-Arctic shipping. If

---

264 See National Oceanic and Atmospheric Administration, "Arctic Change," at https://www.pmel.noaa.gov/arctic- zone/detect/land-road.shtml?page=land.

265 This section prepared by Jonathan L. Ramseur, Specialist in Environmental Policy, Resources, Science, and Industry Division.

266 Arctic Council, Emergency Prevention, Preparedness and Response Working Group, *Guide to Oil Spill Response in Snow and Ice Conditions*, 2015, at https://oaarchive.arctic-council.org/handle/11374/403.

267 Arctic Monitoring and Assessment Programme (AMAP), Arctic Oil and Gas 2007 (2008).

268 See this report's section "Implications for Sea Transportation," by John Frittelli.

trans-Arctic shipping were to occur on a frequent basis, it would represent a considerable portion of the overall risk of oil pollution in the region. In recent decades, many of the world's largest oil spills have been from oil tankers, which can carry millions of gallons of oil.[269]

Although the level of trans-Arctic shipping is uncertain, many expect oil exploration and extraction activities to intensify in the region.[270] Oil well blowouts from offshore oil extraction operations have been a source of major oil spills, eclipsing the largest tanker spills. The largest unintentional oil spill in recent history was from the 2010 *Deepwater Horizon* incident in the Gulf of Mexico.[271] During that incident, the uncontrolled well released (over an 87-day period) approximately 200 million gallons of crude oil.[272] The second-largest unintentional oil spill in recent history—the *IXTOC I*, estimated at 140 million gallons—was due to an oil well blowout in Mexican Gulf Coast waters in 1979.[273]

Until the 2010 *Deepwater Horizon* incident, the spill record for offshore platforms in U.S. federal waters had shown improvement from prior years.[274] A 2003 National Research Council (NRC) study of oil and gas activities on Alaska's North Slope stated "blowouts that result in large spills are unlikely."[275] Similar conclusions were made in federal agency documents regarding deepwater drilling in the Gulf of Mexico before the

---

[269] For example, the *Exxon Valdez* spilled approximately 11 million gallons of oil, but its carrying capacity was approximately 60 million gallons.

[270] See this report's section "Implication of Changes in the Arctic for Oil, Gas, and Mineral Exploration and Development," by Peter Folger and Marc Humphries.

[271] Larger oil spills occurred during the 1991 Iraq War, but many of those spills were deliberate. A 1910-1911 onshore oil blowout in the California San Joaquin Valley is reported to have spilled 9.4 million barrels of crude oil (almost 400 million gallons).

[272] An estimated 17% of this oil did not enter the Gulf environment but was directly recovered from the wellhead by the responsible party (British Petroleum, BP). See the Federal Interagency Solutions Group, Oil Budget Calculator Science and Engineering Team, *Oil Budget Calculator: Deepwater Horizon-Technical Documentation*, November 2010; and CRS Report R42942, *Deepwater Horizon Oil Spill: Recent Activities and Ongoing Developments*.

[273] National Research Council (NRC) of the National Academies of Science, *Oil in the Sea III: Inputs, Fates, and Effects* (2003).

[274] See CRS Report RL33705, *Oil Spills: Background and Governance*; and Dagmar Etkin (Environmental Research Consulting), Analysis of U.S. Oil Spillage, Prepared for American Petroleum Institute, August 2009.

[275] National Research Council of the National Academies of Science, *Cumulative Environmental Effects of Oil and Gas Activities on Alaska's North Slope* (2003).

2010 *Deepwater Horizon* event.[276] Some would likely contend that the underlying analyses behind these conclusions should be adjusted to account for the 2010 Gulf oil spill. However, others may argue that the proposed activities in U.S. Arctic waters present less risk of an oil well blowout than was encountered by the *Deepwater Horizon* drill rig, because the proposed U.S. Arctic operations would be in shallower waters (150 feet) than the deepwater well (approximately 5,000 feet) that was involved in the 2010 Gulf oil spill. In addition, Shell Oil has stated that the pressures in the Chukchi Sea (the location of Shell's recent interest) would be two to three times less than they were in well involved in the 2010 Gulf oil spill.[277] Regardless of these differences, even under the most stringent control systems, some oil spills and other accidents are likely to occur from equipment failure or human error.

Potential Impacts

No oil spill is entirely benign. Even a relatively minor spill, depending on the timing and location, can cause significant harm to individual organisms and entire populations. Regarding aquatic spills, marine mammals, birds, bottom-dwelling and intertidal species, and organisms in early developmental stages—eggs or larvae—are especially vulnerable. However, the effects of oil spills can vary greatly. Oil spills can cause impacts over a range of time scales, from only a few days to several years, or even decades in some cases.

Conditions in the Arctic may have implications for toxicological effects that are not yet understood. For example, oil spills on permafrost may persist in an ecosystem for relatively long periods of time, potentially harming plant life through their root systems. Moreover, little is known about the effects of oil spills on species that are unique to the Arctic,

---

[276] See, for example, Minerals Management Service (MMS), Outer Continental Shelf Oil & Gas Leasing Program: 2007-2012, Final Environmental Impact Statement, April 2007, chapter 4; MMS, Proposed Gulf of Mexico OCS Oil and Gas Lease Sale 206, Central Planning Area, Environmental Assessment, October 2007.

[277] Letter from Marvin E. Odum, President, Shell Oil Company to S. Elizabeth Birnbaum, Minerals Management Service (May 14, 2010). Cited in a staff paper from the National Commission on the BP Deepwater Horizon Oil Spill and Offshore Drilling ("The Challenges of Oil Spill Response in the Arctic," January 2011).

particularly, species' abilities to thrive in a cold environment and the effect temperature has on toxicity.[278]

The effects of oil spills in high-latitude, cold-ocean environments may last longer and cause greater damage than expected. Some recent studies have found that oil spills in lower latitudes have persisted for longer than initially expected, thus raising the concern that the persistence of oil in the Arctic may be understated. In terms of wildlife, population recovery may take longer in the Arctic because many of the species have longer life spans and reproduce at a slower rate.[279]

Response and Cleanup Challenges in the Arctic Region

Climate changes in the Arctic are expected to increase human activities in the region, many of which impose a risk of oil pollution, particularly from oil spills. Conditions in the Arctic region impose unique challenges for personnel charged with (1) oil spill response, the process of getting people and equipment to the incident, and (2) cleanup duties, either recovering the spilled oil or mitigating the contamination so that it poses less harm to the ecosystem. These challenges may play a role in the policy development for economic activities in the Arctic.

Spill Response Challenges

Response time is a critical factor for oil spill recovery. With each hour, spilled oil becomes more difficult to track, contain, and recover, particularly in icy conditions, where oil can migrate under or mix with surrounding ice.[280] Most response techniques call for quick action, which may pose logistical challenges in areas without prior staging equipment or trained response professionals. Many stakeholders are concerned about a "response gap" for oil spills in the Arctic region.[281] A response gap is a period of time in which oil spill response activities would be unsafe or

---

278 AMAP, *Arctic Oil and Gas 2007* (2008).

279 AMAP, *Arctic Oil and Gas 2007* (2008).

280 World Wildlife Fund, *Oil Spill: Response Challenges in Arctic Waters* (2007).

281 Coastal Response Research Center, *Opening the Arctic Seas: Envisioning Disasters and Framing Solutions* (2009), partnership between the National Oceanic and Atmospheric Administration and the University of New Hampshire.

infeasible. The response gap for the northern Arctic latitudes is likely to be extremely high compared to other regions.[282]

According to a 2014 National Research Council (NRC) report, "the lack of infrastructure in the Arctic would be a significant liability in the event of a large oil."[283] The Coast Guard has no designated air stations north of Kodiak, AK, which is almost 1,000 miles from the northernmost point of land along the Alaskan coast in Point Barrow, AK.[284] Although some of the communities have airstrips capable of landing cargo planes, no roads connect these communities.[285] Vessel infrastructure is also limited. The nearest major port is in the Aleutian Islands, approximately 1,300 miles from Point Barrow. Two of the major nonmechanical recovery methods—in situ burning and dispersant application—may be limited (or "precluded") by the Arctic conditions and lack of logistical support: aircraft, vessels, and other infrastructure.[286]

A 2010 Government Accountability Office (GAO) report identified further logistical obstacles that would hinder an oil spill response in the region, including "inadequate" ocean and weather information for the Arctic and technological problems with communications.[287] A 2014 GAO report highlighted steps taken by some groups (e.g., the National Oceanic

[282] Although the response gap in the Arctic has not been quantified, a recent estimate of Prince William Sound (PWS) may be instructive. A 2007 study found a response gap for PWS of 38% for the time of the study period (65% during the winter season). Note that PWS has existing infrastructure for response, while the more remote Arctic areas do not. Nuka Research and Planning Group, LLC, *Response Gap Estimate for Two Operating Areas in Prince William Sound, Alaska* (2007), Report to Prince William Sound Regional Citizens' Advisory Council.

[283] National Research Council (NRC) of the National Academies of Science, *Responding to Oil Spills in the U.S. Arctic Marine Environment*, 2014.

[284] U.S. Coast Guard, *Report to Congress: U.S. Coast Guard Polar Operations*, December 2008.

[285] A single road connects Deadhorse, Alaska and the Prudhoe Bay with central Alaska (Fairbanks). For more details and maps of the area, see Nuka Research and Planning Group, *Oil Spill Prevention and Response in the U.S. Arctic Ocean: Unexamined Risks, Unacceptable Consequences*, Commissioned by Pew Environment Group, November 2010.

[286] World Wildlife Fund, Oil Spill: Response Challenges in Arctic Waters (2007). For further discussion of issues relating to oil spills, see CRS Report RL33705, *Oil Spills: Background and Governance*.

[287] Government Accountability Office, *Coast Guard: Efforts to Identify Arctic Requirements Are Ongoing, but More Communication about Agency Planning Efforts Would Be Beneficial*, GAO-10-870, September 2010.

and Atmospheric Administration) to improve some of these logistical elements.[288]

Oil Spill Cleanup Challenges

The history of oil spill response in the Aleutian Islands highlights the challenges and concerns for potential spills in the Arctic region:

> The past 20 years of data on response to spills in the Aleutians has also shown that almost no oil has been recovered during events where attempts have been made by the responsible parties or government agencies, and that in many cases, weather and other conditions have prevented any response at all.[289]

The behavior of oil spills in cold and icy waters is not as well understood as oil spills in more temperate climates.[290] The 2014 NRC report highlights some recent advancements in understanding oil spill behavior in arctic climates. At the same time, the report recommends further study in multiple areas.

The 2014 NRC report states that in colder water temperatures or sea ice, "the processes that control oil weathering—such as spreading, evaporation, photo-oxidation, emulsification, and natural dispersion—are slowed down or eliminated for extended periods of time."[291] In some respects, the slower weathering processes may provide more time for response strategies, such as in situ burning or skimming. On the other hand, the longer the oil remains in an ecosystem, the more opportunity there is for exposure.

In addition, the 2014 report states the following:

---

288 Government Accountability Office, *Key Issues Related to Commercial Activity in the U.S. Arctic over the Next Decade*, GAO-14-299, March 2014.

289 Transportation Research Board of the National Academy of Sciences, *Risk of Vessel Accidents and Spills in the Aleutian Islands: Designing a Comprehensive Risk Assessment* (2008), Special Report 293, National Academies Press. Washington, DC.

290 National Research Council (NRC) of the National Academies of Science, *Responding to Oil Spills in the U.S. Arctic Marine Environment*, 2014.

291 National Research Council (NRC) of the National Academies of Science, *Responding to Oil Spills in the U.S. Arctic Marine Environment*, 2014.

> Arctic conditions impose many challenges for oil spill response—low temperatures and extended periods of darkness in the winter, oil that is encapsulated under ice or trapped in ridges and leads, oil spreading due to sea ice drift and surface currents, reduced effectiveness of conventional containment and recovery systems in measurable ice concentrations, and issues of life and safety of responders.

Existing Policy Framework

Considering both the recent increase in human activity in the region (and expectation of further interest) and the response and recovery challenges that an oil spill would impose in Arctic waters, many would assert that the region warrants particular attention in terms of governance. However, the existing framework for international governance of maritime operations in the Arctic region lacks legally binding requirements. While the Safety of Life at Sea Convention (SOLAS) and other International Maritime Organization (IMO) conventions include provisions regarding ships in icy waters, the provisions are not specific to the polar regions. Although the IMO has "Guidelines for Ships Operating in Arctic," a 2009 NOAA report described the nonbinding IMO provisions as "inconsistent with the hazards of Arctic navigation and the potential for environmental damage from such an incident."[292]

In 2013, the member states of the Arctic Council signed an Agreement on Cooperation on Marine Oil Pollution Preparedness and Response in the Arctic.[293] The agreement's objective is to "strengthen cooperation, coordination, and mutual assistance ... on oil pollution preparedness and response in the Arctic."

In addition, the United States has separate bilateral agreements with Canada and Russia that address oil spill response operations. The agreement with Canada was established in 1974 for the Great Lakes and has been amended several times to add more geographic areas, including Arctic waters. According to the 2014 NRC report: "Formal contingency planning and exercises with Canada have enabled both the United States

---

292 Coastal Response Research Center, *Opening the Arctic Seas: Envisioning Disasters and Framing Solutions*, (2009), partnership between the National Oceanic and Atmospheric Administration and the University of New Hampshire.

293 Available at http://www.arctic-council.org.

and Canada to refine procedures and legal requirements for cross-border movement of technical experts and equipment in the event of an emergency."

The U.S.-Russian agreement was made in 1989 and applies to oil spills in Arctic waters. However, the 2014 NRC report asserts that the agreement has not been tested to the same extent as the U.S.-Canada agreement.

Fisheries[294]

The effects of climate change such as increasing sea surface temperatures and decreasing permanent sea ice are altering the composition of marine ecosystems in the Arctic. These changes are likely to affect the ranges and productivity of living marine resources including species that support marine fisheries. Furthermore, as a greater portion of the waters in the central Arctic Ocean become open for longer periods, the region's resources will become more accessible to commercial fishing. Large commercial fisheries already exist in the Arctic, including in the Barents and Norwegian Seas north of Europe, the Central North Atlantic off Greenland and Iceland, the Bering Sea off Russia and the United States (Alaska), and the Newfoundland and Labrador Seas off northeastern Canada.[295] As environmental changes occur, fisheries managers will be challenged to adjust management measures for existing fisheries. Uncertainties related to these changes and potential new fisheries in the central Arctic Ocean have prompted many fishery managers to support precautionary approaches to fisheries management in the region.

On June 1, 2008, Congress passed a joint resolution (P.L. 110-243) that directed "the United States to initiate international discussions and take necessary steps with other nations to negotiate an agreement for managing migratory and transboundary fish stocks in the Arctic Ocean." The joint resolution also supported establishment of a new international

---

[294] This section was prepared by Harold Upton, Analyst in Natural Resources Policy, Resources, Science, and Industry Division.

[295] Erik J. Molenaar and Robert Corell, *Arctic Fisheries*, Arctic Transform, February 9, 2009; available at http://arctic- transform.org/download/FishBP.pdf.

fisheries management organization or organizations for the region. International cooperation is necessary to manage Arctic resources because fish stocks are shared to some degree among the five adjacent jurisdictional zones of the Arctic rim nations. Further, a large portion of the central Arctic Ocean lies outside the Exclusive Economic Zones (EEZ) of these nations. Ideally, regional management would recognize the need to coordinate management for those fish populations that move among these national jurisdictional zones and high seas.

For waters under U.S. jurisdiction, in 2009, the National Marine Fisheries Service in the Department of Commerce's National Oceanic and Atmospheric Administration implemented the North Pacific Council's Fishery Management Plan for Fish Resources of the Arctic Management Area.[296] The management area includes marine waters in the U.S. EEZ of the Chukchi and Beaufort Seas.[297] The plan initially prohibits commercial fishing in the Arctic Management Area and moves the northern boundary of the Bering Sea/Aleutian Islands king and tanner crab fishery management plan out of the Arctic Management Area south to the Bering Strait. The plan takes a precautionary approach by requiring the collection of more information before developing commercial fisheries in the region.

On July 16, 2015, the five nations that surround the Arctic Ocean signed a declaration to prevent unregulated commercial fishing in the high seas portion of the central Arctic Ocean.[298] The five nations agree that a precautionary approach to fishing is needed because there is limited scientific knowledge of marine resources in the region. Currently, there is no commercial fishing in central Arctic Ocean and it is questionable whether existing fisheries resources could sustain a fishery.

---

[296] 74 *Federal Register* 56734-56746, November 3, 2009.

[297] The state of Alaska has jurisdiction over waters from 0-3 nautical miles from the baseline. The baseline generally follows the shoreline.

[298] The five nations include Canada, Denmark, Norway, the Russian Federation, and the United States. See https://www.regjeringen.no/globalassets/departementene/ud/vedlegg/folkerett/declaration-on-arctic-fisheries-16-july- 2015.pdf.

The declaration includes the following interim measures:

- to authorize our vessels to conduct commercial fishing in the high seas area only pursuant to one or more marine regional or subregional fisheries management organizations or arrangements that are or may be established to manage such fishing in accordance with recognized international standards;
- to establish a joint program of scientific research with the aim of improving understanding of the ecosystems of this area and promote cooperation with relevant scientific bodies;
- to promote compliance with these interim measures and with relevant international law, including by coordinating our monitoring, control, and surveillance activities in this area; and
- to ensure that any noncommercial fishing in this area does not undermine the purpose of the interim measures, is based on scientific advice and is monitored, and that data obtained through any such fishing is shared.

The declaration also recognizes the interests of indigenous peoples and the need to encourage other countries to take actions that are consistent with the interim measures. It appears that future management arrangements may include China, the EU, Iceland, Japan, and South Korea. Iceland has stated it regrets that although it has repeatedly asked to participate in the collaboration, the five states decided to keep Iceland outside consultations on the declaration.[299] It remains an open question as to whether an Arctic Ocean regional fishery management organization will be established, which countries would be included in such an arrangement, and if commercial fisheries will be developed in the central Arctic Ocean.

---

[299] Ministry of Foreign Affairs, Iceland, "The Position of Iceland on the Consultation on Arctic Fishing," press release, July 24, 2015, http://arcticportal.org/library/press-releases/1524-the-position-of-iceland-on-the-consultation-on-arctic-fishing.

Protected Species[300]

Concern over development of the Arctic relates to how such development might affect threatened and endangered species. Under the Endangered Species Act (ESA, 16 U.S.C. §§1531-1543), the polar bear was listed as threatened on May 15, 2008. The failure by the Fish and Wildlife Service (FWS) to make a 90-day finding on a 2008 petition to list Pacific walrus led to submission of 60- days' notice of a future citizen suit. However, eventually walruses were listed as candidate species under ESA;[301] this status means that federal agencies carrying out actions that may affect the species must confer with FWS though they are not necessarily obliged to modify their actions. Both polar bears and walruses are heavily dependent during their life cycles on thick sea ice, making them especially susceptible to the shrinking Arctic ice cap.

On December 30, 2008, the National Marine Fisheries Service (NMFS) determined that a listing of ribbon seal as threatened or endangered was not warranted.[302] On October 22, 2010, NMFS listed the southern distinct population segment (DPS) of spotted seals as threatened.[303] Listing of two other DPS (Okhotsk and Bering Sea) had earlier been determined to not be warranted.[304] On December 10, 2010, NMFS proposed that (1) four subspecies of ringed seal be listed as threatened,[305] and (2) that two DPS of one subspecies of bearded seal be listed as threatened.[306]

In either terrestrial or marine environments, the extreme pace of change makes a biological response many times more difficult. For species with adaptations for a specific optimum temperature for egg development,

---

[300] Prepared by Lynne Corn and Eugene Buck, Specialists in Natural Resources Policy, Resources, Science, and Industry Division.

[301] Fish and Wildlife Service, "Endangered and Threatened Wildlife and Plants; Review of Native Species That Are Candidates for Listing as Endangered or Threatened; Annual Notice of Findings on Resubmitted Petitions; Annual Description of Progress on Listing Actions; Proposed Rule," 76 *Federal Register* 66370-66439, October 26, 2001.

[302] 73 *Federal Register* 79822-79828.

[303] 75 *Federal Register* 65239-65248.

[304] 74 *Federal Register* 53683-58696, October 20, 2009.

[305] 75 *Federal Register* 77476-77495.

[306] 75 *Federal Register* 77496-77515.

or production of young timed to match the availability of a favored prey species, or seed dispersal in predictable fire regimes, etc., evolutionary responses may well not keep pace with the rate of change.[307] While species of plants and animals farther south might migrate, drift, or be transplanted from warming habitats to more northerly sites that may continue to be suitable,[308] once a terrestrial species reaches the Arctic Ocean, it is very literally at the end of the line. No more northern or colder habitat is available.

The Marine Mammal Protection Act (MMPA; 16 U.S.C. §§1361 et seq.) protects whales, seals, walruses, and polar bears. The MMPA established a moratorium on the "taking" of marine mammals in U.S. waters and by U.S. nationals on the high seas, including the Arctic. The MMPA protects marine mammals from "clubbing, mutilation, poisoning, capture in nets, and other human actions that lead to extinction." Under the MMPA, the Secretary of Commerce, acting through National Marine Fisheries Service, is responsible for the conservation and management of whales and seals. The Secretary of the Interior, acting through the Fish and Wildlife Service, is responsible for walruses and polar bears.[309] Despite the MMPA's general moratorium on taking, the MMPA allows U.S. citizens to apply for and obtain authorization for taking small numbers of mammals incidental to activities other than commercial fishing (e.g., offshore oil and gas exploration and development) if the taking would have only a negligible impact on any marine mammal species or stock, provided that monitoring requirements and other conditions are met.

---

307 Among biologists, it is traditionally said that a species faced with extreme change can respond in three basic ways: "migrate, mutate, or die." When change is rapid enough, mutation (accompanied by natural selection of individuals within the population more suited to the changed environment) may not be able to occur fast enough, leaving migration and death as the only options. The problem of response rate is more severe for species that reproduce slowly (e.g., polar bears) and less severe for species that reproduce rapidly (e.g., algae).

308 The efficacy and the effect of this tactic is often questioned, since natural migration is unlikely to involve the entire suite of species in an ecosystem (e.g., host plants might not move north (or up) as fast as their moth herbivores, nor as fast as the birds that depend on the moths). Moreover, the southerners will not find a land of sterile bare dirt—the species that are already there may be threatened themselves by the competition from the new arrivals, perhaps tipping the balance and pushing still more species toward extinction.

309 Under the MMPA, both NMFS and FWS have responsibility for additional marine mammal species (e.g., manatees, sea otters, dolphins) which are not currently found in the Arctic.

Indigenous People Living in the Arctic[310]

People have been living in the Arctic for thousands of years, and indigenous peoples developed highly specialized cultures and economies based on the physical and biological conditions of the long-isolated region. However, with trade, the influx of additional populations especially since the 19th century, and ongoing physical changes in the Arctic, indigenous populations have already experienced substantial change in their lifestyles and economies. Over the past two decades, greater political organization across indigenous populations has increased their demands for international recognition and broader rights, as well as attention to the economic, health, and safety implications of climate change in the North.

Background

Seven of the eight Arctic nations have indigenous peoples,[311] whose predecessors were present in parts of the Arctic over 10,000 years ago, well before the arrival of peoples with European backgrounds.[312] Current Arctic indigenous peoples comprise dozens of diverse cultures and speak dozens of languages from eight or more non-Indo-European language families.[313]

Before the arrival of Europeans, Arctic indigenous peoples lived in economies that were chiefly dependent, in varying proportions, on hunting land and marine mammals, catching salt- and fresh-water fish, herding reindeer (in Eurasia), and gathering, for their food, clothing, and other

---

310 This section was originally prepared by Roger Walke, who was a Specialist in American Indian Policy, Domestic Social Policy Division, until his retirement from CRS in October 2010. It has been updated by Jane A. Leggett, Specialist in Environmental and Energy Policy in CRS's Resources, Science and Industry Division.

311 *Arctic Human Development Report*, ed. Joan Nymand Larsen et al. (Akureyri, Iceland: Stefansson Arctic Institute, 2004), p. 47; this report is subsequently cited in this section as *AHDR*. The seven countries are Canada, Denmark- Greenland, Finland, Norway, Russia, Sweden, and the United States.

312 John F. Hoffecker, *A Prehistory of the North: Human Settlement of the Higher Latitudes* (New Brunswick, NJ: Rutgers University Press, 2005), pp. 8, 81, 112-115.

313 *AHDR*, pp. 47, 53; David Crystal, *Cambridge Encyclopedia of Language*, 2nd ed. (Cambridge: Cambridge University Press, 1997), chap. 50; *Ethnologue: Languages of the World*, 16th ed., ed. M. Paul Lewis (Dallas: SIL International, 2009), available at http://www.ethnologue.com/. The number of languages and language families varies not only with definitions of the Arctic but with definitions of languages and language families.

products.[314] Indigenous peoples' interaction with and knowledge of Arctic wildlife and environments has developed over millennia and is the foundation of their cultures.[315]

The length of time that Arctic indigenous peoples were in contact with Europeans varied across the Arctic. As recorded by Europeans, contact began as early as the 9th century CE, if not before, in Fennoscandia[316] and northwestern Russia, chiefly for reasons of commerce (especially furs); it progressed mostly west-to-east across northern Asia, reaching northeastern Arctic Asia by the 17th century.[317] North American Arctic indigenous peoples' contact with Europeans started in Labrador in the 16th century and in Alaska in the 18th century, and was not completed until the early 20th century.[318] Greenland's indigenous peoples first saw European-origin peoples in the late 10th century, but those Europeans died out during the 15th or 16th century and Europeans did not return permanently until the 18th century.[319]

Contact led to significant changes in Arctic indigenous economies, political structures, foods, cultures, and populations, starting especially in the 20th century. For example, life expectancy among Alaska Natives has

---

314 Jim Berner et al., *Arctic Climate Impact Assessment* (Cambridge: Cambridge University Press, 2005), chapter 12; this report is subsequently cited in this section as *ACIA*.

315 *ACIA*, pp. 654-655.

316 Fennoscandia refers to the Scandinavian Peninsula, Finland, the Kola Peninsula of Russia, and certain parts of Russia bordering on Finland.

317 Janet Martin, *Treasure in the Land of Darkness: The Fur Trade and Its Significance for Medieval Russia* (Cambridge: Cambridge University Press, 1986), pp. 41-42; James Forsyth, *A History of the Peoples of Siberia: Russia's North Asian Colony, 1581-1990* (Cambridge: Cambridge University Press, 1992), pp. 69-83, 102; Lassi K. Heininen, "Different Images of the Arctic and the Circumpolar North in World Politics," in *Knowledge and Power in the Arctic*, Proceedings at a Conference in Rovaniemi, April 16-18, 2007, Arctic Centre Reports 48, ed. Paula Kankaanpaa et al. (Rovaniemi, Finland: University of Lapland, Arctic Centre, 2007), p. 125.

318 James W. VanStone, "Exploration and Contact History of Western Alaska," and David Damas, "Copper Eskimo," and J. Garth Taylor, "Historical Ethnography of the Labrador Coast," in *Handbook of North American Indian: Vol. 5, Arctic,* vol. ed. David Damas, gen. ed. William C. Sturtevant (Washington: Smithsonian, 1984), pp. 149-155, 408, 509- 510.

319 Inge Kleivan, "History of Norse Greenland," in *Handbook, Vol. 5, Arctic,* op. cit., pp. 549-555; Finn Gad, "Danish Greenland Policies," in *Handbook of North American Indians: Vol. 4, History of Indian-White Relations*, vol. ed. Wilcomb E. Washburn, gen. ed. William C. Sturtevant (Washington: Smithsonian, 1988), p. 110.

increased from 47 years in 1950 to over 69 years in 2000 (though it still lags behind that of U.S. residents overall, at 77 years).[320]

Also, at present, most Arctic indigenous peoples have become minorities in their countries' Arctic areas, except in Greenland and Canada. (One source estimates that, around 2003, about 10% of an estimated 3.7 million people in the Arctic were indigenous.)[321] While many Arctic indigenous communities remain heavily dependent on hunting, fishing, and herding and are more likely to depend on traditional foods than nonindigenous Arctic inhabitants,[322] there is much variation.

Most Arctic indigenous people may no longer consume traditional foods as their chief sources of energy and nutrition.[323] Major economic change is also relatively recent but ongoing.[324] Many Arctic indigenous communities have developed a mixture of traditional economic activities and wage employment.[325] The economics of subsistence and globalization will be key factors in the effects of climate change on Arctic indigenous peoples, and on their reactions to Arctic climate change.

Arctic indigenous peoples' current political structures vary, as do their relationships with their national governments. Some indigenous groups govern their own unique land areas within the national structure, as in the United States and Canada; others have special representative bodies, such as the Saami parliaments in Norway, Finland, and Sweden;[326] a few areas have general governments with indigenous majorities, such as Greenland

---

[320] Parkinson, Alan J. The Arctic Human Health Initiative. Washington, DC: U.S. Department of Health and Human Services, 2006.

[321] *AHDR*, pp. 19, 29. Estimates of Arctic indigenous populations are complicated by varying definitions not only of the Arctic but also of indigenous peoples; for instance, Russia does not count some non-European Arctic ethnic groups, such as the Yakut, as "indigenous minorities" (see "Peoples of the Arctic: Characteristics of Human Populations Relevant to Pollution Issues," in *AMAP Assessment Report: Arctic Pollution Issues*, ed. Simon J. Wilson et al. (Oslo: Arctic Monitoring and Assessment Programme, 1998), pp. 167-169; this report is subsequently cited in this section as *AMAP 1998.*

[322] *AMAP 1998*, chapter 5; see also Birger Poppel et al., *SLiCA Results*, Survey of Living Conditions in the Arctic (Anchorage: Institute of Social and Economic Research, University of Alaska Anchorage, 2007), pp. 4-7, http://www.arcticlivingconditions.org.

[323] Annika E. Nilson and Henry P. Huntington, *Arctic Pollution 2009* (Oslo: Arctic Monitoring and Assessment Programme, 2009), pp. 39-41; this report is subsequently cited in this section as *AMAP 2009.*

[324] *ACIA*, p. 1000.

[325] *SLiCA Results,* op. cit., pp. v, 4-8.

[326] *AHDR*, p. 232.

(a member country of Denmark), Nunavut territory in Canada, and the North Slope and Northwest Arctic boroughs in Alaska.[327] Control of land, through claims and ownership, also varies among Arctic indigenous peoples, as do rights to fishing, hunting, and resources.[328] Arctic indigenous peoples' political relationships to their national and local governments, and their ownership or claims regarding land, are also significant factors in the responses to Arctic climate change by the indigenous peoples and by Arctic nations' governments.

Effects of Climate Change

Arctic climate change is expected to affect the economies, population, subsistence, health, infrastructure, societies, and cultures of Arctic indigenous peoples. Changes in sea ice and sea level, permafrost, tundra, weather, and vegetation distributions, as well as increased commercial shipping, mineral extraction, and tourism, will affect the distribution of land and sea mammals, of freshwater and marine fish, and of forage for reindeer. These will in turn affect traditional subsistence activities and related indigenous lifestyles.[329] Arctic indigenous peoples' harvesting of animals is likely to become riskier and less predictable, which may increase food insecurity, change diets, and increase dependency on outside, nontraditional foods.[330] Food cellars in many locations have thawed during summers, threatening food safety. Related health risks of diabetes, obesity, and mental illness have been associated with these changes.[331]

Sea, shoreline ice, and permafrost changes have damaged infrastructure and increased coastal and inland erosion, especially in Alaska, where GAO found in 2003 that "coastal villages are becoming more susceptible to flooding and erosion caused in part by rising

[327] *AHDR*, chapter 4, and pp. 232-233.
[328] *AHDR*, chapters 6-7, and pp. 232-233.
[329] *ACIA*, pp. 1000-1001, 1004.
[330] *ACIA*, pp. 1000-1001, 1004.
[331] Parkinson, Alan J. *The Arctic Human Health Initiative*. Washington, DC: U.S. Department of Health and Human Services, 2006; Brubaker, Michael, James Berner, Raj Chavan, and John Warren. "Climate Change and Health Effects in Northwest Alaska." *Global Health Action* 4 (October 18, 2011).

temperatures."[332] In response, Congress funded the U.S. Army Corps of Engineers to conduct a Baseline Erosion Assessment that identified and prioritized among the 178 communities identified at risk from erosion.[333] (Risks from flooding were not examined.) GAO concluded in 2009 that many Native villages must relocate, but even those facing imminent threats have been impeded by various barriers, including difficulties identifying appropriate new sites, piecemeal programs for state and federal assistance, and obstacles to eligibility for certain federal programs.[334] The Alaska Federation of Natives placed among its 2010 federal priorities a request to Congress to mitigate flooding and erosion in Alaska Native villages and to fund relocation of villages where necessary.[335] However, "the cost is extraordinary," acknowledges Senator Lisa Murkowski.[336]

Oil, gas, and mineral exploration and development are expected to increase, as are other economic activities, such as forestry and tourism, and these are expected to increase economic opportunities for all Arctic residents, including indigenous peoples.[337] Pressures to increase participation in the wage economy, however, may speed up changes in indigenous cultures.

---

[332] U.S. Government Accountability Office (GAO), *Alaska Native Villages: Villages Affected by Flooding and Erosion Have Difficulty Qualifying for Federal Assistance*, GAO-04-895T, June 29, 2004, p. i, http://www.gao.gov/new.items/ d04895t.pdf. See also, Government Accountability Office, *Alaska Native Villages: Most Are Affected by Flooding and Erosion, but Few Qualify for Federal Assistance*, GAO-04-142, December 12, 2003, http://www.gao.gov/new.items/ d04142.pdf.

[333] U.S. Army Corps of Engineers. Alaska Baseline Erosion Assessment (BEA), March 2009.

[334] GAO, *Alaska Native Villages: Limited Progress Has Been Made on Relocating Villages Threatened by Flooding and Erosion*, June 3, 2009.

[335] Alaska Federation of Natives, Human Resources Committee, *2010 Federal Priorities* (Anchorage: Alaska Federation of Natives, 2010), pp. 22-23, available at https://www.nativefederation.org/wp-content/uploads/2018/07/ 2010-afn-federal-priorities.pdf. See also, Government Accountability Office, *Alaska Native Villages: Limited Progress Has Been Made on Relocating Villages Threatened by Flooding and Erosion*, GAO-09-551, June 3, 2009, http://www.gao.gov/new.items/d09551.pdf. For a more detailed anecdote, see a 2012 interview with Brice Eningowuk, Shishmaref Erosion and Relocation Committee member at Shearer, Christine. "Climate Crisis: Alaskan Village Shishmaref Sinking Into the Sea." *Imagined Magazine*, July 8, 2012, http://imaginedmag.com/2012/07/climate-crisis-alaskan-village-shishmaref-sinking-into-the-sea/.

[336] Press Office of Senator Lisa Murkowski. "Murkowski Urges Greater Tribal Consultation from Administration; Senator Spotlights 'Monumental' Climate Struggles Faced by Alaska Natives, Coastal Communities," July 19, 2012.

[337] *ACIA*, pp. 1001, 1004.

Increased economic opportunities may also lead to a rise in the nonindigenous population, which may further change the circumstances of indigenous cultures. Some representatives of Arctic indigenous people have related a "conflicting desire between combating climate change and embracing the potential for economic growth through foreign investment."[338]

Although important advances in public health have occurred in indigenous communities over past decades, some health problems may increase with continued Arctic climate change. Economic development may exacerbate Arctic pollution problems, including higher exposure to mercury, air pollution, and food contamination. The influx and redistribution of contaminants in the air, oceans, and land may change in ways that are now poorly understood.[339] Warmer temperatures and longer warm seasons may increase insect- and wildlife-borne diseases.[340] Climate change may lead to damage to water and sanitation systems, reducing protection against waterborne diseases.[341] Changes in Arctic indigenous cultures may increase mental stress and behavioral problems.[342]

The response to climate change by Arctic indigenous peoples has included international activities by Arctic indigenous organizations and advocacy before their national governments. As one report noted, "the rise of solidarity among indigenous peoples organizations in the region is

---

[338] Aqqaluk Lynge, Chair, Inuit Circumpolar Council, quoted in Stephanie McFeeters, "Lynge talks future of Inuit people," *The Dartmouth*, February 8, 2012.

[339] See, for example, "Health: Increased Bacterial Loads in Potable Water Could Have Significant Health Effects on Indigenous People From the Arctic to Uganda, Says Vanier Scholar." National Aboriginal Health Organization (NAHO), February 17, 2012. http://www.naho.ca/blog/2012/02/17/health-increased-bacterial-loads-in-potable-water-could-have-significant-health-effects-on-indigenous-people-from-the-arctic-to-uganda-says-vanier-scholar/; or, Kallenborn et al., *Combined Effects of Selected Pollutants and Climate Change in the Arctic Environment*. Oslo, Norway: Arctic Monitoring and Assessment Programme (AMAP), Arctic Council, 2011.

[340] *AMAP Assessment 2009: Human Health in the Arctic*, ed. Simon J. Wilson and Carolyn Symon (Oslo: Arctic Monitoring and Assessment Programme, 2009), pp. 4-6, 143.

[341] Brubaker, Michael, James Berner, Raj Chavan, and John Warren. "Climate Change and Health Effects in Northwest Alaska." *Global Health Action* 4 (October 18, 2011); John Warren, "Climate change could affect human health," *Mukluk Telegraph*, January/February 2005, pp. 5-6.

[342] John Warren, "Climate change could affect human health," *Mukluk Telegraph*, January/February 2005, pp. 5-6.

surely a development to be reckoned with by all those interested in policy issues in the Arctic."[343]

Six national or international indigenous organizations are permanent participants of the Arctic Council, the regional intergovernmental forum.[344] Due in part to advocacy by Arctic indigenous people, the United Nations General Assembly adopted in 2007 the Declaration on the Rights of Indigenous Peoples.[345] In April 2009, the Inuit Circumpolar Council (an organization of Inuit in the Arctic regions of Alaska, Canada, Greenland, and Russia) hosted in Alaska the worldwide "Indigenous Peoples Global Summit on Climate Change."[346] The conference report, forwarded to the Copenhagen Conference of the Parties of the U.N. Framework Convention on Climate Change (December 2009), noted "accelerating" climate change caused by "unsustainable development" and, among several recommendations, called for a greater indigenous role in national and international decisions on climate change, including a greater role for indigenous knowledge in climate change research, monitoring, and mitigation.[347]

APPENDIX A. ARCTIC RESEARCH AND POLICY ACT (ARPA) OF 1984 (TITLE I OF P.L. 98-373)

The text of the Arctic Research and Policy Act (ARPA) of 1984 (Title I of P.L. 98-373 of July 31, 1984)[348] is as follows:

---

[343] *AHDR*, p. 235.

[344] See http://www.arctic-council.org/. The six organizations are the Aleut International Association, Arctic Athabaskan Council, Gwich'in Council International, Inuit Circumpolar Council, RAIPON (Russian Association of Indigenous Peoples of the North), and Saami Council.

[345] United Nations, Declaration on the Rights of Indigenous Peoples, Resolution A/RES/61/295, at http://www.un- documents.net/a61r295.htm.

[346] See http://www.indigenoussummit.com/servlet/content/home.html.

[347] K. Galloway-McLean et al., *Report of the Indigenous Peoples' Global Summit on Climate Change: 20-24 April 2009, Anchorage, Alaska* (Darwin, Australia: United Nations University—Traditional Knowledge Initiative, 2009), pp. 5-7; available at http://www.indigenoussummit.com/servlet/content/home.html.

[348] Title II of P.L. 98-373 is the National Critical Materials Act of 1984.

TITLE I – ARCTIC RESEARCH AND POLICY

SHORT TITLE

SEC. 101. This title may be cited as the "Arctic Research and Policy Act of 1984."

FINDINGS AND PURPOSES

SEC. 102. (a) The Congress finds and declares that-

(1) the Arctic, onshore and offshore, contains vital energy resources that can reduce the Nation's dependence on foreign oil and improve the national balance of payments;

(2) as the Nation's only common border with the Soviet Union, the Arctic is critical to national defense;

(3) the renewable resources of the Arctic, specifically fish and other seafood, represent one of the Nation's greatest commercial assets;

(4) Arctic conditions directly affect global weather patterns and must be understood in order to promote better agricultural management throughout the United States;

(5) industrial pollution not originating in the Arctic region collects in the polar air mass, has the potential to disrupt global weather patterns, and must be controlled through international cooperation and consultation;

(6) the Arctic is a natural laboratory for research into human health and adaptation, physical and psychological, to climates of extreme cold and isolation and may provide information crucial for future defense needs;

(7) atmospheric conditions peculiar to the Arctic make the Arctic a unique testing ground for research into high latitude communications, which is likely to be crucial for future defense needs;

(8) Arctic marine technology is critical to cost-effective recovery and transportation of energy resources and to the national defense;

(9) the United States has important security, economic, and environmental interests in developing and maintaining a fleet of icebreaking vessels capable of operating effectively in the heavy ice regions of the Arctic;

(10) most Arctic-rim countries, particularly the Soviet Union, possess Arctic technologies far more advanced than those currently available in the United States;

(11) Federal Arctic research is fragmented and uncoordinated at the present time, leading to the neglect of certain areas of research and to unnecessary duplication of effort in other areas of research;

(12) improved logistical coordination and support for Arctic research and better dissemination of research data and information is necessary to increase the efficiency and utility of national Arctic research efforts;

(13) a comprehensive national policy and program plan to organize and fund currently neglected scientific research with respect to the Arctic is necessary to fulfill national objectives in Arctic research;

(14) the Federal Government, in cooperation with State and local governments, should focus its efforts on the collection and characterization of basic data related to biological, materials, geophysical, social, and behavioral phenomena in the Arctic;

(15) research into the long-range health, environmental, and social effects of development in the Arctic is necessary to mitigate the adverse consequences of that development to the land and its residents;

(16) Arctic research expands knowledge of the Arctic, which can enhance the lives of Arctic residents, increase opportunities for international cooperation among Arctic-rim countries, and facilitate the formulation of national policy for the Arctic; and

(17) the Alaskan Arctic provides an essential habitat for marine mammals, migratory waterfowl, and other forms of wildlife which are important to the Nation and which are essential to Arctic residents.

(b) The purposes of this title are-

(1) to establish national policy, priorities, and goals and to provide a Federal program plan for basic and applied scientific research with respect to the Arctic, including natural resources and materials, physical, biological and health sciences, and social and behavioral sciences;

(2) to establish an Arctic Research Commission to promote Arctic research and to recommend Arctic research policy;

(3) to designate the National Science Foundation as the lead agency responsible for implementing Arctic research policy; and

(4) to establish an Interagency Arctic Research Policy Committee to develop a national Arctic research policy and a five year plan to implement that policy.

ARCTIC RESEARCH COMMISSION

SEC. 103. (a) The President shall establish an Arctic Research Commission (hereafter referred to as the "Commission").

(b)(1) The Commission shall be composed of five members appointed by the President, with the Director of the National Science Foundation serving as a nonvoting, ex officio member. The members appointed by the President shall include-

(A) three members appointed from among individuals from academic or other research institutions with expertise in areas of research relating to the Arctic, including the physical, biological, health, environmental, social, and behavioral sciences;

(B) one member appointed from among indigenous residents of the Arctic who are representative of the needs and interests of Arctic residents and who live in areas directly affected by Arctic resource development; and

(C) one member appointed from among individuals familiar with the Arctic and representative of the needs and interests of private industry undertaking resource development in the Arctic.

(2) The President shall designate one of the appointed members of the Commission to be chairperson of the Commission.

(c)(1) Except as provided in paragraph (2) of this subsection, the term of office of each member of the Commission appointed under subsection (b)(1) shall be four years.

(2) Of the members of the Commission originally appointed under subsection (b)(1)-

(A) one shall be appointed for a term of two years;

(B) two shall be appointed for a term of three years; and

(C) two shall be appointed for a term of four years.

(3) Any vacancy occurring in the membership of the Commission shall be filled, after notice of the vacancy is published in the Federal Register, in the manner provided by the preceding provisions of this section, for the remainder of the unexpired term.

(4) A member may serve after the expiration of the member's term of office until the President appoints a successor.

(5) A member may serve consecutive terms beyond the member's original appointment.

(d)(1) Members of the Commission may be allowed travel expenses, including per diem in lieu of subsistence, as authorized by section 5703 of title 5, United States Code. A member of the Commission not presently employed for compensation shall be compensated at a rate equal to the daily equivalent of the rate for GS-16 of the General Schedule under section 5332 of title 5, United States Code, for each day the member is engaged in the actual performance of his duties as a member of the Commission, not to exceed 90 days of service each year. Except for the purposes of chapter 81 of title 5 (relating to compensation for work injuries) and chapter 171 of title 28 (relating to tort claims), a member of the Commission shall not be considered an employee of the United States for any purpose.

(2) The Commission shall meet at the call of its Chairman or a majority of its members.

(3) Each Federal agency referred to in section 107(b) may designate a representative to participate as an observer with the Commission.

These representatives shall report to and advise the Commission on the activities relating to Arctic research of their agencies.

(4) The Commission shall conduct at least one public meeting in the State of Alaska annually.

DUTIES OF COMMISSION

SEC. 104. (a) The Commission shall-

(1) develop and recommend an integrated national Arctic research policy;

(2) in cooperation with the Interagency Arctic Research Policy Committee established under section 107, assist in establishing a national Arctic research program plan to implement the Arctic research policy;

(3) facilitate cooperation between the Federal Government and State and local governments with respect to Arctic research;

(4) review Federal research programs in the Arctic and suggest improvements in coordination among programs;

(5) recommend methods to improve logistical planning and support for Arctic research as may be appropriate and in accordance with the findings and purposes of this title;

(6) suggest methods for improving efficient sharing and dissemination of data and information on the Arctic among interested public and private institutions;

(7) offer other recommendations and advice to the Interagency Committee established under section 107 as it may find appropriate; and

(8) cooperate with the Governor of the State of Alaska and with agencies and organizations of that State which the Governor may designate with respect to the formulation of Arctic research policy.

(b) Not later than January 31 of each year, the Commission shall-

(1) publish a statement of goals and objectives with respect to Arctic research to guide the Interagency Committee established under section 107 in the performance of its duties; and

(2) submit to the President and to the Congress a report describing the activities and accomplishments of the Commission during the immediately preceding fiscal year.

COOPERATION WITH THE COMMISSION

SEC. 105. (a)(1) The Commission may acquire from the head of any Federal agency unclassified data, reports, and other nonproprietary information with respect to Arctic research in the possession of the agency which the Commission considers useful in the discharge of its duties.

(2) Each agency shall cooperate with the Commission and furnish all data, reports, and other information requested by the Commission to the extent permitted by law; except that no agency need furnish any information which it is permitted to withhold under section 552 of title 5, United States Code.

(b) With the consent of the appropriate agency head, the Commission may utilize the facilities and services of any Federal agency to the extent that the facilities and services are needed for the establishment and development of an Arctic research policy, upon reimbursement to be agreed upon by the Commission and the agency head and taking every feasible step to avoid duplication of effort.

(c) All Federal agencies shall consult with the Commission before undertaking major Federal actions relating to Arctic research.

ADMINISTRATION OF THE COMMISSION

SEC. 106. The Commission may-

(1) in accordance with the civil service laws and subchapter III of chapter 53 of title 5, United States Code, appoint and fix the compensation of an Executive Director and necessary additional staff personnel, but not to exceed a total of seven compensated personnel;

(2) procure temporary and intermittent services as authorized by section 3109 of title 5, United States Code;

(3) enter into contracts and procure supplies, services, and personal property; and

(4) enter into agreements with the General Services Administration for the procurement of necessary financial and administrative services, for which payment shall be made by reimbursement from funds of the Commission in amounts to be agreed upon by the Commission and the Administrator of the General Services Administration.

LEAD AGENCY AND INTERAGENCY ARCTIC RESEARCH POLICY COMMITTEE

SEC. 107. (a) The National Science Foundation is designated as the lead agency responsible for implementing Arctic research policy, and the Director of the National Science Foundation shall insure that the requirements of section 108 are fulfilled.

(b)(1) The President shall establish an Interagency Arctic Research Policy Committee (hereinafter referred to as the "Interagency Committee").

(2) The Interagency Committee shall be composed of representatives of the following Federal agencies or offices:

(A) the National Science Foundation;

(B) the Department of Commerce;

(C) the Department of Defense;

(D) the Department of Energy;

(E) the Department of the Interior;

(F) the Department of State;

(G) the Department of Transportation;

(H) the Department of Health and Human Services;

(I) the National Aeronautics and Space Administration;

(J) the Environmental Protection Agency; and

(K) any other agency or office deemed appropriate.

(3) The representative of the National Science Foundation shall serve as the Chairperson of the Interagency Committee.

DUTIES OF THE INTERAGENCY COMMITTEE

SEC. 108. (a) The Interagency Committee shall-

(1) survey Arctic research conducted by Federal, State, and local agencies, universities, and other public and private institutions to help determine priorities for future Arctic research, including natural resources and materials, physical and biological sciences, and social and behavioral sciences;

(2) work with the Commission to develop and establish an integrated national Arctic research policy that will guide Federal agencies in developing and implementing their research programs in the Arctic;

(3) consult with the Commission on-

(A) the development of the national Arctic research policy and the 5-year plan implementing the policy;

(B) Arctic research programs of Federal agencies;

(C) recommendations of the Commission on future Arctic research; and

(D) guidelines for Federal agencies for awarding and administering Arctic research grants;

(4) develop a 5-year plan to implement the national policy, as provided for in section 109;

(5) provide the necessary coordination, data, and assistance for the preparation of a single integrated, coherent, and multiagency budget request for Arctic research as provided for in section 110;

(6) facilitate cooperation between the Federal Government and State and local governments in Arctic research, and recommend the undertaking of neglected areas of research in accordance with the findings and purposes of this title;

(7) coordinate and promote cooperative Arctic scientific research programs with other nations, subject to the foreign policy guidance of the Secretary of State;

(8) cooperate with the Governor of the State of Alaska in fulfilling its responsibilities under this title;

(9) promote Federal interagency coordination of all Arctic research activities, including-

(A) logistical planning and coordination; and

(B) the sharing of data and information associated with Arctic research, subject to section 552 of title 5, United States Code; and

(10) provide public notice of its meetings and an opportunity for the public to participate in the development and implementation of national Arctic research policy.

(b) Not later than January 31, 1986, and biennially thereafter, the Interagency Committee shall submit to the Congress through the President, a brief, concise report containing-

(1) a statement of the activities and accomplishments of the Interagency Committee since its last report; and

(2) a description of the activities of the Commission, detailing with particularity the recommendations of the Commission with respect to Federal activities in Arctic research.

5-YEAR ARCTIC RESEARCH PLAN

SEC. 109. (a) The Interagency Committee, in consultation with the Commission, the Governor of the State of Alaska, the residents of the Arctic, the private sector, and public interest groups, shall prepare a comprehensive 5-year program plan (hereinafter referred to as the "Plan") for the overall Federal effort in Arctic research. The Plan shall be prepared and submitted to the President for transmittal to the Congress within one year after the enactment of this Act and shall be revised biennially thereafter.

(b) The Plan shall contain but need not be limited to the following elements:

(1) an assessment of national needs and problems regarding the Arctic and the research necessary to address those needs or problems;

(2) a statement of the goals and objectives of the Interagency Committee for national Arctic research;

(3) a detailed listing of all existing Federal programs relating to Arctic research, including the existing goals, funding levels for each of the 5 following fiscal years, and the funds currently being expended to conduct the programs;

(4) recommendations for necessary program changes and other proposals to meet the requirements of the policy and goals as set forth by the Commission and in the Plan as currently in effect; and

(5) a description of the actions taken by the Interagency Committee to coordinate the budget review process in order to ensure interagency coordination and cooperation in (A) carrying out Federal Arctic research programs, and (B) eliminating unnecessary duplication of effort among these programs.

COORDINATION AND REVIEW OF BUDGET REQUESTS

SEC. 110. (a) The Office of Science and Technology Policy shall-

(1) review all agency and department budget requests related to the Arctic transmitted pursuant to section 108(a)(5), in accordance with the national Arctic research policy and the 5-year program under section 108(a)(2) and section 109, respectively; and

(2) consult closely with the Interagency Committee and the Commission to guide the Office of Science and Technology Policy's efforts.

(b)(1) The Office of Management and Budget shall consider all Federal agency requests for research related to the Arctic as one integrated, coherent, and multiagency request which shall be reviewed by the Office of Management and Budget prior to submission of the

President's annual budget request for its adherence to the Plan. The Commission shall, after submission of the President's annual budget request, review the request and report to Congress on adherence to the Plan.

(2) The Office of Management and Budget shall seek to facilitate planning for the design, procurement, maintenance, deployment, and operations of icebreakers needed to provide a platform for Arctic research by allocating all funds necessary to support icebreaking operations, except for recurring incremental costs associated with specific projects, to the Coast Guard.

AUTHORIZATION OF APPROPRIATIONS; NEW SPENDING AUTHORITY

SEC. 111. (a) There are authorized to be appropriated such sums as may be necessary for carrying out this title.

(b) Any new spending authority (within the meaning of section 401 of the Congressional Budget Act of 1974) which is provided under this title shall be effective for any fiscal year only to such extent or in such amounts as may be provided in appropriation Acts.

DEFINITION

SEC. 112. As used in this title, the term "Arctic" means all United States and foreign territory north of the Arctic Circle and all United States territory north and west of the boundary formed by the Porcupine, Yukon, and Kuskokwim Rivers; all contiguous seas, including the Arctic Ocean and the Beaufort, Bering, and Chukchi Seas; and the Aleutian chain.

APPENDIX B. P.L. 101-609 OF 1990, AMENDING ARCTIC RESEARCH AND POLICY ACT (ARPA) OF 1984

The Arctic Research and Policy Act (ARPA) of 1984 (see Appendix A) was amended by P.L. 101-609 of November 16, 1990. The text of P.L. 101-609 is as follows:

SECTION 1. Except as specifically provided in this Act, whenever in this Act an amendment or repeal is expressed as an amendment to, or repeal of a provision, the reference shall be deemed to be made to the Arctic Research and Policy Act of 1984.

SEC. 2. Section 103(b)(1) (15 U.S.C. 4102(b)(1)) is amended—

(1) in the text above clause (A), by striking out 'five' and inserting in lieu thereof 'seven';

(2) in clause (A), by striking out 'three' and inserting in lieu thereof 'four'; and

(3) in clause (C), by striking out 'one member' and inserting in lieu thereof 'two members'.

SEC. 3. Section 103(d)(1) (15 U.S.C. 4102(d)(1)) is amended by striking out 'GS-16' and inserting in lieu thereof 'GS-18'.

SEC. 4. (a) Section 104(a) (15 U.S.C. 4102(a)) is amended—

(1) in paragraph (4), by striking out `suggest' and inserting in lieu thereof `recommend';

(2) in paragraph (6), by striking out `suggest' and inserting in lieu thereof `recommend';

(3) in paragraph (7), by striking out `and' at the end thereof;

(4) in paragraph (8), by striking out the period and inserting in lieu thereof a semicolon; and

(5) by adding at the end thereof the following new paragraphs:

'(9) recommend to the Interagency Committee the means for developing international scientific cooperation in the Arctic; and

'(10) not later than January 31, 1991, and every 2 years thereafter, publish a statement of goals and objectives with respect to Arctic research to guide the Interagency Committee established under section 107 in the performance of its duties.'.

Section 104(b) is amended to read as follows:

'(b) Not later than January 31 of each year, the Commission shall submit to the President and to the Congress a report describing the activities and accomplishments of the Commission during the immediately preceding fiscal year.'.

SEC. 5. Section 106 (15 U.S.C. 4105) is amended—

(1) in paragraph (3), by striking out 'and' at the end thereof;

(2) in paragraph (4), by striking out the period at the end thereof and inserting in lieu thereof; and'; and

(3) by adding at the end thereof the following new paragraph:

'(5) appoint, and accept without compensation the services of, scientists and engineering specialists to be advisors to the Commission. Each advisor may be allowed travel expenses, including per diem in lieu of subsistence, as authorized by section 5703 of title 5, United States Code. Except for the purposes of chapter 81 of title 5 (relating to compensation for work injuries) and chapter 171 of title 28 (relating to tort claims) of the United States Code, an advisor appointed under this paragraph shall not be considered an employee of the United States for any purpose.'

SEC. 6. Subsection (b)(2) of section 108 (15 U.S.C. 4107(b)(2)) is amended to read as follows:

'(2) a statement detailing with particularity the recommendations of the Commission with respect to Federal interagency activities in Arctic research and the disposition and responses to those recommendations.'

APPENDIX C. JANUARY 2009 ARCTIC POLICY DIRECTIVE (NSPD 66/HSPD 25)

On January 12, 2009, the George W. Bush Administration released a presidential directive establishing a new U.S. policy for the Arctic region. The directive, dated January 9, 2009, was issued as National Security Presidential Directive 66/Homeland Security Presidential Directive 25 (NSPD 66/HSPD 25). The text of NSPD 66/HSPD 25 is as follows:

SUBJECT: Arctic Region Policy

I. PURPOSE

A. This directive establishes the policy of the United States with respect to the Arctic region and directs related implementation actions. This directive supersedes Presidential Decision Directive/NSC-26 (PDD-26; issued 1994) with respect to Arctic policy but not Antarctic policy; PDD-26 remains in effect for Antarctic policy only.

B. This directive shall be implemented in a manner consistent with the Constitution and laws of the United States, with the obligations of the United States under the treaties and other international agreements to which the United States is a party, and with customary international law as recognized by the United States, including with respect to the law of the sea.

II. BACKGROUND

A. The United States is an Arctic nation, with varied and compelling interests in that region. This directive takes into account several developments, including, among others:

1. Altered national policies on homeland security and defense;

2. The effects of climate change and increasing human activity in the Arctic region;

3. The establishment and ongoing work of the Arctic Council; and

4. A growing awareness that the Arctic region is both fragile and rich in resources.

III. POLICY

A. It is the policy of the United States to:

1. Meet national security and homeland security needs relevant to the Arctic region;

2. Protect the Arctic environment and conserve its biological resources;

3. Ensure that natural resource management and economic development in the region are environmentally sustainable;

4. Strengthen institutions for cooperation among the eight Arctic nations (the United States, Canada, Denmark, Finland, Iceland, Norway, the Russian Federation, and Sweden);

5. Involve the Arctic's indigenous communities in decisions that affect them; and

6. Enhance scientific monitoring and research into local, regional, and global environmental issues.

B. National Security and Homeland Security Interests in the Arctic

1. The United States has broad and fundamental national security interests in the Arctic region and is prepared to operate either independently or in conjunction with other states to safeguard these interests. These interests include such matters as missile defense and early warning; deployment of sea and air systems for strategic sealift, strategic deterrence, maritime presence, and maritime security operations; and ensuring freedom of navigation and overflight.

2. The United States also has fundamental homeland security interests in preventing terrorist attacks and mitigating those criminal or hostile acts that could increase the United States vulnerability to terrorism in the Arctic region.

3. The Arctic region is primarily a maritime domain; as such, existing policies and authorities relating to maritime areas continue to apply, including those relating to law enforcement.[1] Human activity in the Arctic region is increasing and is projected to increase further in coming years. This requires the United States to assert a more active and influential national presence to protect its Arctic interests and to project sea power throughout the region.

4. The United States exercises authority in accordance with lawful claims of United States sovereignty, sovereign rights, and jurisdiction in the Arctic region, including sovereignty within the territorial sea, sovereign rights and jurisdiction within the United States exclusive economic zone and on the continental shelf, and appropriate control in the United States contiguous zone.

5. Freedom of the seas is a top national priority. The Northwest Passage is a strait used for international navigation, and the Northern Sea Route includes straits used for international navigation; the regime of transit passage applies to passage through those straits. Preserving the rights and duties relating to navigation and overflight in the Arctic region supports our ability to exercise these rights throughout the world, including through strategic straits.

6. Implementation: In carrying out this policy as it relates to national security and homeland security interests in the Arctic, the Secretaries of State, Defense, and Homeland Security, in coordination with heads of other relevant executive departments and agencies, shall:

a. Develop greater capabilities and capacity, as necessary, to protect United States air, land, and sea borders in the Arctic region;

b. Increase Arctic maritime domain awareness in order to protect maritime commerce, critical infrastructure, and key resources;

c. Preserve the global mobility of United States military and civilian vessels and aircraft throughout the Arctic region;

d. Project a sovereign United States maritime presence in the Arctic in support of essential United States interests; and

e. Encourage the peaceful resolution of disputes in the Arctic region.

C. International Governance

1. The United States participates in a variety of fora, international organizations, and bilateral contacts that promote United States interests in the Arctic. These include the Arctic Council, the International Maritime Organization (IMO), wildlife conservation and management agreements, and many other mechanisms. As the Arctic changes and human activity in the region increases, the United States and other governments should consider, as appropriate, new international arrangements or enhancements to existing arrangements.

2. The Arctic Council has produced positive results for the United States by working within its limited mandate of environmental protection and sustainable development. Its subsidiary bodies, with help from many United States agencies, have developed and undertaken projects on a wide range of topics. The Council also provides a beneficial venue for interaction with indigenous groups. It is the position of the United States that the Arctic Council should remain a high-level forum devoted to issues within its current mandate and not be transformed into a formal international organization, particularly one with assessed contributions. The United States is nevertheless open to updating the structure of the Council, including consolidation of, or making operational changes to, its subsidiary bodies, to the extent such changes can clearly improve the Council's work and are consistent with the general mandate of the Council.

3. The geopolitical circumstances of the Arctic region differ sufficiently from those of the Antarctic region such that an "Arctic Treaty" of broad scope—along the lines of the Antarctic Treaty—is not appropriate or necessary.

4. The Senate should act favorably on U.S. accession to the U.N. Convention on the Law of the Sea promptly, to protect and advance U.S. interests, including with respect to the Arctic. Joining will serve the national security interests of the United States, including the maritime mobility of our Armed Forces worldwide. It will secure U.S. sovereign rights over extensive marine areas, including the valuable natural resources they contain. Accession will promote U.S. interests in the environmental health of the oceans. And it will give the United States a seat at the table when the rights that are vital to our interests are debated and interpreted.

5. Implementation: In carrying out this policy as it relates to international governance, the Secretary of State, in coordination with heads of other relevant executive departments and agencies, shall:

a. Continue to cooperate with other countries on Arctic issues through the United Nations (U.N.) and its specialized agencies, as well as through treaties such as the U.N. Framework Convention on Climate Change, the Convention on International Trade in Endangered Species of Wild Fauna and Flora, the Convention on Long Range Transboundary Air Pollution and its protocols, and the Montreal Protocol on Substances that Deplete the Ozone Layer;

b. Consider, as appropriate, new or enhanced international arrangements for the Arctic to address issues likely to arise from expected increases in human activity in that region, including shipping, local development and subsistence, exploitation of living marine resources, development of energy and other resources, and tourism;

c. Review Arctic Council policy recommendations developed within the ambit of the Council's scientific reviews and ensure the policy recommendations are subject to review by Arctic governments; and

d. Continue to seek advice and consent of the United States Senate to accede to the 1982 Law of the Sea Convention.

D. Extended Continental Shelf and Boundary Issues

1. Defining with certainty the area of the Arctic seabed and subsoil in which the United States may exercise its sovereign rights over natural resources such as oil, natural gas, methane hydrates, minerals, and living marine species is critical to our national interests in energy security, resource management, and environmental protection. The most effective way to achieve international recognition and legal certainty for our extended continental shelf is through the procedure available to States Parties to the U.N. Convention on the Law of the Sea.

2. The United States and Canada have an unresolved boundary in the Beaufort Sea. United States policy recognizes a boundary in this area based on equidistance. The United States recognizes that the boundary area may contain oil, natural gas, and other resources.

3. The United States and Russia are abiding by the terms of a maritime boundary treaty concluded in 1990, pending its entry into force. The United States is prepared to enter the agreement into force once ratified by the Russian Federation.

4. Implementation: In carrying out this policy as it relates to extended continental shelf and boundary issues, the Secretary of State, in coordination with heads of other relevant executive departments and agencies, shall:

a. Take all actions necessary to establish the outer limit of the continental shelf appertaining to the United States, in the Arctic and in other regions, to the fullest extent permitted under international law;

b. Consider the conservation and management of natural resources during the process of delimiting the extended continental shelf; and

c. Continue to urge the Russian Federation to ratify the 1990 United States-Russia maritime boundary agreement.

E. Promoting International Scientific Cooperation

1. Scientific research is vital for the promotion of United States interests in the Arctic region. Successful conduct of U.S. research in the Arctic region requires access throughout the Arctic Ocean and to terrestrial sites, as well as viable international mechanisms for sharing access to research platforms and timely exchange of samples, data, and analyses. Better coordination with the Russian Federation, facilitating access to its domain, is particularly important.

2. The United States promotes the sharing of Arctic research platforms with other countries in support of collaborative research that advances fundamental understanding of the Arctic region in general and potential Arctic change in particular. This could include collaboration with bodies such as the Nordic Council and the European Polar Consortium, as well as with individual nations.

3. Accurate prediction of future environmental and climate change on a regional basis, and the delivery of near real-time information to end-users, requires obtaining, analyzing, and disseminating accurate data from the entire Arctic region, including both paleoclimatic data and observational data. The United States has made significant investments in the infrastructure needed to collect environmental data in the Arctic region, including the establishment of portions of an Arctic circumpolar observing network through a partnership among United States agencies, academic collaborators, and Arctic residents. The United States promotes active involvement of all Arctic nations in these efforts in order to advance scientific understanding that could provide the basis for assessing future impacts and proposed response strategies.

4. United States platforms capable of supporting forefront research in the Arctic Ocean, including portions expected to be ice-covered for the foreseeable future, as well as seasonally ice-free regions, should work with those of other nations through the establishment of an Arctic circumpolar observing network. All Arctic nations are members of the Group on Earth Observations partnership, which provides a framework for organizing an international approach to environmental observations in the region. In addition, the United States recognizes that academic and research institutions are vital partners in promoting and conducting Arctic research.

5. Implementation: In carrying out this policy as it relates to promoting scientific international cooperation, the Secretaries of State, the Interior, and Commerce and the Director of the National Science Foundation, in coordination with heads of other relevant executive departments and agencies, shall:

a. Continue to play a leadership role in research throughout the Arctic region;

b. Actively promote full and appropriate access by scientists to Arctic research sites through bilateral and multilateral measures and by other means;

c. Lead the effort to establish an effective Arctic circumpolar observing network with broad partnership from other relevant nations;

d. Promote regular meetings of Arctic science ministers or research council heads to share information concerning scientific research opportunities and to improve coordination of international Arctic research programs;

e. Work with the Interagency Arctic Research Policy Committee (IARPC) to promote research that is strategically linked to U.S. policies articulated in this directive, with input from the Arctic Research Commission; and

f. Strengthen partnerships with academic and research institutions and build upon the relationships these institutions have with their counterparts in other nations.

F. Maritime Transportation in the Arctic Region

1. The United States priorities for maritime transportation in the Arctic region are:

a. To facilitate safe, secure, and reliable navigation;

b. To protect maritime commerce; and

c. To protect the environment.

2. Safe, secure, and environmentally sound maritime commerce in the Arctic region depends on infrastructure to support shipping activity, search and rescue capabilities, short- and long-range aids to navigation, high-risk area vessel-traffic management, iceberg warnings and other sea ice information, effective shipping standards, and measures to protect the marine environment. In addition, effective search and rescue in the Arctic will require local, State, Federal, tribal, commercial, volunteer, scientific, and multinational cooperation.

3. Working through the International Maritime Organization (IMO), the United States promotes strengthening existing measures and, as necessary, developing new measures to improve the safety and security of maritime transportation, as well as to protect the marine environment in the Arctic region. These measures may include ship routing and reporting systems, such as traffic separation and vessel traffic management schemes in Arctic chokepoints; updating and strengthening of the Guidelines for Ships Operating in Arctic Ice-Covered Waters; underwater noise standards for commercial shipping; a review of shipping insurance issues; oil and other hazardous material pollution response agreements; and environmental standards.

4. Implementation: In carrying out this policy as it relates to maritime transportation in the Arctic region, the Secretaries of State,

Defense, Transportation, Commerce, and Homeland Security, in coordination with heads of other relevant executive departments and agencies, shall:

a. Develop additional measures, in cooperation with other nations, to address issues that are likely to arise from expected increases in shipping into, out of, and through the Arctic region;

b. Commensurate with the level of human activity in the region, establish a risk-based capability to address hazards in the Arctic environment. Such efforts shall advance work on pollution prevention and response standards; determine basing and logistics support requirements, including necessary airlift and icebreaking capabilities; and improve plans and cooperative agreements for search and rescue;

c. Develop Arctic waterways management regimes in accordance with accepted international standards, including vessel traffic-monitoring and routing; safe navigation standards; accurate and standardized charts; and accurate and timely environmental and navigational information; and

d. Evaluate the feasibility of using access through the Arctic for strategic sealift and humanitarian aid and disaster relief.

G. Economic Issues, Including Energy

1. Sustainable development in the Arctic region poses particular challenges. Stakeholder input will inform key decisions as the United States seeks to promote economic and energy security. Climate change and other factors are significantly affecting the lives of Arctic inhabitants, particularly indigenous communities. The United States affirms the importance to Arctic communities of adapting to climate change, given their particular vulnerabilities.

2. Energy development in the Arctic region will play an important role in meeting growing global energy demand as the area is thought to contain a substantial portion of the world's undiscovered energy resources. The United States seeks to ensure that energy development throughout the Arctic occurs in an environmentally sound manner, taking into account the interests of indigenous and local communities, as well as open and transparent market principles. The United States seeks to balance access to, and development of, energy and other natural resources with the protection of the Arctic environment by ensuring that continental shelf resources are managed in a responsible manner and by continuing to work closely with other Arctic nations.

3. The United States recognizes the value and effectiveness of existing fora, such as the Arctic Council, the International Regulators Forum, and the International Standards Organization.

4. Implementation: In carrying out this policy as it relates to economic issues, including energy, the Secretaries of State, the Interior, Commerce, and Energy, in coordination with heads of other relevant executive departments and agencies, shall:

a. Seek to increase efforts, including those in the Arctic Council, to study changing climate conditions, with a view to preserving and enhancing economic opportunity in the Arctic region. Such efforts shall include inventories and assessments of villages, indigenous communities, subsistence opportunities, public facilities, infrastructure, oil and gas development projects, alternative energy development opportunities, forestry, cultural and other sites, living marine resources, and other elements of the Arctic's socioeconomic composition;

b. Work with other Arctic nations to ensure that hydrocarbon and other development in the Arctic region is carried out in accordance with accepted best practices and internationally recognized standards and the 2006 Group of Eight (G-8) Global Energy Security Principles;

c. Consult with other Arctic nations to discuss issues related to exploration, production, environmental and socioeconomic impacts, including drilling conduct, facility sharing, the sharing of environmental data, impact assessments, compatible monitoring programs, and reservoir management in areas with potentially shared resources;

d. Protect United States interests with respect to hydrocarbon reservoirs that may overlap boundaries to mitigate adverse environmental and economic consequences related to their development;

e. Identify opportunities for international cooperation on methane hydrate issues, North Slope hydrology, and other matters;

f. Explore whether there is a need for additional fora for informing decisions on hydrocarbon leasing, exploration, development, production, and transportation, as well as shared support activities, including infrastructure projects; and

g. Continue to emphasize cooperative mechanisms with nations operating in the region to address shared concerns, recognizing that most known Arctic oil and gas resources are located outside of United States jurisdiction.

H. Environmental Protection and Conservation of Natural Resources

1. The Arctic environment is unique and changing. Increased human activity is expected to bring additional stressors to the Arctic environment, with potentially serious consequences for Arctic communities and ecosystems.

2. Despite a growing body of research, the Arctic environment remains poorly understood. Sea ice and glaciers are in retreat. Permafrost is thawing and coasts are eroding. Pollutants from within and outside the Arctic are contaminating the region. Basic data are lacking in many fields. High levels of uncertainty remain concerning the effects of climate change and increased human activity in the Arctic. Given the need for decisions to be based on sound scientific and socioeconomic information, Arctic environmental research, monitoring, and vulnerability assessments are top priorities. For example, an understanding of the probable consequences of global climate variability and change on Arctic

ecosystems is essential to guide the effective long-term management of Arctic natural resources and to address socioeconomic impacts of changing patterns in the use of natural resources.

3. Taking into account the limitations in existing data, United States efforts to protect the Arctic environment and to conserve its natural resources must be risk-based and proceed on the basis of the best available information.

4. The United States supports the application in the Arctic region of the general principles of international fisheries management outlined in the 1995 Agreement for the Implementation of the Provisions of the United Nations Convention on the Law of the Sea of December 10, 1982, relating to the Conservation and Management of Straddling Fish Stocks and Highly Migratory Fish Stocks and similar instruments. The United States endorses the protection of vulnerable marine ecosystems in the Arctic from destructive fishing practices and seeks to ensure an adequate enforcement presence to safeguard Arctic living marine resources.

5. With temperature increases in the Arctic region, contaminants currently locked in the ice and soils will be released into the air, water, and land. This trend, along with increased human activity within and below the Arctic, will result in increased introduction of contaminants into the Arctic, including both persistent pollutants (e.g., persistent organic pollutants and mercury) and airborne pollutants (e.g., soot).

6. Implementation: In carrying out this policy as it relates to environmental protection and conservation of natural resources, the Secretaries of State, the Interior, Commerce, and Homeland Security and the Administrator of the Environmental Protection Agency, in coordination with heads of other relevant executive departments and agencies, shall:

a. In cooperation with other nations, respond effectively to increased pollutants and other environmental challenges;

b. Continue to identify ways to conserve, protect, and sustainably manage Arctic species and ensure adequate enforcement presence to safeguard living marine resources, taking account of the changing ranges or distribution of some species in the Arctic. For species whose range includes areas both within and beyond United States jurisdiction, the United States shall continue to collaborate with other governments to ensure effective conservation and management;

c. Seek to develop ways to address changing and expanding commercial fisheries in the Arctic, including through consideration of international agreements or organizations to govern future Arctic fisheries;

d. Pursue marine ecosystem-based management in the Arctic; and

e. Intensify efforts to develop scientific information on the adverse effects of pollutants on human health and the environment and work with other nations to reduce the introduction of key pollutants into the Arctic.

IV. Resources and Assets

A. Implementing a number of the policy elements directed above will require appropriate resources and assets. These elements shall be implemented consistent with applicable law and authorities of agencies, or heads of agencies, vested by law, and subject to the availability of appropriations. The heads of executive departments and agencies with responsibilities relating to the Arctic region shall work to identify future budget, administrative, personnel, or legislative proposal requirements to implement the elements of this directive.

[1] These policies and authorities include Freedom of Navigation (PDD/NSC-32), the U.S. Policy on Protecting the Ocean Environment (PDD/NSC-36), Maritime Security Policy (NSPD-41/HSPD-13), and the National Strategy for Maritime Security (NSMS).[349]

[349] Source for text: http://fas.org/irp/offdocs/nspd/nspd-66.htm. The text is also available at http://www.nsf.gov/od/opp/ opp_advisory/briefings/may2009/nspd66_hspd25.pdf.

INDEX

A

B

C

D

E

F

I

J

L

M

N

O

P

R

S

T

U